MODERN MORPHOMETRICS IN PHYSICAL ANTHROPOLOGY

DEVELOPMENTS IN PRIMATOLOGY: PROGRESS AND PROSPECTS

Series Editor:
Russell H. Tuttle
University of Chicago, Chicago, Illinois

This peer-reviewed book series will meld the facts of organic diversity with the continuity of the evolutionary process. The volumes in this series will exemplify the diversity of theoretical perspectives and methodological approaches currently employed by primatologists and physical anthropologists. Specific coverage includes: primate behavior in natural habitats and captive settings: primate ecology and conservation; functional morphology and developmental biology of primates; primate systematics; genetic and phenotypic differences among living primates; and paleoprimatology.

ALL APES GREAT AND SMALL
VOLUME 1: AFRICAN APES
Edited by Birute M. F. Galdikas, Nancy Erickson Briggs, Lori K. Sheeran, Gary L. Shapiro and Jane Goodall

THE GUENONS: DIVERSITY AND ADAPTATION IN AFRICAN MONKEYS
Edited by Mary E. Glenn and Marina Cords

ANIMAL BODIES, HUMAN MINDS: APE, DOLPHIN, AND PARROT LANGUAGE SKILLS
William A. Hillix and Duane M. Rumbaugh

COMPARATIVE VERTEBRATE COGNITION: ARE PRIMATES SUPERIOR TO NON-PRIMATES
Lesley J. Rogers and Gisela Kaplan

ANTHROPOID ORIGINS: NEW VISIONS
Callum F. Ross and Richard F. Kay

MODERN MORPHOMETRICS IN PHYSICAL ANTHROPOLOGY
Edited by Dennis E. Slice

MODERN MORPHOMETRICS IN PHYSICAL ANTHROPOLOGY

Edited by

Dennis E. Slice

Institute for Anthropology
University of Vienna
Vienna, Austria

KLUWER ACADEMIC / PLENUM PUBLISHERS
New York, Boston, Dordrecht, London, Moscow

Library of Congress Cataloging-in-Publication Data

Dennis E. Slice.
    Modern morphometrics in physical anthropology / edited by
    Dennis E. Slice.
        p.   cm. — (Developments in primatology)
    Includes bibliographical references and index.
    ISBN 0-306-48697-0 (hardback) — ISBN 0-306-48698-9 (ebook)
      1. Physical anthropology—Methodology.
      2. Physical anthropology—Mathematics.
      3. Physical anthropology—Computer simulation.
      4. Morphology—Mathematics. 5. Morphology—Statistical methods.
      6. Morphology—Computer simulation. 7. Anthropometry.
      8. Craniometry.   I. Slice, Dennis E.   II. Series.

GN50.8.M63 2005
599.9—dc22

                                          2004054594

ISBN: 0-306-48697-0 (hardback)
ISBN: 0-306-48698-9 (ebook)

©2005 Kluwer Academic / Plenum Publishers, New York
233 Spring Street, New York, New York 10013

http://www.wkap.nl/

10 9 8 7 6 5 4 3 2 1

A C.I.P. record for this book is available from the Library of Congress

Permissions for books published in Europe: *permissions@wkap.nl*
Permissions for books published in the United States of America:*permissions@wkap.com*

Leslie F. Marcus (1930–2002)

On February 22, 2002, the field of morphometrics lost a great friend, teacher, and colleague with the death of Dr. Leslie F. Marcus. At the time of his death, Les was a research associate in paleontology at the American Museum of Natural History and Professor of Biology Emeritus at Queens College and the Graduate School of the City University of New York. From the start of his career, he worked at the leading edge of quantitative biology, as evidenced by his earning an M.A. in Statistics in 1959 prior to starting work on his Ph.D. in paleontology that was completed in 1962. In the ensuing years, he worked to integrate computers into biological research in areas such as database development, museum cataloging, and data acquisition, and he was commited to the proper use of multivariate statistical techniques in ecological, evolutionary, and systematic research. Generations of scientists benefited from his teaching and his always constructively critical eye.

Les was particularly concerned with methods for the quantification of biological shape. This led him to continuously evaluate the most recent techniques in this area and, ultimately, to the modern morphometric methods that are the foundation of the work presented here. Having identified state-of-the-art approaches that would benefit so many researchers, he used his considerable international reputation and extensive collegial network to introduce these methods to scientists and students throughout the world. Every chapter in this volume has been influenced in one way or another by his contributions, and as the field continues to grow in the years to come, his absence will be felt even more greatly. For these and so many other reasons, we dedicate this volume to the lasting contributions and memory of Dr. Leslie F. Marcus.

PREFACE

The explosive growth in computing capability and accessibility in recent years has had a profound impact on all areas of biology concerned with the analysis of shape variation. Today, almost every student has easy access to computing resources that exceed those available in the entire world just a few decades ago. Such computational capacity has enabled mathematicians, statisticians, and others to not only reassess the analyses that could be applied to linear and angular measurements that would have been familiar to generations of anthropologists, but also to reexamine the nature of the data being analyzed. As described in more detail in the introductory chapter, traditional sets of measurement data that completely encode their relative spatial relationships are difficult to construct. The Cartesian coordinates of the individual points that define traditional measurements, on the other hand, concisely encode all the available geometric information in those points. It is the development of new methods of collection, processing, analysis, and visualization based on Cartesian coordinates, made possible by relatively recent technological and theoretical advances, that has led to a major shift in the way biological shape analysis — that is, morphometrics — is conducted.

Physical anthropologists and their colleagues in related fields have been early adopters of this new approach to shape analysis, and a number of investigators quickly grasped the potential of the modern morphometric methods to address shape-related questions in such areas as growth and development, evolution, ecology, and functional morphology. As a result, ever more publications and meeting presentations in anthropology incorporate modern morphometric techniques into their research methods. The relatively new and somewhat specialized nature of these analyses, however, has, to a degree, limited their use to research groups with training from specialized morphometrics workshops or with sources of local expertise.

To address this situation, a symposium dedicated to modern morphometrics in physical anthropology was proposed and accepted for the 2002 AAPA meeting in Buffalo, New York. The goal of the symposium was to introduce researchers to the fundamental elements of modern morphometric analysis, illustrate how these methods compare with traditional approaches and how they have contributed to major research projects, and provide a glimpse of new ideas and methods that may further contribute to research in the not-too-distant future. The symposium was well attended and judged to be a great success. The next step seemed obvious, and that was to provide the same in book form to reach a much larger audience.

This is that book. Most of the chapters in this volume are derived directly from presentations in the AAPA symposium. A few participants were not able to contribute as their material was already scheduled for publication elsewhere, but we were able to invite other contributors who could not present in the original symposium. The introductory chapter is designed to expose readers new to the field to most of the terms, concepts, and methods and provide references through which they can gain the prerequisite knowledge to apply these methods in their own work. Subsequent chapters speculate as to the future of morphometric analysis and extend basic methods to embrace even more sophisticated analyses and data sets. Some contributions compare the results of modern and traditional analyses in an applied context, while others focus on specific research questions addressed through the use of modern morphometric methods. It is hoped that at least some part of this book will appeal and be of use to most researchers interested in problems of shape analysis.

This book is obviously the product of the efforts of a great many people. I would like to extend my sincerest appreciation to all of the contributors for their hard work in providing such outstanding material, their encouragement throughout the project, and their patience in explaining to me unfamiliar aspects of their own research. All contributions were critically examined by at least two reviewers—one experienced in morphometrics and the other in the particular subject matter of the chapter, and their efforts are gratefully acknowledged. Philip L. Walker provided suggestions and support for the organization of the original symposium for the American Association of Physical Anthropologists, and Andrea Macaluso, Krista Zimmer, and Felix Portnoy of Kluwer Academic Publishers were sources of limitless encouragement and technical help from the start of this undertaking. The series editor, Russell H. Tuttle, played an important role in identifying and addressing some of the rough edges. To all of these individuals, I extend my sincerest thanks. My own work on this project would not have been possible without the generous support of Dr. Edward G. Hill and members of the Winston-Salem community and

Prof. Dr. Horst Seidler and the faculty and students of the Institute for Anthropology at the University of Vienna.

And finally, I would like to express my appreciation for the mentorship, encouragement, and friendship offered to me by the late Leslie F. Marcus to whom this volume is dedicated. Thanks, Les!

Dennis E. Slice

Winston-Salem, North Carolina, 2004

CONTENTS

CONTRIBUTORS

Markus Bastir, Hull York Medical School, The University of York, Helsington, York YO10 5DD, United Kingdom, and Atapuerca Research Team, Department of Paleobiology, Museo Nacional de Ciencias Naturales, CSIC, José Abascál 2, 28006 Madrid, Spain. Email: mbastir@mncn.csic.es

Michel Baylac, Muséum National d'Histoire Naturelle, Département Systématique et Evolution, 45, rue Buffon F-75005 Paris. Email: baylac@cimrs1.mnhn.fr

Fred L. Bookstein, Institute for Anthropology, University of Vienna, Althanstrasse 14, A-1091 Vienna, Austria and Biophysics Research Division, University of Michigan, Ann Arbor, MI. Email: fred@brainmap.med.umich.edu

Theodore M. Cole, III, Department of Basic Medical Science, School of Medicine, University of Missouri—Kansas City, 2411 Holmes St., Kansas City, MO 64108. Email: ColeT@umkc.edu

Martin Friess, Anthrotech Inc., 503 Xenia Ave, Yellow Springs, OH 45387. Email: martin@anthrotech.net or mfriess@amnh.org

Waleed Gharaibeh, Department of Ecology and Evolution, State University of New York at Stony Brook, Stony Brook, NY 11794-5245. Email: Waleed@life.bio.sunysb.edu

Philipp Gunz, Institute for Anthropology, University of Vienna, Althanstrasse 14, A-1091 Vienna, Austria. Email: philipp.gunz@univie.ac.at

Katerina Harvati, Department of Anthropology, New York University, New York, NY. Current address: Max-Planck-Institute for Evolutionary Anthropology, Deutscher Platz 6, D-04103 Leipzig, Germany. Email: harvati@eva.mpg.de

Richard L. Jantz, Department of Anthropology, 250 South Stadium Hall, University of Tennessee, Knoxville, TN 37996. Email: rjantz@utk.edu

Johann Kim, Vertebrate Paleontology, American Museum of Natural History, 79th St. at Central Park West, New York, NY 10024. Email: johann.kim@univie.ac.at

Erin H. Kimmerle, Department of Anthropology, 250 South Stadium Hall, University of Tennessee, Knoxville, TN 37996. Email: ehk121@utk.edu

Subhash R. Lele, Department of Mathematical and Statistical Sciences, University of Alberta, Edmonton, Alberta, Canada T6G 2G1. Email: slele@ualberta.ca

Ashley H. McKeown, Department of Anthropology, 250 South Stadium Hall, University of Tennessee, Knoxville, TN 37996. Email: amckeown@utk.edu

Kieran P. McNulty, Department of Sociology & Anthropology, Baylor University, One Bear Place #97326, Waco, TX 76798–7326. Email: kieran_mcnulty@baylor.edu

Philipp Mitteroecker, Institute for Anthropology, University of Vienna, Althanstrasse 14, A-1091 Vienna, Austria. Email: philipp.mitteroecker@univie.ac.at

Wesley Allan Niewoehner, Department of Anthropology, California State University—San Bernardino, 5500 University Parkway, San Bernardino, CA 92407. Email: wniewoeh@csusb.edu

Hermann Prossinger, Institute for Anthropology, University of Vienna, Vienna, Austria. Email: hermann.prossinger@univie.ac.at

David Paul Reddy, Radio Logic, Incorporated, PO Box 9665, New Haven, CT 06536-0665. Email: reddy@amnh.org

Joan T. Richtsmeier, Department of Anthropology, The Pennsylvania State University, University Park, PA 16802 and Center for Craniofacial Development and Disorders, The Johns Hopkins University, Baltimore, MD 21205. Email: jta10@psu.edu

Antonio Rosas, Atapuerca Research Team, Department of Paleobiology, Museo Nacional de Ciencias Naturales, José Abascál 2, 28006 Madrid, Spain. Email: arosas@mncn.csic.es

H. David Sheets, Department of Physics, Canisius College, 2001 Main St., Buffalo, NY 14208. Email: sheets@canisius.edu

Michelle Singleton, Department of Anatomy, Midwestern University, 555 31st Street, Downers Grove, IL 60515. Email: msingl@midwestern.edu

Dennis E. Slice, Institute for Anthropology, University of Vienna, Althanstrasse 14, A-1091 Vienna, Austria. Email: dslice@morphometrics.org

Andrea B. Taylor, Doctor of Physical Therapy Division, Department of Community and Family Medicine, and Department of Biological Anthropology and Anatomy, Duke University Medical Center, Box 3907 Durham, NC 27710. Email: andrea.taylor@duke.edu

Daniel J. Wescott, University of Missouri—Columbia, Institute for Anthropology, 107 Swallow Hall, Columbia, MO 65211. Email: WescottD@missouri.edu

CHAPTER ONE

Modern Morphometrics

Dennis E. Slice

INTRODUCTION

The quantification of human proportions has a long history. As far back as the Middle Kingdom (*c.* 1986–1633 BC), Egyptian artisans used square grids and standard proportions to produce consistent depictions of human (and other) figures, even establishing different formulae for males and females (Robins, 1994) (Figure 1). The German anatomist Johann Sigismund Elsholtz formalized the scientific measurement of living individuals, "anthropometry," in his 1654 Doctoral dissertation (Kolar and Salter, 1996), and his particular interest in symmetry would appeal to many present-day anthropologists and general biologists. From the 19th century to the present day, the measurement and analysis of human beings and their skeletal remains have been a central theme in anthropology, though not always with beneficent motivation (e.g., Gould, 1981). During this time, anthropologists have often taken advantage of the state-of-the-art in statistical methodology, but they have not been just passive consumers of technological innovation. Indeed, pervasive interest in our own species, its artifacts, and our closest relatives has motivated and contributed much to the development of statistical methods that are now taken for granted in areas far afield from anthropology. The early work of the biometric laboratory established by Galton and Pearson bears witness to the vital

Dennis E. Slice • Institute for Anthropology, University of Vienna, Vienna, Austria.

Modern Morphometrics in Physical Anthropology, edited by Dennis E. Slice.
Kluwer Academic/Plenum Publishers, New York, 2005.

1

Dennis E. Slice

Figure 1. Nakht and wife (New Kingdom: *c.* 1570–1070 BC). Note same number of gridlines (18) from soles of feet to hairline and different waist heights for male and female figures. Grid completed from surviving traces. Drawing by Ann S. Fowler in Robins (1994). Reproduced by permission of the author.

interplay between the development of statistical methodology and anthropological research (e.g., Mahalanobis, 1928, 1930; Morant, 1928, 1939; Pearson, 1903, 1933).

This dynamic interaction between physical anthropology and statistical development continues today as new methods of shape analysis are inspired by anthropological problems. In turn, the availability of new morphometric tools opens new avenues of research or offers more powerful alternatives to traditional methods. The current volume provides a snapshot of this state of affairs in the early 21st century. Contributions include speculations on new directions in morphometrics, the development and extension of tools for shape analysis, and illustrations of how the latest morphometric methods have provided better and more powerful means to address basic research questions.

In this introductory chapter, I have tried to provide an overview of the basic terminology, concepts, and methods relating to the vast field that is modern morphometrics. My intention is to both relieve individual authors of having to reiterate methodological summaries and to provide an accessible introduction to students and researchers new to the field. Additional information to any level of technical detail can be found through the cited literature. Adams et al. (2004) provide a similar summary from a slightly different perspective. The reviews by Bookstein (1993, 1996) and Reyment (1996) include interesting and valuable historic components in addition to useful technical information.

We begin with some definitions.

DEFINITIONS

The field of morphometrics brings with it a plethora of terms and concepts that are seldom part of a biological, or even mathematical, curriculum. Slice et al. (1996) and the updated online version available at the Stony Brook website (see below) provide definitions of many of these. Here we present only the most fundamental terms necessary for an appreciation of modern morphometric methods.

Shape—the geometric properties of an object that are invariant to location, scale, and orientation.

Shape is the property about which we are most concerned here, and the definition contains two important points. The first is that we are interested in the geometric properties of an object. By focusing on geometry, we exclude properties, such as color and texture, that would otherwise meet the invariance requirements set out above (but see the Chapter 2 by Bookstein and Chapter 7 by Prossinger, this volume).

The second point is the invariance to location, scale, and orientation. By this we confine ourselves to geometric properties that do not change if the position or orientation of the specimen changes and, furthermore, would not change with the magnification or reduction of the object. This can be achieved either by the use of invariant measures, such as distance ratios or angles, or through methods that register all data into a common coordinate system, for example, the Procrustes superimpositions. In the latter case, the parameters for location, orientation, and scale built into the superimposition models are referred to as *nuisance* parameters. This is, in fact, only a technical designation as the relationship between shape and these other sources of variation, especially and usually size, may be of scientific interest and not just sources of annoyance. Morphometricians use the term *form* to refer to data containing only size and shape.

Size measure—any positive, real-valued measure of an object that scales as a positive power of the geometric scale of the form.

For the most part, we are concerned here with linear measures of size, say $g(\mathbf{X})$, that can be characterized by:

$$g(a\mathbf{X}) = ag(\mathbf{X})$$

$\mathbf{X}$ is our data, $g(\mathbf{X})$ is our size measure, and a is some magnification factor. This equation means that if we compute our size measure for our original data and for the same data scaled by some factor, a, the size measure for the scaled data will be a times that of the original data. In more concrete terms, if we multiply our data by a factor of two, a proper, linear size measure will be doubled.

Size has long been recognized as an important component of the comparison of structures (e.g., Burnaby, 1966; Huxley, 1932; Mosimann, 1970). It tends to dominate the variability between sexes, populations, species, and even individuals, and the researcher is interested in methods for separating size variation from that due to other factors. Even though the above provides a precise definition for what is a proper size variable, there are any number of measures that are consistent with that definition, but each may behave differently in the presence of shape variation. For instance, the distance between two well-defined points on an object is a proper size measure, but for the same data set, different distances could indicate no changes, increases, or decreases in size when individuals or groups differ in shape. The question of which size measure to use is only partially answerable in that under certain circumstances some size measures have optimal properties that can be used to argue for their use, for example, centroid size described below. In other situations, or depending upon the ultimate research focus, a case can be made for other measures, for example, the cube root of body weight in allometry studies.

Shape variable—any geometric measure of an object that is invariant to the location, scale, and orientation of the object.

Shape variables are the grist for the analytical mill that will be used to answer research questions. Coordinates of well-defined points, sufficient sets of distances between such points, the coordinates of points used to sample an outline, and angular differences used to encode the arc of a curve are all proper shape variables so long as they possess, or have been processed to achieve, the requisite invariances and capture geometric information about the structures for which they have been defined.

Geometric morphometrics—the suite of methods for the acquisition, processing, and analysis of shape variables that retain *all* of the geometric information contained within the data.

Geometric morphometrics brings together all of the acquisition, processing, analysis, and display methods for the study of shape that is characteristic of modern morphometric methods. Generally attributed to Les Marcus, to whom this volume is dedicated, and first used in print, perhaps, by Rohlf and Marcus (1993), this term is specifically meant to represent those methods that rigorously adhere to the exhaustive acquisition and analysis of shape information as defined above. This distinguishes these methods from what have been referred to as "traditional" morphometric methods that do not necessarily capture or retain sufficient information to reconstruct the spatial relationships among structures by which the measurements are defined. A key benefit to the use of geometric morphometric methods is that since all geometric information is retained throughout a study, results of high-dimensional multivariate analyses can be mapped back into physical space to achieve appealing and informative visualizations that are frequently not possible with alternative methods. The current volume is filled with examples.

MORPHOMETRIC DATA

The specific variables used in a morphometric analysis are chosen based on the question being investigated, the material under study, the equipment available for data acquisition, and to a greater or lesser extent the biases or experience of the researcher. However, there are several general classes of variables that are most frequently used for shape analysis. These classes each have their own benefits and/or limitations and admit different types of processing, analysis, and interpretation. Many of these are illustrated in Figure 2 and discussed below.

One thing that these different types of variables have in common is the assumption of the identity of data recorded for each individual, for example, the width of my head is the "same" variable as the width of your head. This implicit sameness of measurements may apply to individual variables, like head width,

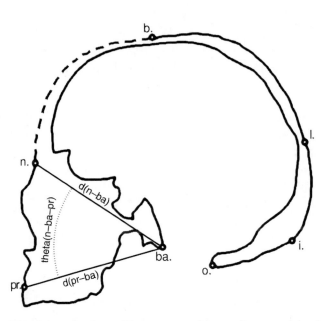

Figure 2. Morphometric data. Distances, angles, outlines, and landmarks: b.—bregma, n.—nasion, pr.—prosthion, ba.— basion, o.—opisthion, i.—inion, l.—lambda. See text for details.

but must sometimes be extended to suites of variables, such as an ordered series of points recorded along a curve where several variables (the coordinates of the points) are used to characterize a single structure of interest. This distinction has implications in the processing of data and the interpretation of results. This issue is discussed in greater detail as it arises in subsequent sections.

Distances, Ratios, and Angles

Distances are perhaps the oldest and most familiar variables used for morphometric analysis. They are measured by ruler, caliper, or other specialized device either between well-defined points, such as nasion, basion, and prosthion illustrated in Figure 2, or according to some rule, such as maximum cranial breadth (Howells, 1973). They may be linear, straight-line distances, or they may be curvilinear as, for example, the arclength of the frontal bone indicated by the dashed line in the figure.

Distances have the advantage of being inherently independent of orientation and position. Size is the only information that must be removed from a set of distances to achieve an analysis of shape. One way to meet this invariance is through the use of ratios or indices that encode the *relative* magnitudes of two distances. In Figure 2, one could use as a variable for shape analysis the ratio of the distance between nasion and basion, "d(n-ba)," to the distance between prosthion and basion, "d(pr-ba)," ($\times 100$). This is called the gnathic, or alveolar, index (e.g., Howells, 1989; Hanihara, 2000) and encodes some information about the shape of the face. However, it also introduces the statistical shortcomings associated with ratios (Atchley and Anderson, 1978; Atchley et al., 1976; Pearson, 1897).

A significant problem with distances is that unless they are carefully selected, one may not be able to determine the relative locations of all the endpoints of the measurements, and thus, they can omit information about the shape of the structure. In the current example, the distances between nasion and basion and between prosthion and basion (and their ratio) lack information about the positions of nasion and prosthion relative to each other. In fact, there are any number of face shapes that could have same nasion–basion and prosthion–basion distances. A way to address this deficiency is to add to our variables the angle formed by nasion–basion–prosthion, as shown in the figure as "theta(n-ba-pr)." Angles have the quite desirable property of being invariant not only to location and orientation as are distances, but also are invariant to size. The combination of the two distances and the angle fixes the relative positions of the three points up to a reflection, but mixing variables of different units may cause problems in multivariate analyses (like principal components analysis discussed below) that utilize information about the variances and covariances of variables. In such cases, one must resort to the standardization of the data and the analysis of correlations (not covariances). This introduces another level of abstraction between the analysis and the original specimens.

A better solution would be the addition of another distance variable—the distance between nasion and prosthion. This new set of three distances completely fixes the relative positions of the three points up to a reflection, is invariant to position and orientation, but still contains size information.

A shortcoming of both these solutions is that the selection of a sufficient set of variables to fix the shape of a structure becomes more difficult and tedious as the number of anatomical points or distance definitions increases. There is also the additional, unfortunate situation that such data (even if carefully selected to fix geometry) might yield multivariate summaries (sets of distances) of anatomy that are impossible to realize in space of the original data, that is, in two or three dimensions (Lele and Richtsmeier, 2001; Rao and Suryawanshi, 1996).

Landmark Coordinates

Cartesian coordinates are another type of data that can address many of the problems of distances and angles, but they introduce new ones of their own. They are, in fact, a special set of distances—signed distances of specified points from a set of mutually perpendicular axes. When these points are anatomical structures on a specimen, they are often called landmarks and are frequently the same points used to define traditional distances or angles.

The collection of landmark coordinates can proceed either by the direct recording of point locations on the specimen using specialized digitizing hardware or by the use of software operating on representations of the specimen like digital images (two dimensional) or medical imaging data (three dimensional).

Figure 2 shows a number of landmarks familiar to physical anthropologists, but as presented their coordinates are undefined. We must first establish a coordinate system with respect to which we can record the positions of the landmarks. Once recorded, the advantages and efficiency of coordinates become apparent as every possible distance and every possible angle that could be defined using these landmarks can be computed using classical geometry and elementary trigonometry.

The analysis of landmark coordinates has its own problems, however. These arise because it is difficult, if not impossible, to define a biologically meaningful set of axes with which to record them. Attempting to define such axes with respect to anatomy, such as in the use of the Frankfurt orientation, nonlinearly transfers variability in the anatomical structures used to define the axes to the landmark coordinates. Instead, coordinate data are usually collected with respect to some convenient, but arbitrary, axes, and these axes are unique with respect to individual specimens. Coordinates so obtained thus have encoded in them orientation and location both with respect to the different axes used in their collection, and, equivalently, the different positioning of the specimens during data collection. The estimation and extraction of these nuisance parameters are important steps in modern, coordinate-based morphometrics, and various approaches to this problem are discussed below.

Besides the fundamental problem of coordinate comparability, landmark coordinates also differ in the quality of the information they encode. This has been codified to a degree by Bookstein's (1991) classification of landmarks as Type I, II, or III. Type I landmarks are those defined with respect to discrete juxtapositions of tissues, such as triple points of suture intersections, Type II landmarks are curvature maxima associated with local structures usually with biomechanical implications, and Type III landmarks are extremal points, like the endpoints of maximum length, breadth, etc., defined with respect to some distant structure. Of these, the two- or three-dimensional locations of Types I and II are most often fully defined with respect to local morphology, and all dimensions are more-or-less biologically informative. Type III landmarks, however, are "deficient" in that they contain meaningful information only in line with the remote defining structure. Variation orthogonal to this direction has a substantial arbitrary component. A similar situation arises in the analysis of outlines discussed later.

Landmark coordinates afford us the opportunity to examine two topics that are more difficult to consider when dealing with other types of morphometric data. The first is the importance of triangles. Triangles, triplets of points, are the simplest geometric structure to have shape. A single point has only location, and a line segment defined by the two endpoints is completely described by its location, orientation, and length (size). Triangles have all of these attributes plus an additional component that is shape.

The second topic made more accessible by landmark coordinates is the consideration of the dimensionality of shape variation, that is, the number of dimensions (degrees of freedom) necessary to represent shapes. The triangle's simplicity makes it a good starting point. Any triangle in a plane requires only six numbers (the coordinates) for its complete geometric description. One degree of freedom is attributable to translation along each axis, one to scaling, and another to orientation. This results in $6 - 2 - 1 - 1 = 2$ degrees of freedom for variation in shape. Note that with only two dimensions left to encode shape we have a chance to graphically explore the structure of this space. This is one reason triangles are such an important part of research into shape theory.

Since the number of nuisance parameters are fixed for planar configurations, the general formula for the dimensionality of shape variation for p points in two dimensions is $2p - 2 - 1 - 1 = 2p - 4$. For three dimensional data the formula is $3p - 3 - 3 - 1 = 3p - 7$, where we have three dimensions in which to translate, three angles of rotation, but still one scale parameter to estimate. To be completely general, the dimensionality of shape variation for an arbitrary number of points, p, in any number of dimensions, k, is $pk - k - k(k-1)/2 - 1$. We return to this topic in our discussion of shape spaces.

Outlines

Some anatomical structures, like brow ridges, orbital rims, or the foramen magnum, do not lend themselves to characterization by well-defined, discrete landmarks. They are, instead, partial or complete boundaries of another structure or traces of local maximum surface curvature that are continuous, one-dimensional features.

Such curve or outline data are usually represented by ordered sets of discrete point coordinates. These can be superficially similar to landmark data, but they are conceptually quite different. It is the entire underlying continuous structure that is to be compared across specimens and not the individual points used to characterize the outline. This deceptive similarity is even more reinforced when equal numbers of points are used to sample individual outlines. In fact, the coordinates of outline points only contain one piece of useful information—the position of the outline in the region around the sample point relative to its position at similar surrounding points.

This distinction usually requires special methods of analysis, and different methods are available for different types of outlines. In general, outlines can be classified as simple or complex, where simple outlines can be expressed as a single-valued function of some other variable, say $y = f(x)$, and complex outlines cannot. Outlines can also be closed or open, where closed outlines have no beginning or end and can be traced repeatedly without lifting the pencil or reversing direction on the outline, while open outlines have distinct starting and ending points. Rohlf (1990) provides a good survey of outline data and earlier methods for their analysis.

Two types of outlines are shown in Figure 2. The arc between bregma and nasion is the mid-sagittal profile of the frontal bone and could be treated as an open, simple curve. The mid-sagittal outline of the entire cranium in the figure would be a complex, closed outline.

Surfaces

Surfaces, two-dimensional regions within some defined boundaries, are not simple extensions of outlines. An immediate problem introduced in the transition from the analysis of outlines to that of surfaces is that the concept of the one-dimensional ordering of sample points is lost. Thus, the analysis of surfaces requires unique morphometric methods that are few. Niewoehner (Chapter 13, this volume) uses projected grids to construct sets of points with which to sample surfaces, and the chapter by Gunz et al. in this volume (Chapter 3) provides some new possibilities for analyzing surfaces within the superimposition framework established for landmarks.

TRADITIONAL METHODS

The bulk of biological literature dealing with shape analysis has used methods that are today called "traditional" morphometrics (Marcus, 1990). These methods are characterized by the application of multivariate statistical procedures to collections of distances, distance ratios, and/or angles gathered to sample the shape of an object. Thus, such approaches are also known by the appellation "multivariate morphometrics" (Blackith and Reyment, 1971).

As described earlier, the distances, ratios, and angles used in traditional, or multivariate, morphometrics more often than not fail to encode all of the geometric information about the biological structures by which they are defined. Without recording and maintaining this geometry, morphometric analyses cannot provide an exhaustive assessment of shape variability or differences and may unnecessarily neglect important, but unanticipated, geometric relationships among the structures under investigation. Furthermore, such incomplete analyses make it difficult to produce graphical depictions of results that can be related to the actual physical specimens.

Bookstein et al. (1985) attempted to remedy these limitations with the development of the truss—a systematic series of measurements designed to fix the geometry (up to a reflection) of the anatomical landmarks. Another method for dealing with the same problem is to simply analyze all possible distances between the landmarks of interest. This is the basis for the Euclidean Distance Matrix Analysis (EDMA) methods described in the section on coordinate-free methods.

SUPERIMPOSITION-BASED MORPHOMETRICS

The deficiencies of distance- and angle-based morphometrics can be addressed by the direct analysis of the coordinates of the landmarks by which traditional measurements are, or could be, defined. Raw coordinates, however, also contain information about the location, orientation, and size of the configuration of landmarks that must be factored out or subtracted off to achieve an analysis of shape. Proper geometric morphometric methods do not simply discard the information in these nuisance parameters, but, rather, sequester it (a phrase due to Bookstein) into a separate suite of non-shape variables available for later consideration.

One way of partitioning the total variation of raw coordinate data into shape and non-shape components is by superimposing all of the configurations within a common reference system and scaling them to a common size. Various ways to do this are described in the following sections.

The data used to illustrate some of the methods are presented in Figure 3 (left) that shows a set of five landmarks on a gorilla scapula whose two-dimensional coordinates were recorded from scanned photographs (provided by Andrea Taylor) using tpsDig (Rohlf, 2001). The complete data set consists of coordinates for 52 male and 42 female adult, west African lowland gorillas, and we seek to compare sexes with respect to scapular shape. Taylor and Slice (Chapter 14, this volume) use similar data (though not exactly the same landmarks) to investigate biomechanical predictions of scapular shape in *Pan* and *Gorilla*.

Two-Point Registration

There are any number of ways one could define measures of size, location, and orientation for a particular set of landmark coordinate data. For two-dimensional data one could simply specify the coordinates of one landmark to define location and the length and direction of a line segment, or baseline, between that point and another to define orientation and scale. This is the "two-point registration" extensively developed by Bookstein (1986, 1991). It is also often called base-line registration or edge-matching.

The operations involved in the two-point registration of planar configurations can be expressed quite concisely in complex notation (Bookstein, 1991), but here and throughout this chapter I use the more general matrix notation that is readily extended to three- and higher-dimensional data. Let

$$\mathbf{X} = \begin{bmatrix} x_1 & y_1 \\ x_2 & y_2 \\ x_3 & y_3 \end{bmatrix}$$

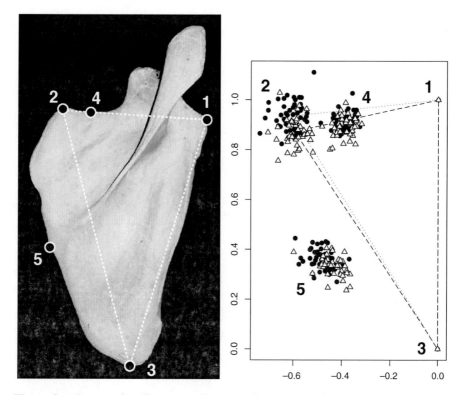

Figure 3. Data used to illustrate various morphometric methods. Left, gorilla scapula and landmarks. Right, boths males (circles, $n = 52$) and females (triangles, $n = 42$) superimposed using two-point registration with landmarks 1 and 3 as the baseline. Mean triangles between extremal angles shown for males (dotted) and females (dashed). Superimposition done with Morpheus et al. (Slice, 1998). Plot generated with R (http://www.r-project.org).

be the x and y coordinates of the three labeled vertices of a triangle, and let us specify that we will use points in the first and second position for our registration. Subtracting off location as encoded in the coordinates of the first point we get

$$\mathbf{X'} = \mathbf{X} - \mathbf{1t}$$

$$= \begin{bmatrix} x_1 & y_1 \\ x_2 & y_2 \\ x_3 & y_3 \end{bmatrix} - \begin{bmatrix} 1 \\ 1 \\ 1 \end{bmatrix} \begin{bmatrix} x_1 & y_1 \end{bmatrix}$$

$$= \begin{bmatrix} x_1 & y_1 \\ x_2 & y_2 \\ x_3 & y_3 \end{bmatrix} - \begin{bmatrix} x_1 & y_1 \\ x_1 & y_1 \\ x_1 & y_1 \end{bmatrix}$$

$$= \begin{bmatrix} 0 & 0 \\ x_2 - x_1 = x_2' & y_2 - y_1 = y_2' \\ x_3 - x_1 = x_3' & y_3 - y_1 = y_3' \end{bmatrix}$$

Next, we rotate the baseline so that it is in some standard alignment that is usually taken to be coincident with the positive x-axis:

$$\mathbf{X}'' = \mathbf{X}'\mathbf{H}^t = \begin{bmatrix} 0 & 0 \\ x_2'' & 0 \\ x_3'' & y_3'' \end{bmatrix}$$

$$\mathbf{H} = \begin{bmatrix} \cos\theta & \sin\theta \\ -\sin\theta & \cos\theta \end{bmatrix}$$

$$\theta = \cos^{-1}\left((\mathbf{x}_2' \cdot \mathbf{e}_1) / \left(\sqrt{\mathbf{x}_2' \cdot \mathbf{x}_2'} \sqrt{\mathbf{e}_1 \cdot \mathbf{e}_1} \right) \right)$$

$$\mathbf{x}_2' = \begin{bmatrix} x_2' & y_2' \end{bmatrix}^t$$

$$\mathbf{e}_1 = \begin{bmatrix} 1 & 0 \end{bmatrix}^t$$

Finally, we divide by the length of the baseline, which because of the standardizations so far, is simply x_2'':

$$\mathbf{X}''' = \frac{1}{x_2''}\mathbf{X}'' = \begin{bmatrix} 0 & 0 \\ 1 & 0 \\ x_3''/x_2'' = x_3''' & y_3''/y_2'' = y_3''' \end{bmatrix}$$

These operations are applied separately to all triangles in a sample. Notice that after these transformations, the coordinates of the first two landmarks are fixed at (0, 0) and (0, 1) for any triangle in the sample. Thus, all information about the shape of a triangle is encoded in the coordinates of the third landmark (x_3''', y_3'''). These are known as Bookstein coordinates or Bookstein shape coordinates for triangles. However, this same method was used by Galton (1907) to characterize facial profiles.

For more than three landmarks, one simply extends the rows of the $\mathbf{X}$ and $\mathbf{1}$ matrix and applies the same transformations to any additional landmarks. For each additional landmark, the dimensionality of the shape variation is increased by two, since the number of dimensions lost due to the standardizations is fixed.

One might ask, how does the choice of baseline effect the resulting shape coordinates and can such effects impact the findings of the analysis? As Bookstein (1991) points out, for small shape variation the use of different baselines effects mainly translations, rotations, and rescalings of the scatter of shape coordinates that will not effect multivariate statistical analysis. The magnitude of the rescaling effect is a function of the relative lengths of alternative baselines, and problems do occur for baselines approaching zero length. Rohlf (2000) showed, for instance, that the power of statistical tests for group differences is severely reduced when the landmarks defining the baseline are nearly coincident. This is not a serious problem in practice, since investigators are likely to make more reasonable choices of baseline that span the object under consideration.

It is somewhat difficult to extend the Bookstein coordinates to three-dimensional data. One possibility was implemented by Slice (1994) and described by Dryden and Mardia (1998). This involves the specification of a third point to establish a baseplane. The data are then transformed, as before, so that the first point is situated on the origin and the vector between first and second points is coincident with the positive x axis. Finally, the configurations are rotated such that the plane defined by the three base points is coincident with the x, y plane with the third point on the positive side of the y axis. The algebra is a bit more complicated than in the two-dimensional case, but the principles remain the same. These operations result in the variation at the third landmark not being completely removed, yet it is not fully three dimensional, either. There can be up to two dimensions of variation at this point, and the interpretation of such results becomes much more difficult. As a result, this method of three-dimensional shape variable construction has been little used.

An example of using two-point registration to construct Bookstein shape coordinates is illustrated in Figure 3 (right), that shows the registration of the sample of male and female gorilla scapula data mentioned above. The triangle between the extreme angles is highlighted, but the entire five-landmark suite of data can be analyzed through the coordinates of the three non-baseline points. Apparent differences between male and female lowland gorillas can be seen in the differences in the locations of sex-specific scatter at the various landmarks.

Procrustes Superimposition

Instead of using just one or two landmarks to estimate the requisite parameters, one might consider methods to incorporate information from all of the landmarks in a configuration. This is the basis for the most widely used method in geometric morphometrics today—the Procrustes superimposition. Major theoretical investigations into Procrustes-based methods are due to David Kendall (1984, 1985, 1989), who was motivated, in part, by questions in archeology (Kendall and Kendall, 1980). Specifically, it was proposed that megalithic sites in England were linearly situated. The statistical question, then, is how can one tell if sets of points are more linearly arranged than one would expect from random placement, and Kendall's approach was to develop the theoretical constructs to test if triangles formed by triplets of these points (sites) were flatter than expected by chance. Since there is no obvious correspondence between sites like there is in anatomical landmarks, Kendall's investigations included a component allowing for the permutation of vertices.

Kendall's work resulted in the deep and elegant mathematical results that form the basis of much of modern shape theory. Kendall's and Bookstein's research intersected in Bookstein's (1986) *Statistical Science* paper for which Kendall was a discussant. Goodall (1991) provides an extensive treatment of the practical and theoretical aspects of the Procrustes methods, and Small (1996) and Dryden and Mardia (1998) are recent statistical texts.

Much theoretical work on Procrustes methods is due to Kendall, but much applied work preceded his endeavors. The earliest known matrix formulation of the two object Procrustes method is due to Mosier (1939) for psychometric application. The matrix formulation of Generalized Procrustes Analysis (see below) for superimposing samples was set forth by Gower (1975), who was concerned with comparing the multivariate scoring of carcasses by meat inspectors. Similar algebraic and geometric comparisons of landmark configurations for anthropological purposes were used by Sneath (1967), and Cole (1996) points out that the earliest use of this technique was, in fact, suggested by the eminent anthropologist Franz Boas (1905) to address shortcomings of the Frankfurt orientation.

Procrustes superimposition is a least-squares method that estimates the parameters for location and orientation that minimize the sum of squared distances between corresponding points on two configurations. A least-squares estimate for scale is also available, but its use does not lead to symmetric results between configurations of different sizes, so all specimens are most often scaled to a standard size. When the least-squares estimate of scale is used, the analysis is called a full Procrustes analysis. It is termed a partial Procrustes analysis in the usual case when configurations are scaled to a common size. The square root of the sum of squared coordinate differences after superimposition is a measure of the shape difference between configurations.

For a particular configuration of p landmarks in k dimensions written as a $p \times k$ matrix, $\mathbf{X}_1$, we model difference in coordinate values relative to a mean-centered reference configuration of corresponding landmarks, $\mathbf{X}_0$, by

$$\mathbf{X}_1 = \frac{1}{r_1}(\mathbf{X}_0 + \mathbf{E}_1)\mathbf{H}_1^t + \mathbf{1t}_1,$$

indicating that numerical values in $\mathbf{X}_1$ differ from those in $\mathbf{X}_0$ by scaling, $1/r_1$, rotation, $\mathbf{H}_1^t$, translation, $\mathbf{t}_1$, and actual shape differences, $\mathbf{E}_1$, which would include measurement error. $\mathbf{1}$ is a $p \times 1$ matrix of ones. Some rearrangement leads to

$$r_1(\mathbf{X}_1 - \mathbf{1t}_1)\mathbf{H}_1 = \mathbf{X}_0 + \mathbf{E}_1$$

that exposes the shape differences in $\mathbf{E}_1$. The estimates of the requisite parameters are

$$\mathbf{t}_1^t = \frac{1}{p}\mathbf{X}_1^t\mathbf{1} = (\bar{x}_1, \bar{y}_1)^t,$$

$$\mathbf{H}_1 = \mathbf{V}_1\mathbf{\Sigma}_1\mathbf{U}_1^t, \quad \text{where } \mathbf{X}_1^t\mathbf{X}_0 = \mathbf{U}_1\mathbf{D}_1\mathbf{V}_1^t,$$

and

$$r_1 = \frac{1}{\sqrt{\text{tr}\left((\mathbf{X}_1 - \mathbf{1t}_1)^1(\mathbf{X}_1 - \mathbf{1t}_1)\right)}} = \frac{1}{\text{CS}_1}$$

It can be shown fairly easily (Boas, 1905; Sneath, 1967) that to superimpose two configurations to minimize the sum of squared distances between landmarks one need only translate the configurations so that the average coordinate in each dimension is the same for both specimens. The exact coordinates of the average location are irrelevant, so both configurations are centered on the origin as indicated in the $\mathbf{t}_1$ vector above.

$\mathbf{H}_1$ is an orthogonal matrix that rigidly rotates the $\mathbf{X}_1$ configuration about the origin to minimize the sum of squared distances between its landmarks and the corresponding landmarks of $\mathbf{X}_0$. It is computed as the product of three matrices, of which $\mathbf{U}$ and $\mathbf{V}$ are derived from the singular value decomposition of the product of the transpose of $\mathbf{X}_1$ and $\mathbf{X}_0$. The matrix $\mathbf{\Sigma}$ is a diagonal matrix with positive and negative ones along the diagonal, each having the same sign as the corresponding element of $\mathbf{D}_1$, which is also diagonal. This substitution is to prevent any stretching of $\mathbf{X}_1$ to achieve a better (in a least-squares sense) fit to $\mathbf{X}_0$—the non-unity elements of $\mathbf{D}_1$ correspond to stretching or compression. As expressed, the rotation could also effect a reflection to improve fit. If such is not desired, $\mathbf{H}_1$ must be tested to see if it is reflecting (has a negative determinant) and adjusted accordingly (change the sign of the diagonal element of $\mathbf{\Sigma}$ in the same position as that of the smallest value of $\mathbf{D}_1$).

The scalar, r_1, is the scale factor and is most frequently computed as the inverse of the configuration's centroid size (CS). Each configuration involved in an analysis is similarly scaled so that all configurations have a standard size—unit CS. CS is used because small, circular, random variation at individual landmarks does not generate a correlation between shape and this measure of size (Bookstein, 1991). It is also the length of the vector containing all of a centered specimen's landmark coordinates written out as a *pk* row or column.

The above formulae can all be derived using matrix notation, and they apply to two-, three-, or higher-dimensional configurations of landmarks. One simply appends columns to the various $\mathbf{X}_i$ and other matrices to allow for the additional coordinates.

So far, the discussion has focused on the superimposition of one configuration of landmarks onto another, specified configuration. Such fitting of two specimens has been called ordinary Procrustes analysis, OPA (Goodall, 1991). Researchers, though, are usually interested in the analysis of samples of more than two specimens. In such cases, it is usual to relate members of a sample to their mean. The problem is that meaningful mean coordinates cannot be computed prior to superimposition and superimposition requires knowledge of the mean configuration. The solution to this is an iterative process in which any specimen is initially selected to stand for the mean. All of the configurations in the sample are fit to that reference, then a new mean is computed as the arithmetic average location of the individual landmarks in the sample and scaled to unit CS. The process is repeated, fitting the sample to the new estimate, and it is guaranteed to produce monotonically decreasing sum-of-squared deviations of the sample configurations around the estimated mean (Gower, 1975). The procedure is terminated when this sum-of-squares no longer decreases by some critical value or, equivalently, when the change in mean estimate from one iteration to the next is deemed negligible. The term used for the fitting of a sample onto an iteratively computed mean is generalized Procrustes analysis (GPA) (Gower, 1975), and it is the superimposition method used for most geometric morphometric studies.

Figure 4 shows the effects of the separate translation, scaling, and rotation steps of a GPA on the gorilla data. The triangles between the extreme angles of the scapulae are highlighted in the last panel for later reference, but the superimposition was based on all five landmarks. In general, one can see a "condensation" of the original variability into nearly circular, apparently unstructured variation around the mean landmark locations as the superimposition translates, scales, then rotates the configurations to minimize their sum of squared differences to the iteratively computed mean. Also, one can see subtle differences between the males and females in average landmark location that suggests sexual dimorphism in scapula shape (tested and visualized below). Both sexes were fit to the grand mean by generalized Procrustes superimposition, and the sex-specific means computed separately afterwards.

In this volume (Chapter 9), McKeown and Jantz compare the results of coordinate-based, generalized Procrustes analysis with those of traditional, distance-based data in the investigation of spatio-temporal affinities in samples of Native American crania.

Resistant Fit

One criticism of the Procrustes superimposition is of its use of a least-squares criterion in estimating the translation and rotation parameters (Seigel and Bensen, 1982). If one or a few landmarks are greatly displaced relative to the others in one specimen or in one sample, these localized differences would inflate the squared distance between configurations. The least-squares Procrustes methods, therefore, spread such large local differences across all of the landmarks to produce a number of smaller differences. This is called the Pinocchio effect referring to the shape of the puppet's head before and after lying. The only real difference would be in the length of the nose, but a Procrustes superimposition would suggest differences over the entire head.

An approach developed to help identify local differences is resistant fitting (Rohlf and Slice, 1990; Seigel and Bensen, 1982; Slice, 1993a, 1996). These are methods based on the use of medians or medians of medians (repeated medians) to estimate the translation, scale, and rotation parameters for superimposition. Seigel and Benson (1982) describe the method for pairs of two

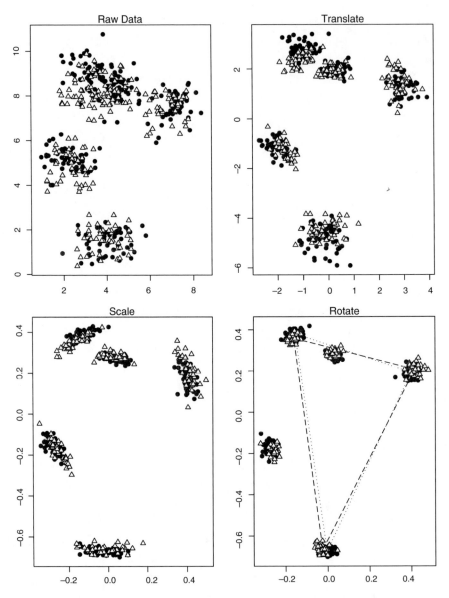

Figure 4. Generalized Procrustes superimposition of the scapula data. Upper left, raw data as digitized. Upper right, data after translation to the origin. Lower left, translated data after scaling to unit CS. Lower right, translated and scaled data after least-squares rotation of individual specimens to their iteratively-computed sample mean. Mean triangles between extremal landmarks shown for males (dotted) and females (dashed). Superimposition done with Morpheus et al. (Slice, 1998). Plots generated with R (http://www.r-project.org).

dimensional configurations, Rohlf and Slice (1990) extend the method to allow for the generalized resistant fitting of two-dimensional data, and Slice (1993a, 1996) develops methods for the generalized resistant fitting of three and higher dimensional samples based on similar work for two configurations by Siegel and Pinkerton (1982). Dryden and Mardia (1998) discuss other alternatives.

Despite its intuitive appeal, the median-based results do no allow nearly as sophisticated a theoretical development as the Procrustes methods (e.g., Dryden Mardia, 1998; Kendall, 1984, 1985; Small, 1996). The current consensus is that the Procrustes methods are to be preferred for use in statistical analysis. The resistant methods, though, can still be useful, especially through the comparison of the results of Procrustes and resistant superimpositions (e.g., Slice, 1996) and, like two-point registration, can be used for suggestive visualizations after statistical analyses based on Procrustes methods have been performed.

SHAPE SPACES

The square root of the sum of squared differences of corresponding landmark coordinates in two (partial) Procrustes-superimposed figures is equivalent to the distance between the tips of the two vectors containing all of the landmark coordinates for each of the two configurations. Given this "Procrustes distance," it is possible to ask about the geometry of a space in which the distance between all points representing shapes is that same distance. Such an inquiry was a major component of the theoretical work of Kendall (1984, 1985) who was able to fully describe the geometry of what is generally referred to as "Kendall's shape space."

Some of Kendall's key results are that the shape space of planar triangles is two-dimensional, but that it is non-Euclidean (curved) and isometric to the surface of a sphere of radius $\frac{1}{2}$. This is consistent with our earlier conclusion about the dimensionality of the shape space for triangles obtained by counting degrees of freedom. Because of the spherical geometry of shape space, Kendall and other mathematicians often prefer to work with great-circle, or Riemannian, distance, ρ, instead of the straight-line, or chord, partial Procrustes distance, d, but there is a simple relationship between the two, $\rho = 2\sin^{-1}(d/2)$ for two-dimensional configurations (Kendall, 1984).

An important corollary of Kendall's results for planar triangles is that it is possible to visualize shape space for such data. Figure 5 shows Kendall's shape

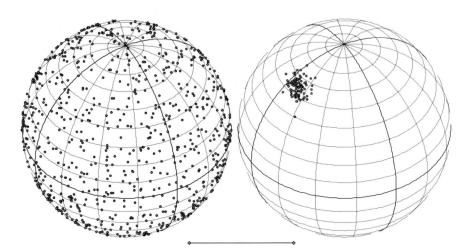

Figure 5. Representation of Kendall's shape space for triangles. Left, 2,000 random triangles generated by normal displacement of vertices from the origin. Right, 94 gorilla scapulae. Males and females not distinguished. Scale bar is 0.5 units. Plots created with tpsTri (Rohlf, 2002).

space for triangles. The plot on the left illustrates one of his other results—that triangles generated by the independent, normal displacement of points from the origin are uniformly distributed in shape space (Kendall, 1984). The positions of 2,000 such points are shown. The right plot illustrates another important point—that the biological variability of interest to researchers is usually concentrated in a relatively small area of shape space. The right panel shows the positions of the triangles formed by the extreme angles of the 94 gorilla scapulae. This is an important feature of biological material for the statistical analysis of shape data. It is also interesting to note that Bookstein shape coordinates are a special, stereographic projection of points in Kendall's shape space (Small, 1996).

Kendall's results are based on the shape distance between two configurations of landmarks. The situation is somewhat different for the generalized Procrustes analysis, where multiple configurations in a sample are fit to in iteratively-computed mean (Slice, 2001). In that case, the geometry of superimposed planar triangles is that of a hemisphere of unit radius. The key difference in the two geometries arises from the fact that in Kendall's shape space the distances between all points are Procrustes distances. With GPA, the specimens are individually superimposed onto the consensus. Only distances between individual specimens and the consensus equal the Procrustes distance on the GPA

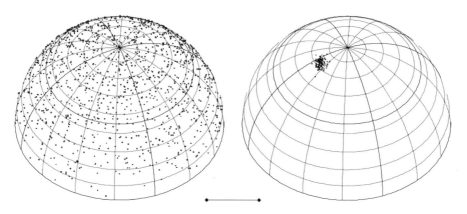

Figure 6. The generalized Procrustes hemisphere for triangles. Data are the same as in the previous figure. Scale bar is 0.5 units. Plots created with tpsTri (Rohlf, 2002).

hemisphere. Distances between specimens do not. For more than three land-marks in two dimensions, the geometry of GPA space is still the surface of a hyper-hemisphere. Like Kendall's shape space, though, the geometry for configurations in three or more dimensions is likely to be much more complicated (Dryden and Mardia, 1993; Small, 1996).

The GPA spaces for the random triangles and gorilla scapulae are illustrated in Figure 6. Rohlf (1999a) and Slice (2001) discuss the simple relationship between the Kendall's shape space, the GPA hemisphere, and various projections into linear tangent spaces (see section on Multivariate analysis). These relationships, however, hold only for planar triangles (Slice, 2001).

MULTIVARIATE ANALYSIS

The purpose of the superimposition methods is to register landmark configurations in a common coordinate system in which the coordinates of the landmarks can be used as shape variables. This is not an end in itself, but provides the researcher with suitable data to explore the structure of shape variation, assess the significance of differences in mean shapes, relate the observed variation to extrinsic factors, and the like. For this, the whole suite of multivariate methods familiar to traditional morphometrics is available for the analysis of differences and variation in superimposed landmark coordinates. For these methods, the student of morphometrics is free to consult standard texts on multivariate statistics, such as Johnson and Wichern (1982) or Krzanowski (1988). One especially useful text is Carroll and Green (1997) that focuses on the geometric interpretation of multivariate analyses. This nicely compliments the geometric theme at the core of modern morphometric analysis.

There is one caveat in the application of parametric multivariate methods to Procrustes-processed data. The theory underlying many multivariate methods assumes a linear, Euclidean space. We have seen, though, that the geometry of Kendall's shape space and that of generalized Procrustes analysis is non-linear, thereby violating this key assumption. One way around this problem is to analyze not the Procrustes coordinates, but their projection into a linear space tangent either to Kendall's shape space or the Procrustes hemisphere. Rohlf (1999a) and Slice (2001) describe and assess various projections (see also Dryden and Mardia, 1998 and Small, 1996). In general, an orthogonal projection from the GPA hemisphere to a linear space tangent at the sample mean seems to best preserve the distances between specimens, though for the relatively small variation found in most biological samples any of the reasonable alternative projections does a fairly decent job and, conversely, using no projection does not violate the assumptions of a linear space too badly.

Singleton in this volume (Chapter 15) uses principal components analysis and regression of Procrustes-superimposed landmarks to investigate allometric, functional, and phylogentic aspects of the shape of masticatory structures in cercopithicines.

An alternative to parametric methods with their restrictive assumptions is the use of nonparametric, randomization tests to effect similar tests (Bookstein, 1997). See, for instance, Manly (1997) for a discussion of randomization and related non-parametric tests. In the case of the gorilla data used so far, differences in mean shape between male and female gorillas are significant for both Bookstein's shape coordinates ($p = 0.001$) and GPA coordinates ($p = 0.001$) when judged by a simple randomization test that compares the observed between-group sum-of-squares to the same value for 999 random shufflings of group membership. Sex accounted for about 12% of total sample variation in Bookstein shape coordinates (Figure 3) and about 11% in GPA coordinates (Figure 4, lower right).

Similar permutation tests are used here by Wescott and Jantz (Chapter 10) to document recent, secular change in the craniofacial shape of Black and White Americans, and Bastir et al. (Chapter 12) use randomized version of partial least-squares analysis (Bookstein, 1991) to examine the integration of the cranium and mandible in hominoids.

VISUALIZATION

Once one has superimposed the configurations, computed means, quantified variation, and/or assessed significant differences or associations, one can take advantage of a geometric approach to morphometrics to generate visualizations of differences, associations, variability, etc. in the space of the original specimens. There are several ways to do this.

Vector Plots

Perhaps the simplest method for visualizing the results of a geometric morphometric analysis is with vector plots. This is done by taking the coordinates of the landmarks of a key configuration, say, a grand mean or the mean of one group in a two-group comparison, and drawing vectors from the landmark locations on that configuration to points specified by the results of your statistical analysis. For instance, say you are comparing two mean shapes, $\bar{X}_1$ and $\bar{X}_2$, which have been computed after fitting the members of both samples to their joint grand mean. The relevant vectors for display might then be the difference vectors between the two groups, $\Delta\bar{X}_{1,2} = \bar{X}_2 - \bar{X}_1$. One would then plot the landmarks of $\bar{X}_1$ and draw vectors from them to the points $\bar{X}_1 + \Delta\bar{X}_{1,2}$. Note that these will be just the locations of the points of $\bar{X}_2$. One can also exaggerate (or diminish) differences by multiplying the displacement matrix by some appropriate factor, for example, $2\Delta\bar{X}_{1,2}$ would double the difference between the two groups, but preserve the direction of the differences. For more than two groups one could plot all pairwise differences as vector differences or plot the differences between group means and the grand mean.

This type of plot is shown in Figure 7 where the shape difference between average male and female gorilla scapula from a five-landmark GPA are shown. The reference configuration is that of the average male, and the vectors (their length multiplied by a factor of two) point in the direction of the shape of the average female.

The machinery of geometric morphometrics and linear statistical analysis make it equally easy to plot results from analyses more complicated than mean comparisons. The results of familiar multivariate analyses like principal components analysis, canonical variates analysis, etc. are expressed as linear combinations of the original variables—the superimposed landmark

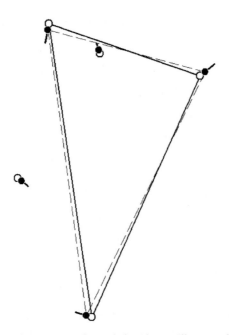

Figure 7. Difference between male and female gorilla scapulae shown as vectors (magnified ×2). Open circles are mean male landmark coordinates. Solid circles are mean female coordinates. Plot created with Morpheus et al. (Slice, 1998).

coordinates. Furthermore, these results are usually in the form of vectors scaled to unit length. To visualize, say, the shape variability captured by the principal component axis associated with the largest amount of variation (usually called the 1st principal component, the one with the largest eigenvalue), one takes the grand mean configuration and adds to it the coefficients of the first PC to generate the positions of the vector tips as described previously. The position of the tips of these vectors corresponds to the shape of a configuration displaced one unit in the positive direction along the first principal component (It is often not appreciated, but the positive/negative directions of individual components are perfectly interchangeable. They are defined only up to reflection). Likewise, the coefficients from a multiple regression of shape onto some other variables can be used to generate the predicted shape for any value of the independent variables, and this can be used as to define the tips in a morphometric vector plot.

One problem with vector plots is that it becomes irresistibly tempting to discuss them in terms of individual points moving or being displaced. Such statements are not justifiable simply on the basis of the plots, and this kind of information is generally unknowable given landmark coordinate data. Consider two triangles of different shapes. Which landmarks in one are in different relative locations compared to the other? One cannot say. The procedures used for shape analysis examine shape differences in their totality, not one landmark position at a time. So, is there a graphical device that can take into account the relative positions of all the landmarks? Yes, the thin-plate spline.

Thin-Plate Splines

The thin-plate spline was adapted for use in morphometrics by Bookstein (1989, 1991). It addresses both the problem of integrating information about the relative locations of all landmarks and the classic problem posed by D'Arcy Thompson (1942) of expressing shape differences between two specimens as a global mapping of the Cartesian space of one specimen into that of another. This latter goal, in fact, has a much longer history with Renaissance artists like Albrecht Dürer using deformed grids to express normal variation and methods of pictorial caricature (Bookstein, 1996b).

The theoretical justification for the method is rather complicated, but the algebra is relatively straightforward. First, one configuration, usually a group or grand mean, is used as a reference and the differences between the landmark locations and those of another specimen, the target, along each coordinate axis are processed separately. That is, one computes the required parameters for the differences in x coordinates between the two specimens, then the y coordinates, and so on. For each coordinate dimension, the differences between the two configurations are treated as displacements at right angles out of the plane of the reference configuration (for the two-dimensional case). The equations so derived are then recombined to express the totality of differences between the two configurations (Figure 8).

To achieve this, we need the coefficients for the equation:

$$f(x,y) = a_1 + a_x x + a_y y + \sum_{i=1}^{p} w_i U\left((x_i, y_i) - (x, y)\right) \tag{1}$$

This function maps a pair of coordinates, (x, y), to a scalar incorporating information about the possesive proximity to each of the p reference landmarks. What we seek are the coefficients such that that scalar at the positions of the landmarks in the reference configuration equals the heights above or below the plane that, in turn, correspond to the coordinate differences between the reference and the

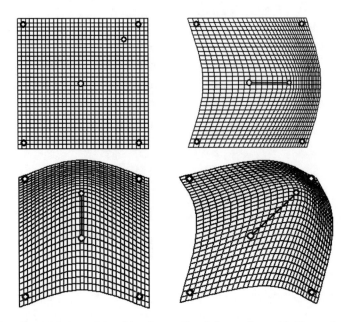

Figure 8. Construction of the thin-plate spline deformation grid. Shown in the upper, left are two five-landmark configurations differing only in the right and upward displacement of the central landmark on the target configuration. To produce a deformation grid for the difference, interpolation formulae are computed separately for the x displacement (upper, right) and the y displacement (lower, left), then combined (lower, right). Note, this construction works even though the configurations are not in Procrustes alignment. Plots created with Morpheus et al. (Slice, 1998).

target. Note that although we set this as a condition for the above equation, the resulting formula can still be applied to any position in the plane of the reference to interpolate heights at points not coincident with reference landmarks. With the addition of one more condition—that the resultant surface be the least bent of any surface passing through the specified heights at the locations of the reference landmarks—the derived surface will be the thin-plate spline. This specification comes from engineering where the equation is used to model the deformation of an infinite, infinitely-thin metal plate, hence the name.

To compute the coefficients for a configuration of p points in $k = 2$ dimensions meeting our requirements, we begin with the construction of a partitioned matrix:

$$\mathbf{L} = \left[\begin{array}{c|c} \mathbf{P}_{p \times p} & \mathbf{Q}_{p \times 3} \\ \hline \mathbf{Q}^t_{p \times 3} & \mathbf{0}_{3 \times 3} \end{array} \right]$$

where $\mathbf{P}$ is symmetric with zeros on the diagonal and off-diagonal elements $p_{i,j} = p_{j,i} = U(r_{i,j}) = r_{i,j}^2 \ln(r_{i,j}^2)$, where $r_{i,j}$ is the Euclidean distance between points i and j of the reference specimen. $\mathbf{Q}$ is a matrix of the landmark coordinates of the reference specimen augmented by an initial column of ones, and $\mathbf{0}$ is a matrix of zeros.

The required coefficients are obtained from the equation:

$$\mathbf{L}^{-1}\mathbf{Y}_{p+3,1} = (\mathbf{w}|a_1, a_x, a_y)^t \tag{2}$$

where $\mathbf{Y}$ is the vector of differences between the reference and the target specimen along the axis currently being considered (the constraints placed on the equation mentioned earlier) augmented by three zeros at the end. The individual elements of $\mathbf{w}$ are the w_i in the earlier equation. Each is associated with one (the ith) landmark on the reference configuration.

We are now free to use Equation (1) and the new coefficients to compute the height of the surface at any point in the plane of the reference. As required, heights at reference landmarks will equal differences between the reference and the target configuration along the coordinate axis under consideration, and heights at other positions will be interpolated so that the resulting surface has minimal bending. For application in morphometric visualization, we then assemble the heights, separately computed with different coefficients for each coordinate axis, into displacement vectors for a given point in the plane of the reference.

To use this information to achieve the thin-plate spline plots seen throughout this book, one constructs a grid of square cells over the reference configuration and computes the interpolated displacement vectors for points on the gridlines. Redrawing the connections between the displaced points results in the thin-plate spline plot (Figure 8). It is important that the initial grid cells be square so that deviations from "squareness" can be interpreted as oriented stretching within the cells of the resulting spline plot. This is not a mathematical requirement. It is just harder to assess how a cell has changed in a plot if you are unsure of its initial shape and distinguishing between initial rectangles and resultant quadrilaterals is more difficult than spotting deviations from squareness.

Note that this construction has no prerequisites about the superimposition of the reference and the target configurations. The construction is also very robust and can represent extreme shape differences beyond any to be encountered in anthropological research. The only exception is that the formulae "blow up" if two points are coincident on the reference, but distinct on the target. This is perceived as a "tearing" of the thin-plate spline model. In biological terms such a situation would represent the genesis of a new biological feature through either evolution or development. Bookstein and Smith (2000) proposes the use of "creases" to model such occurrences.

The w_i used in the thin-plate spline provide the coordinates of an individual specimen with respect to the eigenvectors of the bending energy matrix (see the section on "Warps" below)—the upper, left $p \times p$ submatrix of $\mathbf{L}^{-1}$ (Bookstein, 1991). These eigenvectors are a set of orthonormal axes for local, or non-affine, components of shape differences with respect to the reference configuration. The remainder of the total shape difference between an individual specimen and the reference is the global, affine, or uniform shape difference. These are those differences that can be characterized as stretching or compressing the space in orthogonal directions. Such transformations have the properties that they leave parallel lines parallel and affect the local space precisely the same way everywhere, hence the term global (Figure 9, upper right and lower left). In contrast, the local shape differences represented by the eigenvectors of the bending energy matrix encode different compressions, expansions, displacements, or reorientations of local regions of the space (Figure 9, lower right). The w_i encode the local difference between shapes, but use of such scores in the analysis of total shape difference requires the quantification of the affine component, as well.

Bookstein (1996a) presented a method to compute the affine terms for two-dimensional data based on the Procrustes metric, but this method does not extend easily to higher-dimensional configurations. Recently, Rohlf and Bookstein (2003) have presented more general, complimentary approaches to quantifying affine variation. One is based on the Burnaby-like projection of data (Burnaby, 1966) into a space orthogonal to that of local variation. The other is based on the linear regression of a specimen onto the consensus or reference that is similar to the affine least-squares fitting described in Rohlf and Slice (1990). Regardless of the form of computation, the concatenation of the variables for purely affine shape differences with those for local variation provides a linear space suitable for the application of standard, parametric multivariate statistical tests.

The discussion so far has been modeled differences in two-dimensional configurations as deviations out of the plane of the reference. The situation is a little different, but not by much, for generating thin-plate splines for three-dimensional configurations. Here the model is less-intuitive, with

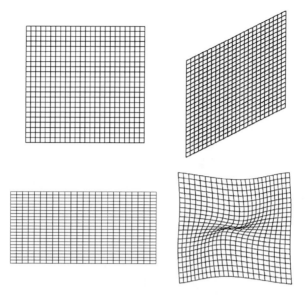

Figure 9. Affine transformations (upper, right and lower, left) are global in the sense that they involve only a simple stretching/compression in orthogonal directions and are the same everywhere in the space. Non-affine, or local, deformations involve twisting, stretching, and shifting of small regions as shown in the lower, right. An exact position must be specified to discuss the effects of such deformations. Plots created with Morpheus et al. (Slice, 1998).

coordinate-wise differences being considered as orthogonal displacements out of the volume of the reference specimen, but it is identical in spirit to the two-dimensional case. Some adjustment is also needed in the U function, which becomes $|r_{i,j}|$ to achieve the requisite minimization (Bookstein, 1991, appendix 1).

Figure 10 shows a thin-plate spline mapping the shape of the average male gorilla scapula landmarks onto those of the average female gorilla. This figure was generated using the average male gorilla landmark locations as the reference in the thin-plate spline equations and the average male–average female difference vectors ($\times 2$) in the **Y** matrix in Equation 2. One can now appreciate the regional expansions, rotations, etc. taking into account the relative positions of all landmarks simultaneously. As with vector plots, any reasonable source of displacements can be used to generate the splines. For instance, one can spline an average configuration using the coefficients of a principal component or the coordinates of a predicted configuration based on a regression analysis.

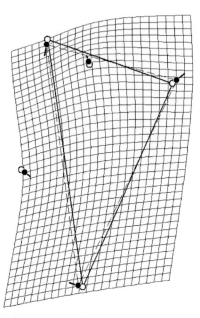

Figure 10. Thin-plate spline deformation grid showing the difference between mean male and female gorilla scapulae (magnified ×2) as a transformation of the male scapula. Vectors are also shown. Compare with Figure 7. Plot created with Morpheus et al. (Slice, 1998).

The thin-plate spline mapping can also be used in other ways. For instance, it can be used to associate other information, such as pixel intensity, in the space around the landmarks of the target configuration with specific locations in the space of the reference. In this way, images associated with individual specimens can be used to construct an average image associated with the reference. This method of image "unwarping" is used by Gharaibeh (Chapter 5, this volume) in his study of the geometric effects of head orientation in the anthropometric analysis of archival photographs.

WARPS, WARPS, AND MORE WARPS

Researchers new to the world of geometric morphometrics are often confused by unfamiliar terminology. One especially noteworthy case is that of the various "warps" that are often referred to in the literature and derive from the thin-plate spline formulation just discussed. There are principal warps, partial warps, relative warps, and singular warps.

Principal warps are the eigenvectors of the *bending energy matrix*, which is the upper, left $p \times p$ submatrix of the L^{-1} matrix used to compute the coefficients for the thin-plate spline. This matrix encodes the local aspects of shape differences, and its eigenvectors are linear combinations of orthogonal displacements of the landmarks of the reference configuration ordered by the energy required to fit the hypothetical metal sheet to those displacements. That is, the first principal warp associated with the largest eigenvalue (bending energy) is the most local deformation of the reference configuration. The second requires the most energy of deformations geometrically orthogonal to the first, and so on. A key feature of the principal warps is that they are functions of the reference configuration alone. They are computed without using any other data configurations, and therefore, carry no information about the sample other than vaguely through the contributions of sample configurations to the mean (if the reference used is the sample mean).

Partial warps are pairs or triplets of principal warps used to encode differences between individual specimens and the reference. The scores on these warps or axes come in multiples since real coordinate axes (x, y, and z) are modeled separately in the thin-plate spline computations. The partial warp scores are the scores for each individual for each coordinate for each principal warp axis, and taken together, the partial warps provide an orthogonal basis for the space of non-affine shape variation. The scores, themselves, are shape variables.

Relative warps are linear combinations of the partial warps and affine components computed to decompose total shape variability into uncorrelated, variance-maximizing variables. In more familiar terms, the relative warps are the principal components of sample variability in shape space with respect to the partial warp and affine scores. Rohlf (1993) describes how this principal component analysis can be tuned to emphasize larger- or smaller-scaled shape variation if the researcher has a reason for doing so.

Recent additions to the morphometric jargon are the *singular warps*. There are pairs of covariance-maximizing linear combinations of two sets of variables observed on individuals. One or both sets can be shape variables, and in the case of the former, the second set can be environmental or other variables the researcher wishes to relate to shape (Bookstein et al., 2003).

EXTENSIONS TO PROCRUSTES ANALYSIS

The combination of multivariate Procrustes analysis, multivariate statistics, and thin-plate spline visualizations is what Bookstein (1993, 1996b) called the "Morphometric synthesis." It also represents the foundation of most coordinate-based analyses to be found in the literature to date. Still, there are a number of current or potential elaborations that are likely to take the approach beyond the now-familiar comparison of mean shapes or the regression of shape on size or extrinsic factors like temperature or epoch. These are mentioned only briefly here, but this economy should not be taken to represent any limitation on the potential of these methods in morphometrics.

One intriguing area of research is that of asymmetry (Palmer, 1996; Palmer and Strobeck, 1986) in which the differences in the relative size of two sides of a bilaterally symmetric structure or bilateral structures of an individual organism are examined across populations. The idea is that in the absence of developmental instability, environmental perturbations, or within-organism substrate competition, organisms should manifest perfect symmetry. Deviations from perfect symmetry, therefore, can give insight into developmental programs, environmental stress, and/or other putative factors. Such variation can be divided into various classes such as: directional asymmetry—where one side always differs in the same way, antisymmetry—where one side differs by some degree, but which side is more-or-less random, and fluctuating asymmetry—where variations are random with respect to the average shape or form. The latter is often interpreted as an indicator of environmental perturbation, but to get to that component, other types of asymmetry must either be partitioned out or discounted.

An early investigation of asymmetry using geometric morphometrics is that of Smith et al. (Bookstein, 1991; Smith et al., 1997), who partition shape variation in right and left honey bee wings into fluctuating and directional components. Klingenberg and McIntyre (1998) and Klingenberg and Zaklan (2000) analyze asymmetry in fly wings using standard MANOVA methods applied to Procrustes coordinates. Mardia et al. (2000) and Kent and Mardia (2001) present a comprehensive methodology for the analysis of asymmetry entirely within the Procrustes geometry, though this has yet to find its way into general morphometric use.

Kimmerle and Jantz (Chapter 11, this volume) use MANOVA methods and regression to examine trends in asymmetry in the crania of both sexes of whites and blacks in the United States. An alternative to the Procrustes framework for the analysis of asymmetry is the use of Euclidean Distance Matrix Analysis (see the section on Coordinate-free methods) presented in the current volume by Richtsmeier et al. (Chapter 8).

The Procrustes methods are restricted by their requirement of named (in most cases) landmarks and the assumption of equipotent information in all coordinate directions. As indicated by Bookstein's landmark classification, this is not always the case. To address this restriction, Bookstein (1991, 1997) proposed sliding such two-dimensional semi-landmarks in the uninformative direction(s) to enhance the relative contribution of informative variability. Bookstein et al. (1999) uncover some rather surprising results using this approach. They find that the inner, midsagittal profile of the frontal bone is invariant with respect to shape between modern and archaic humans. Furthermore, it is nearly indistinguishable in these groups from that of Australopithicines or Pan! In the current volume, Gunz et al. (Chapter 3) extend the method of sliding landmarks both to two-dimensional curves embedded in three-dimensions (space curves) and to the analysis of surfaces. Relevant formulae for these and Bookstein's original two-dimensional case are found there as well.

Quantitative genetics is another area in which the integration of modern morphometric techniques could have a significant impact. Procrustes-derived shape variables have been used by Klingenberg et al. (2001) and Klingenberg and Leary (2001) to explore quantitative trait loci and the relationship between genetic and phenotypic covariance using similar data. Monteiro et al. (2002) investigate the heritability of shape using Procrustes shape variables.

COORDINATE-FREE METHODS

It has been suggested that the Procrustes and other registration-based methods are undesirable since they involve a distinct and unnecessary superimposition step after which analyses involve an inestimable coordinate covariance structure (e.g., Lele and Richtsmeier, 2001). To avoid this, several "coordinate-free" approaches have been suggested that utilize variables that are invariant to orientation and location within a specific coordinate system (size may be factored out separately). Rao and Suryawanshi (1996, 1998) suggested the use of sufficient sets of interior angles between landmarks and combinations of interlandmark distances as variables to quantify and analyze shape. The early truss methods of Bookstein et al. (1985) would also be included in this category.

The most widely used method of this school, however, is based on Euclidean Distance Matrix Analysis, or EDMA (Lele and Cole, 1995, 1996; Lele and Richtsmeier, 1991, 2001). The Euclidean Distance Matrix (EDM), or Form Matrix (FM) is simply a symmetric, $p \times p$ matrix in which the off-diagonal elements, $e_{i,j} = e_{j,i}$, are the Euclidean distances between landmarks i and j. Diagonal elements are zero, and in practice, only one set of off-diagonals is used for statistical testing. These measurements are a highly redundant set of variables that completely fix the geometry of the landmarks up to a reflection and are invariant to any choice of orthogonal basis vectors with respect to which the original coordinates might have been collected. Note that, while invariant to location and orientation, such an assemblage still contains size information that must be partialled out if a pure shape analysis is desired.

Analysis proceeds via the pairwise comparison of EDMs, which might be two configurations or mean EDMs for two groups. Mean form matrices can be computed using the method of moments (Stoyan, 1990; Lele, 1993), but there is a problem in that the resulting average interpoint distances might not represent a physically realizable structure. Lele (1993) suggests one "flattens" the estimates back into image/physical space by using only the first two/three eigenvectors of the mean form matrix scaled by their associated eigenvalues.

The developers have proposed several methods for the statistical assessment of pairs of EDMs. In its first incarnation, EDMA I (Lele and Richtsmeier, 1991), ratios of corresponding distances in the two EDMs are used to construct a relative form difference matrix (rFDM). If the two configurations are precisely the same, these values will be all unity, and if the two differ only by size, the elements will be some constant value. It is seldom the case that configurations will be identical, so the authors proposed a non-parametric test for shape differences. The test statistic, T, is the ratio of the largest ratio in the rFDM to the smallest. The significance of T is assessed relative to confidence limits obtained from bootstrap resampling of the two populations (Richtsmeier and Lele, 1993). Lele and Cole (1995, 1996) proposed another approach for heteroscedastic samples, EDMA II, that uses the test statistic, Z, the maximum absolute value of the arithmetic difference of elements in two EDMs. The significance of Z is tested by a parametric bootstrap procedure in which on generates normally distributed samples with the same mean shape and covariance structure as the original samples and determines if the resulting confidence limits on Z contain zero. To achieve a shape, instead of form, analysis, EDMs in EDMA II may be scaled by some user-specified measure of size, for example, baseline length or CS (Lele and Cole, 1995, 1996).

The relative merit of EDMA vs. the Procrustes approaches has been one of the more contentious issues of modern morphometrics. The proponents of EDMA (Lele, 1993; Lele and McCulloch, 2002; Lele and Richtsmeier, 2001) argue that the use of inestimable landmark covariance matrices and inconsistent mean and covariance estimates are serious problems with Procrustes-based methods ("consistency" is a statistical term meaning that as the sample size goes to infinity, the estimated value converges in probability on the true value). Rohlf (1999b, 2000, 2003) compared a number of morphometric methods including EDMA and GPA using simulations of random triangles generated with independent, isotropic error. He found the EDMA methods introduced artifactual covariance structure into randomly generated samples (Rohlf, 1999b), had a more complicated and structured power surface (Rohlf, 2000), and for realistic sample sizes, produced more biased mean estimates (Rohlf, 2003). EDMA supporters counter that the simplistic models of isotropic error are not representative of real-world data sets (Lele and Richtsmeier, 1990) and reject the power comparisons as being artifactual and lacking valid, analytical support (Richtsmeier, personal communication).

Lele and Richtsmeier (2001) is a comprehensive outline of the case for EDMA and with contributions from Cole, provides extensions of the EDMA approach to the study of growth, classification, clustering, asymmetry, molecular structures and phylogenetics. In this volume, Richtsmeier et al. present an EDMA-based method for the study of asymmetry using a mouse model for Trisomy 21, Down syndrome.

OUTLINE METHODS

The differences between outline data and landmark coordinates usually require special consideration, and different methods are available for different types of outlines (see Rohlf, 1990). One method of particular importance in morphometrics is elliptic Fourier analysis (EFA) (Kuhl and Giardina, 1982) for two-dimensional, closed contours. This method does not require evenly spaced sample points or equal numbers of points across specimens, can handle arbitrarily complex outlines, provides parameterization of the entire outline, and includes optional standardizations for size, location, orientation, and digitizing starting point to support the analysis of shape. In general, though, one would like to seamlessly combine outline and landmark data in a single analysis. An outline might pass through bona fide landmarks, or it could have structurally associated landmarks nearby. In Chapter 6, this volume, Baylac and Frieß combine EFA and GPA to study the effects of cranial deformation using partial and complete cranial outlines and landmarks.

Another possibility is to develop a common framework for the joint analysis of landmarks and outlines. McNulty (Chapter 16, this volume) uses equally spaced sample points between anatomical landmarks on brow ridges superimposed using only the anatomical landmarks to quantify supraorbital morphology in extant and fossil hominoids. The sliding landmarks method developed by Bookstein (e.g., 1997) allows sample points to slide along tangents to the outline in a Procrustes analysis in order to reduce uninformative variation in that direction. Their use in analyzing curves in three-dimension is discussed by Gunz et al. in Chapter 3, this volume. Reddy et al. (Chapter 4) present a modification of the sliding landmark method to take advantage of the usually high density of initial sample points collected when quantifying such curves, and apply the method to the study the Neanderthal "bun."

FRACTALS

The concept of a fractal is an interesting one with many potential applications in biology (Slice, 1993b), including physical anthropology. In general, a fractal is "a set for which the Hausdorff-Besicovitch dimension strictly exceeds the topological dimension" (Mandelbrot, 1983). Such sets can be point sets or outlines or surfaces or volumes, and all possess a distinguishing property call self-similarity (see Feder, 1988; Peitgen and Saupe, 1988). This means that some geometric aspect of the set is repeated at different scales. The similarity may be exact, with larger-scale features being composed of smaller-scale copies of a fundamental structure, or statistical, where scale-adjusted variability at all scales is similar. The implication of such self-similarity for morphometrics is great. One finds, for instance, that genuinely fractal closed outlines have an infinite perimeter since no matter how small a measurement scale is used the perimeter never "smoothes out" into a two-dimensional Euclidean curve (Slice, 1993b). The characterization of a biological structure in the form of the fractal dimensions, d, can be used to quantify the complexity of a shape over some finite range of measurement scales.

Structures amenable to such analysis include complex sutures like those in ammonite shells and cervid skulls examined by Long (1985). The comparison of the fractal dimensions of sutures from a more accurate tracing was used by Palmqvist (1997) to challenge the previous hominid affinities of the Orce skull fragment supported by Gilbert and Palmqvist (1995). In this volume,

Prossinger (Chapter 7) argues for the fractal nature of frontal sinuses and presents a plausible model for the development and modeling of such structures.

FINAL COMMENTS

This chapter has been a rather fast-paced and necessarily superficial overview of the large and growing field of modern shape analysis with a slight bias toward applications in physical anthropology. It is hoped that this will provide readers having relatively little exposure to morphometric methods with sufficient information to follow and appreciate subsequent chapters. Newcomers are not expected to master, or necessarily fully grasp, all of the subtleties and mathematical details summarized here, but at least they will have been exposed to the material. If, as is likely, these same readers want to learn more about the methods and begin to use them in their own research, the references cited herein will provide a good place to start. In addition, a number of user-friendly resources are available to enhance their education. One is the morphometrics website at the State University of New York at Stony Brook (http://life.bio.sunysb.edu/morph) developed and maintained by F. James Rohlf. The site contains free, downloadable software for data acquisition and morphometric analysis, downloadable data sets, and a morphometrics bulletin board, bibliographies, book reviews, and links to other sites. A second valuable resource is the morphometric mailing list, morphmet, started by Les Marcus. Over four hundred people from around the world are subscribers and are inevitably willing to assist others. Subscription information can be found at the Stony Brook morphometrics site, http://life.bio.sunysb.edu/morph/morphmet.html.

ACKNOWLEDGMENTS

The comments and criticisms of Fred Bookstein, F. James Rohlf, Andrea B. Taylor, and Katherine Whitcome were most helpful in the preparation of this manuscript. Any errors or other shortcomings are clearly my own. This work was supported, in part, by the Austrian Ministry of Culture, Science, and Education, and the Austrian Council for Science and Technology (grant numbers: GZ 200.049/3-VI/I2001 and GZ 200.093/1-VI/I2004 to Horst Seidler) and Dr. Edward G. Hill and members of the Winston-Salem community, to whom the author is especially grateful.

REFERENCES

Adams, D. C., Rohlf, F. J., and Slice, D. E., 2004, Geometric morphometrics: Ten years of progress after the "revolution," *Italian Journal of Zoology.* 71(1):5–16.

Atchley, W. R. and Anderson, D., 1978, Ratios and the Statistical Analysis of Biological Data, *Syst. Zool.* 27(1):71–78.

Atchley, W. R., Gaskins, C. T., and Anderson, D., 1976, Statistical properties of ratios. I. Empirical results. *Syst. Zool.* 25(2):137–148.

Blackith R. and Reyment R. A., 1971, *Multivariate Morphometrics*, Academic Press, New York.

Boas, F., 1905, The horizontal plane of the skull and the general problem of the comparison of variable forms, *Sci.* 21(544):862–863.

Bookstein, F. L., 1986, Size and shape spaces for landmark data in two dimensions (with discussion), *Stat. Sci.* 1:181–242.

Bookstein, F. L., 1989, Principal warps: thin-plate splines and the decomposition of deformations, *IEEE Trans. on Pattern Anal. and Machine Intell.* 11:567–585.

Bookstein F. L., 1991, *Morphometric Tools for Landmark Data: Geometry and Biology*, Cambridge University Press, Cambridge.

Bookstein, F. L., 1993, A brief history of the morphometric synthesis, in: *Contributions to Morphometrics*, L. F. Marcus, E. Bello, and Garciá-Valdecasas, eds., Monografias del Museo Nacional de Ciencias Naturales 8, Madrid, pp. 15–40.

Bookstein, F. L., 1996a, A standard formula for the uniform shape component in landmark data, in: *Advances in Morphometrics*, L. F. Marcus, M. Corti, A. Loy, G. J. P. Naylor, and D. E. Slice, eds., Plenum Press, New York, pp. 153–168.

Bookstein, F. L., 1996b, Biometrics, biomathematics and the morphometric synthesis, *Bull. of Math. Biol.* 58:313–365.

Bookstein, F. L., 1997, Landmark methods for forms without landmarks: Localizing group differences in outline shape, *Med. Image. Anal.* 1:225–243.

Bookstein F. L. and Smith, B. R., 2000, Inverting development: geometric singularity theory in embryology, in: *Mathematical Modeling, Estimation, and Imaging*, D. Wilson, H. Tagare, F. Bookstein, F. Préteaux, and E. Dougherty, eds., Proc. SPIE, vol. 4121, pp. 139–174.

Bookstein, F. L., Chernoff, B., Elder R. L., Humphries, J. M. Jr., Smith, G. R., and Strauss, R. E., 1985, *Morphometrics in Evolutionary Biology*, Special Publication 15, Academy of Natural Sciences Press, Philadelphia.

Bookstein, F. L., Gunz, P., Mitteröcker, P., Prossinger, H., Schäfer, K., and Seidler, H., 2003, Cranial integration in *Homo*: Singular warps analysis of the midsagittal plane in ontogeny and evolution, *J. Hum. Evol.* 44(2):167–187.

Bookstein, F. L., Schäfer, K., Prossinger, H., Seidler, H., Fieder, M., Stringer, C. et al., 1999, Comparing frontal cranial profiles in archaic and modern Homo by morphometric analysis, *Anatomi. Rec. (New Anatomist)* 257:217–224.

Burnaby, T. P., 1966, Growth-invariant discriminant functions and generalized distances, *Biometrics* 22:96–110.

Carroll, J. D. and Green, P. E., 1997, *Mathematical Tools for Applied Multivariate Analysis*, Academic Press, San Diego.

Cole, T. M. III., 1996, Historical note: early anthropological contributions to "geometric morphometrics," *Am. J. Phys. Anthropol.* 101(2):291.

Dryden, I. L. and Mardia, K. V., 1993, Multivariate shape analysis, *Sankhya* 55:460–480.

Dryden, I. L. and Mardia K. V., 1998, *Statistical Shape Analysis*, John Wiley & Sons, New York.

Feder, J., 1988, *Fractals*, Plenum Press, New York.

Galton, F., 1907, Classification of portraits, *Nature* 76:617–619.

Gilbert, J. and Palmqvist, P., 1995, Fractal analysis of the Orce skull sutures, *J. Hum. Evol.* 28:561–575.

Goodall, C. R., 1991, Procrustes methods in the statistical analysis of shape, *J. Roy. Statistical Society, Series B*, 53:285–339.

Gould, S. J., 1981, *The Mismeasure of Man*, W.W. Norton & Company, New York.

Gower, J. C., 1975, Generalized Procrustes analysis, *Psychometrika*, 40:33–51.

Hanihara, T., 2000, Frontal and facial flatness of major human populations, *Am. J. Phy. Anthropol.* 111(1):105–134.

Howells, W. W., 1973, *Cranial Variation in Man: A Study by Multivariate Analysis of Difference Among Recent Human Populations*, Papers of the Peabody Museum of Archaeology and Ethnology, No. 67. Harvard University, Cambridge, MA.

Howells, W. W., 1989, Skull shapes and the map: craniometric analyses in the dispersion of modern *Homo*, Papers of the Peabody Museum of Archaeology and Ethnology, No. 79. Harvard University, Cambridge, MA.

Huxley, J. S., 1932, *Problems of Relative Growth*, Methuen, London. Reprinted 1972, Dover Publications, New York.

Johnson, R. A. and Wichern, D. W., 1982, *Applied Multivariate Statistical Analysis*, Prentice-Hall Inc., Inglewood Cliffs, New Jersey.

Kendall, D. G., 1984, Shape-manifolds, Procrustean metrics and complex projective spaces, *Bulletin of the London Mathematical Society* 16:81–121.

Kendall, D. G., 1985, Exact distributions for shapes of random triangles in convex sets, *Advan. Appl. Prob.* 17:308–329.

Kendall, D. G., 1989, A survey of the statistical theory of shape, *Stat. Sci.* 4:87–120.

Kendall, D. G. and Kendall, W. S., 1980, Alignments in two dimensional random sets of points, *Advan. Appl. Prob.* 12:380–424.

Kent, J. T. and Mardia, K. V., 2001, Shape, Procrustes tangent projections and bilateral symmetry, *Biometrika* 88:469–485.

Klingenberg C. P. and Leamy, L. J., 2001, Quantitative genetics of geometric shape in the mouse mandible, *Evolution* 55:2342–2352.

Klingenberg, C. P. and McIntyre, G. S., 1998, Geometric morphometrics of developmental instability: Analyzing patterns of fluctuating asymmetry with Procrustes methods, *Evolution* 52:1363–1375.

Klingenberg, C. P. and Zaklan, S. D., 2000, Morphological integration between developmental compartments in the *Drosophila* wing, *Evolution* 54(4):1273–1285.

Klingenberg C. P., Leamy L. J., Routman E. J., and Cheverud J. M., 2001, Genetic architecture of mandible shape in mice: Effects of quantitative trait loci analyzed by geometric morphometrics, *Genetics* 157:785–802.

Kolar, J. C. and Salter, E. M., 1996, *Craniofacial Anthropometry*, Charles C. Thomas Publishers Ltd., Springfield.

Krzanowski, W. J., 1988, *Principles of Multivariate Analysis*, Claredon Press, Oxford.

Kuhl, F. P. and Giardina, C. R., 1982, Elliptic Fourier features of a closed contour, *Comp. Graph. Imag. Process.* 18:236–258.

Lele, S. R., 1993, Euclidean distance matrix analysis: estimation of mean form and form difference, *Math. Geol.* 25:573–602.

Lele, S. R. and Cole T. M. III, 1995, Euclidean distance matrix analysis: a statistical review, in: *Current Issues in Statistical Shape Analysis, Volume 3*, University of Leeds, Leeds, pp. 49–53.

Lele, S. R. and Cole, T. M. III, 1996, A new test for shape differences when variance-covariance matrices are unequal, *J. Hum. Evol.* 31:193–212.

Lele, S. R. and McCulloch, C., 2002, Invariance and morphometrics, *J. Am. Stat. Assoc.* 97(459):796–806.

Lele, S. R. and Richtsmeier, J. T., 1990, Statistical models in morphometrics: Are they realistic? *Syst. Zool.* 39(1):60–69.

Lele S. R. and Richtsmeier, J. T., 1991, Euclidean distance matrix analysis: A coordinate free approach for comparing biological shapes using landmark data, *Am. J. Phy. Anthropol.* 86:415–427.

Lele, S. R. and Richtsmeier, J. T., 2001, *An Invariant Approach to Statistical Analysis of Shapes*, Chapman & Hall/CRC, New York.

Long, C. A., 1985, Intricate sutures as fractal curves, *J. Morph.* 185:285–295.

Mahalanobis, P. C., 1928, On the need for standardization in measurements on the living, *Biometrika* 20A(1/2):1–31.

Mahalanobis, P. C., 1930, A statistical study of certain anthropometric measurements from Sweden, *Biometrika* 22(1/2):94–108.

Mandelbrot, B. B., 1983, *The Fractal Geometry of Nature, 2nd Edition*, W. H. Freeman and Company, New York.

Manly, B. F. J., 1997, *Randomization, Bootstrap and Monte Carlo Methods in Biology*, *2nd Edition*, Chapman & Hall/CRC, New York.

Marcus, L. F., 1990, Traditional morphometrics, in: *Proceedings of the Michigan Morphometrics Workshop*, F. J. Rohlf and F. L. Bookstein, eds., Special Publication Number 2, University of Michigan Museum of Zoology, Ann Arbor, MI, pp. 77–122.

Mardia, K. V., Bookstein, F. L., and I. J. Moreton, I. J., 2000, Statistical assessment of bilateral symmetry of shapes, *Biometrika* 87:285–300.

Monteiro, L. R., Diniz-Filho, J. A. F., dos Reis, S. F., and Araújo, E. D., 2002, Geometric estimates of heritability in biological shape, *Evolution* 56:563–572.

Morant, G. M., 1928, A preliminary classification of European races based on cranial measurements, *Biometrika* 20B(3/4):301–375.

Morant, G. M., 1939, The use of statistical methods in the investigation of problems of classification in anthropology: Part I. The general nature of the material and the form of intraracial distributions of metrical characters, *Biometricka* 31(1/2): 72–98.

Mosier, C. I., 1939, Determining a simple structure when loadings for certain tests are known, *Psychometrika* 4:149–162.

Mosimann, J. E., 1970, Size allometry: size and shape variables with characterization of the log-normal and generalized gamma distributions, *Jour. Amer. Stat. Assoc.* 65:930–945.

Palmer, A. R., 1996, Waltzing with asymmetry, *BioScience* 46:518–532.

Palmer, A. R. and Strobeck, C., 1986, Fluctuating asymmetry: measurement, analysis, patterns, *Annu. Rev. Ecol. Syst.* 17:391–421.

Palmqvist, P., 1997, A critical re-evaluation of the evidence for the presence of hominids in Lower Pleistocene times at Venta Micena, Southern Spain, *J. Hum. Evol.* 33: 83–89.

Pearson, K., 1897, Mathematical contributions to the theory of evolution—on a form of spurious correlation which may arise when indices are used in the measurement of organs, *Proceedings of the Royal Society of London* 60:489–498.

Pearson, K., 1903, Craniological notes: Professor Aurel von Torok's attack on the arithmetical mean, *Biometrika* 2(3):339–345.

Pearson, K., 1933, The cranial coordinatograph, the standard planes of the skull, and the value of Cartesian geometry to the craniologist, with some illustrations of the uses of the new methods, *Biometrika* 25(3/4):217–253.

Peitgen, H. and Saupe, D., 1988, *The Science of Fractal Images*, Springer-Verlag, New York.

Rao, C. R. and Suryawanshi, S., 1996, Statistical analysis of shape of objects based on landmark data, *Proceedings of the National Academy of Sciences, U.S.A.* 93:12132–12136.

Rao, C. R. and Suryawanshi, S., 1998, Statistical analysis of shape through triangulation of landmarks: A study of sexual dimorphism in hominids, *Proceedings of the National Academy of Sciences, U.S.A.* 95:4121–4125.

Reyment, R. A., 1996, An idiosyncratic history of early morphometrics, in: L. F. Marcus, M. Corti, A. Loy, G. J. P. Naylor and D. E. Slice, eds., *Advances in Morphometrics*, Plenum Press, New York, pp. 15–22.

Richtsmeier, J. T. and Lele, S. R., 1993, A coordinate-free approach to the analysis of growth-patterns: Models and theoretical considerations, *Biol. Rev.* 68:381–411.

Robins, G., 1994, *Proportion and Style in Ancient Egyptian Art*, University of Texas Press, Austin.

Rohlf, F. J., 1990, Fitting curves to outlines. Proceedings of the Michigan morphometrics workshop, in: *Proceedings of the Michigan Morphometrics Workshop*, F. J. Rohlf and F. L. Bookstein, eds., Special Publication Number 2, University of Michigan Museum of Zoology, Ann Arbor, MI, pp. 167–177.

Rohlf, F. J., 1993, Relative warp analysis and an example of its application to mosquito wings, in: *Contributions to Morphometrics*, L. F. Marcus, E. Bello and Garciá-Valdecasas, eds., Monografias del Museo Nacional de Ciencias Naturales 8, Madrid, pp. 131–159.

Rohlf, F. J., 1999a, Shape statistics: Procrustes superimpositions and tangent spaces, *Journal of Classification* 16:197–223.

Rohlf, F. J., 1999b, On the use of shape spaces to compare morphometric methods, *Hystrix* 11:9–25.

Rohlf, F. J., 2000, Statistical power comparisons among alternative morphometric methods, *Am. J. Phys. Anthropol.* 111:463–478.

Rohlf, F. J., 2001, tpsDig, Version 1.31. Department of Ecology and Evolution, State University of New York at Stony Brook, Stony Brook, New York.

Rohlf, F. J., 2002, tpsTri. Version 1.17. Department of Ecology and Evolution, State University of New York at Stony Brook, New York.

Rohlf, F. J., 2003, Bias and error in estimates of mean shape in morphometrics, *J. Hum. Evol.* 44:665–683.

Rohlf, F. J. and Bookstein, F. L., 2003, Computing the uniform component of shape variation, *Syst. Biol.* 52:66–69.

Rohlf, F. J. and Marcus, L. F., 1993, A revolution in morphometrics, *Trends in Ecology and Evolution* 8:129–132.

Rohlf, F. J. and Slice, D. E., 1990, Extensions of the Procrustes method for the optimal superimposition of landmarks, *Syst. Zool.* 39:40–59.

Siegel, A. F. and Benson, R. H., 1982, A robust comparison of biological shapes, *Biometrics* 38:341–350.

Siegel, A. F. and Pinkerton, J. R., 1982, Robust comparison of three-dimensional shapes with an application to protein molecule configurations, Technical Report No. 217. Series 2, Department of Statistics, Princeton University.

Slice, D. E., 1993a, Extensions, comparisons, and applications of superimposition methods for morphometric analysis, Ph.D. dissertation. Department of Ecology and Evolution, State University of New York, Stony Brook, New York.

Slice, D. E., 1993b, The fractal analysis of shape, in: *Contributions to Morphometrics*, L. F. Marcus, E. Bello, and Garciá-Valdecasas, eds., Monografías del Museo Nacional de Ciencias Naturales 8, Madrid, pp. 161–190.

Slice, D. E., 1994, GRF-ND: Generalized rotational fitting of N-dimensional data, Department of Ecology and Evolution, State University of New York at Stony Brook, New York.

Slice, D. E., 1996, Three-dimensional, generalized resistant fitting and the comparison of least-squares and resistant-fit residuals, in: *Advances in Morphometrics*, L. F. Marcus, M. Corti, A. Loy, G. J. P. Naylor and D. E. Slice, eds., Plenum Press, New York, pp. 179–199.

Slice, D. E., 1998, Morpheus et al.: Software for morphometric research. Revision 01-31-00, Department of Ecology and Evolution, State University of New York at Stony Brook, New York.

Slice, D. E., 2001, Landmark coordinates aligned by Procrustes analysis do not lie in Kendall's shape space, *Syst. Biol.* 50:141–149.

Slice, D. E., Bookstein, F. L., Marcus, L. F., and Rohlf, F. J., 1996, A glossary for morphometrics, in: *Advances in Morphometrics*, L. F. Marcus, M. Corti, A. Loy, G. J. P. Naylor and D. E. Slice, eds., Plenum Press, New York, pp. 531–551.

Small, C. G., 1996, *The Statistical Theory of Shape*, Springer-Verlag, New York.

Smith, D., Crespi, B., and Bookstein, F. L., 1997, Asymmetry and morphological abnormality in the honey bee, *Apis mellifer*: Effects of ploidy and hybridization, *Evolutionary Biology* 10:551–574.

Sneath, P. H. A., 1967, Trend surface analysis of transformation grids, *J. Zool., London* 151:65–122.

Stoyan, D., 1990, Estimation of Distances and Variances in Booksteins landmark model, *Biometrical J.* 32:843–849.

Thompson, D'A. W., 1942, *On Growth and Form*, Cambridge University Press, Cambridge.

PART ONE

Theory and Methods

CHAPTER TWO

After Landmarks

Fred L. Bookstein

y argument here springs principally out of the fact that the morphometrics of named location data is essentially complete. By "named location data" I mean not mere digitized landmark points but the Cartesian coordinates of *simplicial complexes*—general assemblages of named points, smooth curves, and smooth open or closed surfaces.[1] By an "essentially complete" morphometrics I mean that for point–curve–surface data, Procrustes tangent space coordinates, classic linear multivariate analysis, and thin-plate spline displays are powerful, robust, and easily combined tools for cogent reporting of biologically meaningful patterns over a huge range of empirical designs. Over a busy decade of development there has emerged a complete multivariate methodology for biometric size-and-shape analysis of named location data, whatever the geometric scheme, that helps to further a wide variety of studies concerned with mean differences, variation and ordination, or covariances of form with its causes or effects. The argument for this proposition will not be reviewed here (but see, for example,

---

[1] There are a few restrictions arising from the mathematical notion of a "simplicial complex" and the way that it has been borrowed for this application. For instance, intersections of curves must be named landmarks, and also sharp corners or centers of curvature of sharply curved arcs of otherwise smooth curves; likewise, intersections of curves with surfaces. Intersections of surfaces must be named curves, and also ridges along which smooth surfaces are particularly sharply curved.

Fred L. Bookstein • Institute of Anthropology, University of Vienna, and Biophysics Research Division, University of Michigan, University of Michigan.

Modern Morphometrics in Physical Anthropology, edited by Dennis E. Slice.
Kluwer Academic/Plenum Publishers, New York, 2005.

the review by Adams, Rohlf, and Slice (2004), or the chapters by Slice and by Mitteroecker et al. elsewhere in this volume).

And so it is not too early to begin speculating on the next toolkit even as our research and teaching communities assimilate this one. Quantitative morphology has long dealt with information beyond the coordinates of named points, curves, or surface structures, and many of these other types of information deserve morphometric methods of their own. Sometimes the underweighted information is at relatively large scale, such as a growth-gradient; sometimes it is at relatively small scale, such as spacing of nearby structures; and sometimes it is multilocal, such as pertains to bilateral symmetry or to patterns of spacing down an axis.

This essay speculates on future extensions of the central formalism under three general headings: changes in characterizing what is to be regarded as "local," changes in the *a-priori* structure of graphic summaries, and extensions of location-related information, the actual data set to be analyzed, beyond the simple notion of digitized coordinates. Under each heading I sketch the limitations of the current methodological toolkit, limitations often cryptic or rarely noted elsewhere, and then show a range of real examples that suggest possible new or newly applied formalisms for data alignment, variance and covariance, linear models, and graphic displays.

SUGGESTION 1: CHANGING THE CHARACTERIZATION OF WHAT IS LOCAL

Today the standard approach to interpreting findings in Kent tangent space begins with the specification of a privileged subspace (cf. Bookstein, 1996, or Rohlf and Bookstein, 2003), the uniform (affine) transformations (of dimension two for two-dimensional three data, five for three-dimensional data). Affine transformations being those that are "the same everywhere," the complement of the uniform space is the space of transformations that have affine part the identity but that nevertheless are *not* the same everywhere, and are thereby susceptible to searches for local features, such as regions that appear to be responsive to spatially limited causes of form-variation. Often it is sufficient to examine the space of bending by eye, but the uniform subspace is treated more formally, by its own specific matrix operations. The division into uniform vs nonuniform subspaces is under the control of a *bending energy* $\int_{\mathbf{R}^m} \sum \sum \sum_{i,j,k=1,m} (\partial^2 y_k / \partial x_i \partial x_j)^2$ that applies to interpolating functions $y(x)$, where x is a vector of length m (the dimension of the data), and y is an m-vector of functions of x. The "uniform" transformations consist precisely of those interpolating functions y for which this bending energy is zero: the functions for which the first derivatives are constant, which is to say, the exactly linear maps taking square grids into grids of parallelograms and cubes into parallelopipeds. Whether or not the map is linear, the integral can be thought of as a summary of the extent to which the resulting deformed grid cells are *not* all the same—their *variation* as little graphical objects on their own.

The standard thin-plate spline achieves the minimum value of the bending energy for interpolations from one (fixed) landmark configuration to another, and thereby the graphic is matched to the multivariate statistics. In other words: the ultimate justification of the usual thin-plate spline is based in a formalism of what kind of information is to be taken as local, that is, a postulate of bio-mathematics. For instance, because the spline minimizes the variation of first derivatives of the map in this specific sense, where those derivatives are seen to be extreme it must be because the data demand those extrema (cf. the method of creases, Bookstein, 2002), and so those loci (along with the associated rates and directions) themselves become useful derived descriptors of patterns of form-change or form-variation. Morphometrics shares these useful functions with several other domains of applied mathematics, including mathematical geology and environmetrics. In these other communities, the part with zero bending is called *drift*. In this chapter, which is in part an attempt to bridge the biometric literature with these other domains, I use the terms "drift" and "trend" interchangeably.

Thus the standard exposition: but the assumption that the concept of "uniformity" pertains to the *first* derivative has nowhere been confronted.

By a startlingly small change of formalism one easily substitutes a criterion according to which Huxley's (1932) classic growth-gradient, plausibly parameterized as a linear trend in directional derivatives of a map, now is considered global rather than local. This is the thin-plate spline that minimizes the alternative bending energy $\int_{\mathbf{R}^m} \sum \sum \sum \sum_{i,j,k,l=1,m} (\partial^3 y_l / \partial x_i \partial x_j \partial x_k)^2$, sum of the squared *third* derivatives of the map. Now any quadratic trend-surface map (cf. Sneath, 1967) is considered bending-free (inasmuch as all of the third derivatives of a quadratic polynomial are identically zero). Just as the first-order affine term of a spline could be reduced to a standard parametric scheme (decomposition into principal strains, Bookstein, 1991, sec. 6.1), so too a second-order (quadratically) uniform term likewise can be reduced to a standard set of axes

(Bookstein, 1991, sec. 7.4). When I am discussing these new splines, I use the words "drift" and "trend," again interchangeably, to refer to what is now a term with uniform *second* derivatives, just as in the context of the more familiar thin-plate spline they referred to a component with uniform *first* derivatives.

The substitute formalism here applies solely to the *report* of a morphometric data set, leaving the labeled coordinate data themselves completely unchanged, except insofar as the "sliding" algorithms for spline relaxation along curves and surfaces would now result in slightly different semilandmark locations.

In view of the cognitive neuropsychology of human vision, a new discipline will be required for interpreting the alternative grids. The familiar thin-plate grid tends toward uniformity at infinity, so that visually interesting landscapes of the grid must necessarily be from the vicinity of the actual structures supplying coordinate data. The quadratic spline usually has at least one catastrophe (edge of folding) at some distance from the data, an extrapolated locus of no particular biological meaning *per se*, and so the viewer's attention must be much more rigorously vignetted to the vicinity of the actual data driving the interpolations.

With the definition of what is "global" thus altered, so is the characterization of what is "local." The new bending-energy formalism gives rise to a complete new spectrum of principal warps and partial warps, each of which is now growth-gradient-free. There is thus a plausible alternative to the method of creases (Bookstein, 2002) as well. In the context of a global term that was linear (affine), the method of creases represented the extension to higher dimensions of a search for inflections (extrema of the first derivative, zeroes of the second derivative) for an interpolation function in one Cartesian dimension. The one-dimensional equivalent for this new construal of drift (a quadratic trend, that is, a map with constant second derivative) is the analogous search for extrema of the *second* derivative of an interpolation.

Detailed lore of this new version of the old thin-plate spline will be published elsewhere. For formulas, see Wahba, 1990, pages 30–34, the case $d = 3$ in her notation (in which d is the degree of the derivatives that are squared and then integrated to give the bending energy). For $m = 2$, the kernel function is $r^4 \log r$; for $m = 3$, it is r^3, the same as for the one-dimensional cubic spline. Notice that for two-dimensional (pictorial) data, the quadratic trend subspace has six more dimensions than the linear trend subspace, a total of eight instead of two. Here I would like to present just one example of how a familiar data set becomes reinterpreted by this substitution, in which the new spline is applied to the classic Vilmann data set of landmark octagons from the midline neural skull of the rodent. From the complete data archive (Bookstein, 1991) we extract age-specific means at seven stages of development, 7 days of age to 90.

Figure 1 shows the analyses of deformations from the age-7 mean to the mean shapes at ages of 14, 21, 30, 40, 60, and 90 days. In the top row is the ordinary (linear-drift) thin-plate spline; in the second row, the new version with quadratic drift term. Now the "global" trend (third row) is a full 90% of the total signal—the combination of becoming trapezoidal with vertical

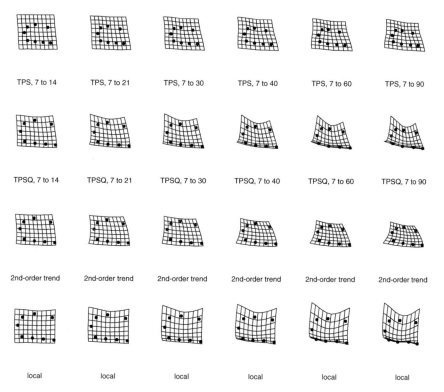

Figure 1.   Thin-plate spline representations for the Vilmann rat midsagittal neurocranial data set, I: analysis of shape change from the age-7 average. Top row: the familiar (affine-trend) thin-plate spline for growth from age 7 to 14, 21, 30, 40, 60, and 90 days of age. Second row: the suggested alternate (quadratic-trend) thin-plate spline (TPSQ) for the same comparisons. Third row: quadratic trend components of the same comparisons. All trend deformations combine vertical compression with a relative shortening of the upper margin of the calva. Fourth row: residuals (local bending) from the quadratic trend. All show a bending along the top of the calva. The data are the eight calvarial landmarks figured in Bookstein (1991), sec. 3.4.1 and listed in Appendix A.4.5. Grid sectors outside the form should not be interpreted.

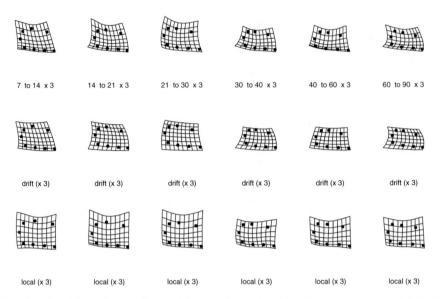

7 to 14 x 3   14 to 21 x 3   21 to 30 x 3   30 to 40 x 3   40 to 60 x 3   60 to 90 x 3

drift (x 3)   drift (x 3)   drift (x 3)   drift (x 3)   drift (x 3)   drift (x 3)

local (x 3)   local (x 3)   local (x 3)   local (x 3)   local (x 3)   local (x 3)

Figure 2. Thin-plate spline representations for the Vilmann rat midsagittal neurocranial data set, II: analysis of shape changes between successive age-specific averages. Upper row: quadratic-trend thin-plate splines. Second row: Quadratic components. Third row: residuals from trend. All deformations are exaggerated threefold.

compression that otherwise required the first partial warp in addition to the linear uniform term. (As is customary [Rohlf and Bookstein, 2003], the trend term is estimated by least-squares in Procrustes distance, not the corresponding term of the exactly interpolating spline.) The bottom row shows the *residuals* from the drift term, that is, the *local* or *multilocal* part of the fit. This points to a new feature, the bending of the upper margin of the calva, that did not emerge as a feature from analyses using the ordinary thin-plate spline.

These were "integral" representations of growth, from the youngest stage in the data to all the later stages. We can investigate more finely in time by studying deformations of each mean stage into the next, Figure 2. Each transformation is exaggerated threefold. From the second row, we see that the quadratic trend, a combination of relative shortening of the upper margin and relative compression of height, characterizes each of the six growth intervals separately.[2]

---

[2] In the estimate of the trend term for this landmark design, the bowing toward the left or right (the y^2 dependence of the transformed x-coordinate) is determined by one single posterior landmark, Opisthion, as it is the only landmark that is found halfway up the form. Issues of appropriate spacing of landmarks in connection with estimates of a quadratic trend term are different from those pertinent to a linear trend.

Residuals from the quadratic trend, bottom row, show a continual infusion of the bending of the upper margin already seen in Figure 1. The successive shape changes are now modest enough in magnitude that none of the interpolations fold in the vicinity of the data, even at threefold extrapolation. Furthermore, they are nearly identical over time—a confirmation that this shape sequence is modeled well by one single relative warp in addition to a global trend term (Bookstein, 1991, sec. 7.6).

The comparison of Figures 1 and 2 is instructive. For instance, we see quite clearly how, as with the ordinary (linear-drift) thin-plate spline, the new quadratic-drift grids are not put forward as models for actual tissue changes, but instead as guideposts for the extraction of biometric features. In Figure 1, the maps over large age intervals induce singularities along the cranial base in the shape changes over larger age intervals, an appearance of "rolling up" that is obviously not realistic. As the quadratic drift shows (third row), there is no such vertical gradient in the landmark data. (Remember these drift terms are fitted by least-squares in shape space, not as the corresponding term of the exactly interpolating spline.) How does this rolling-up arise? Notice the sharp discrepancy across the form between the strong bending of the landmarks of the upper margin and the nearly invariant geometry of those along the lower margin. When there is no bending cost to a quadratic term, as with this particular spline, the exact interpolation represents the ordinate as if it lay on a parabolic cylinder of axis nearly aligned with the line along which landmark locations are hardly changing (the cranial base). The paraboloid can be as bent as it needs to be around that line; hence the line-singularity of the spline map there. According to the least-squares fit to the trend *per se*, however, there is no such trend of vertical derivative in the actual data observed. Figures 1 and 2 are wholly consistent in this respect. Reparametrizing—in effect, *regridding*—in this way often leads to greater constancy of apparent morphogenetic factors. (For an extended discussion in more mathematical language, see Miller et al., 2002.)

An even more local interpretation of these transformations was published previously in Bookstein (2002), where I noted that *as a deviation from an affine trend* each of the 20 rats with complete data had a crease located along this same upper calvarial border. Figure 3 shows the same phenomenon as a property of the quadratic trend rather than the nonaffine bending. (The limitation to twelve of the twenty rats is purely for reasons of legibility.) Alternate columns of this figure represent double the drift and double the local residual characterizing

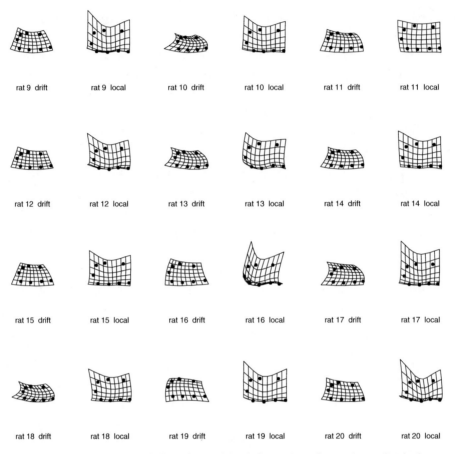

Figure 3. Quadratic-trend thin-plate spline deformations for twelve individual rats, 14 to 40 days of age. Columns 1, 3, 5: global (quadratic) trend. Columns 2, 4, 6: residual bending. Notice the strong resemblance of the individual trend deformations, and the variable bending of the top of the calva in the residual grids. Each deformation is extrapolated twofold for legibility.

the shape change of this octagon for one rat, from age 14 to age 40, for each rat in turn. The doubling is merely for legibility.

There is clearly a very strong family resemblance among these grids that can be quantified by relative warp analysis of this new global term. Ninety percentage of the total Procrustes sum of squares for quadratic trend is carried by one single dimension, the combination of vertical compression and trapezoid formation we have already noted. The remaining variation of the global trend is spherical. The local (bending) part of these transformations is characterized by one single relative warp for which the sample extreme scores belong to animal 11 and animal 20; evidently this component concerns the degree of bending of the upper calvarial margin. The crease noted in Bookstein (2002) for each of these animals actually arises by virtue of the combination of the large-scale trapezoidal process with this bending factor; it is always very near the bottom of that bend. Notice, also, that the extent of this bending along the upper margin is correlated to the appearance of the singularity (here, the rolling-up) of the residual grid along the lower margin already discussed in connection with Figure 1.

The ordinary (affine) uniform term of a shape sample is estimated by its own least-squares projection *a-priori* (Rohlf and Bookstein, 2003), and is thus in a direction that is fixed (up to sampling variation of the grand mean) rather than being computed from the full sphere of directions in Procrustes tangent space in order to optimize some sample criterion, like a mean difference or a relative warp. The same increase in efficiency applies to this new global term. For data in two dimensions, it is estimated using eight degrees of freedom, considerably fewer than the full tangent-space dimensionality (for this example) of 12. Nothing up our sleeve here: We have moved from full shape space to an *a-priori* subspace, just as in the consideration of the more usual linear uniform term itself. The linear affine term explains about 50% of the total variance of this 164-animal data set using two *a-priori* dimensions; the quadratic global term, more than 90% on eight that are likewise *a-priori* (functions of the mean form, not its variation). The additional six dimensions of "growth-gradient" thus explain nearly as much variation as the two dimensions of affine change themselves did, leaving only 9% of the total variance for the remaining four dimensions of local rearrangement. One might expect, however, that in other applications, such as tumor growth, these local terms might have a larger role to play.

SUGGESTION 2: CHANGES IN THE GEOMETRIC STRUCTURE OF GRAPHICAL SUMMARIES

In the standard contemporary toolkit, grids are in essence algebraic formulas (i.e., static objects) depicted as fixed diagrams in some Cartesian coordinate system (perhaps the one aligned with the principal moments of the Procrustes average form). This display was originally developed to suit the case encountered first, the discrete landmark point configuration. As the scope of the new morphometrics was extended, its graphics did not keep pace: the conventional diagram style has not hitherto been extended to exploit the special properties of semilandmarks on curves or surfaces. When the form under comparison is characterized by extended features, it is an impoverishment to report its comparisons by a pictorial grammar making no reference to those features. In this section I sketch three extensions of the current toolkit that might aid the task of biological understanding in particular applications. All the examples in this section pertain to problems of three-dimensional data analysis and display, as for data in two dimensions most of these problems can be handled tacitly by the scientist's retina, which is likewise two-dimensional.

Tumbling the Splined Grid

The components of biological objects rarely are characterized by straight edges or angles of 90°. While the arguments of the three-dimensional thin-plate spline formulas are constant on mutually perpendicular planes, it is not necessary to subordinate the report of a transformation grid to any such algebraic straitjacket. The Edgewarp program package (Bookstein and Green, 2002) includes a display mode that intersects an arbitrary three-dimensional thin-plate spline (incorporating either version, linear or quadratic, of the trend formalism) by an arbitrary sequence of query planes. In effect, the report of a grid is a special case of the report of a solid medical image, for which this sort of "navigation" is proving rather important (Bookstein et al., 2000).

The data for this example, courtesy of Philipp Gunz and Philipp Mitteroecker of the University of Vienna, are part of a larger study of sexual dimorphism and allometry in the anthropoids that is still in preparation. The example involves 533 landmarks and semilandmarks for the comparison of an adult male chimpanzee skull to the skull of a human two-year-old (for the semilandmark scheme, see Mitteroecker et al., this volume). Figure 4 traces the parietal crest of the chimpanzee skull outer surface (black line) and shows one of a series of section planes constructed precisely normal to this ridge curve. The tumbling grid I am calling to your attention is a warped plane over the child's skull that is the deformed image of a square grid moving perpendicular to some curve on the chimp skull. These dynamic displays are difficult to represent on the printed page. From the continuum of deformed squares, Figure 5 shows a selection of eight in three different perpendicular views. In Edgewarp, the actual display also shows them "face-on," not only after foreshortening as demonstrated here.

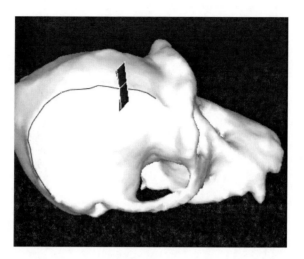

Figure 4. Surface of a chimpanzee skull, with the parietal ridge traced and one normal plane indicated. There are 533 semilandmark points on this surface of 15,909 triangles. Data from Philipp Gunz and Philipp Mitteroecker, University of Vienna (see also chapter by Mitteroecker et al., this volume); display by the `Edgewarp` program package (Bookstein and Green, 2002).

In reality, these multiple grids are displayed as an interactive "movie," with the scientist free to start and stop the navigation at will, rotate any of the sections by 90°, or examine the sequence of local deformations from any point of view.

Bilateral Asymmetry

While the geometric-statistical structure of asymmetry analysis in Procrustes space is thoroughly understood (Mardia et al., 2000), the corresponding graphical methods are not yet adequately explored. The methods apply to any form made up out of some landmarks or curves that are unpaired—that lie along a putative midplane—and also landmarks or curves that are *paired* left and right. In Procrustes space, the description of asymmetry (along with its classical components fluctuating and directional asymmetry) reduces to a comparison of the form or its average with its own mirror-image (in any mirroring plane whatsoever). The set of precisely symmetrical forms of this sort, regardless of the counts of paired and unpaired features, constitutes a mathematical hyperplane, and the mirroring is represented by a vector perpendicular to that hyperplane, for which Procrustes coordinate shifts sum to zero over the paired landmarks

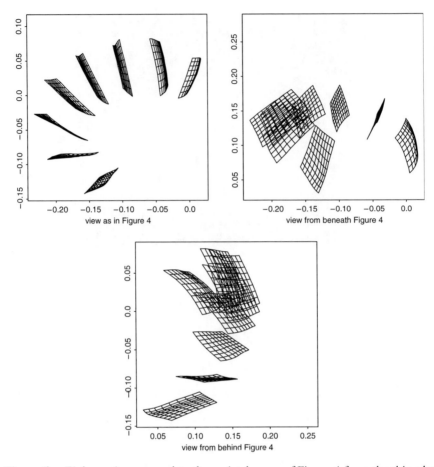

Figure 5. Eight sections normal to the parietal curve of Figure 4 from the thin-plate spline that warps the 533 semilandmarks of the chimp skull onto the homologously placed landmarks for the skull of a human two-year-old. The warped grids are shown in three orthogonal views, of which the first is the view of Figure 4.

as a group (left and right together). In the limit of small deformations, the thin-plate splines for the mirroring deformation, left side vs right side, are inverses.

In the conventional thin-plate spline, the trend for a mirroring deformation like this is limited to a simple shear along the manifold of unpaired structures. It is thereby of hardly any help in the description of asymmetry (since only rarely is asymmetry characterized by such a shear). The spline introduced in the preceding section, with a quadratic trend term, is much better suited to this application. Figure 6 compares the two as applied to a previously published

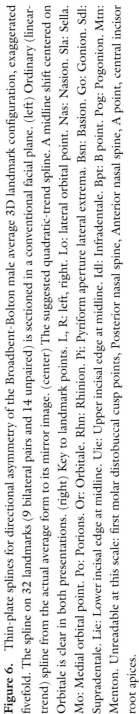
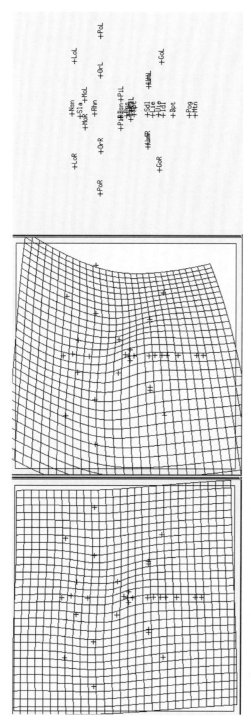

Figure 6. Thin-plate splines for directional asymmetry of the Broadbent-Bolton male average 3D landmark configuration, exaggerated fivefold. The spline on 32 landmarks (9 bilateral pairs and 14 unpaired) is sectioned in a conventional facial plane. (left) Ordinary (linear-trend) spline from the actual average form to its mirror image. (center) The suggested quadratic-trend spline. A midline shift centered on Orbitale is clear in both presentations. (right) Key to landmark points. L, R: left, right. Lo: lateral orbital point. Nas: Nasion. Sla: Sella. Mo: Medial orbital point. Po: Porions. Or: Orbitale. Rhn: Rhinion. Pi: Pyriform aperture lateral extrema. Bsn: Basion. Go: Gonion. Sdl: Supradentale. Lie: Lower incisal edge at midline. Uie: Upper incisal edge at midline. Idl: Infradentale. Bpt: B point. Pog: Pogonion. Mtn: Menton. Unreadable at this scale: first molar distobuccal cusp points, Posterior nasal spine, Anterior nasal spine, A point, central incisor root apices.

threee-dimensional data set, the grand average of all the three-dimensionalized images from the entire Broadbent-Bolton normative male craniofacial sample (see Dean et al., 2000, or Mardia et al., 2000). Figure 6 shows, on the left, the ordinary thin-plate spline for the comparison of the actual mean form to its own mirror-image, extrapolated fivefold for legibility. The sectioning plane to which this grid corresponds is more or less the conventional "facial plane" of orthodontic analysis. The principal feature of the grid is a lateral shift centered on the semilandmark Orbitale. In the center panel is the corresponding grid (i.e., from the identically positioned and scaled sectioning plane) for the quadratic thin-plate spline. The graphics of the midline shift is the same, but now there is a new and more obvious visual feature also, one corresponding very well to the familiar size dominance of one hemifacial complex over the other. The global nature of this asymmetry is even clearer in Figure 7 (left), which shows the quadratic thin-plate spline using a sectioning plane twice as large in every direction. The size contrast is global; the shift of the midline *per se*, merely local. At right in the same figure is the quadratic spline for directional asymmetry when three paired semilandmarks are omitted (both Orbitales, both Medial orbital points, and both lateral extrema of the pyriform aperture—see

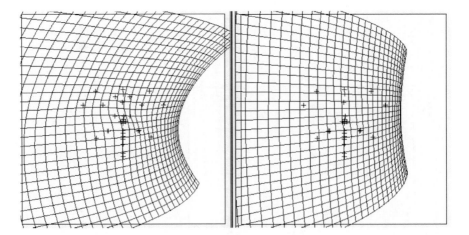

Figure 7.   Large-scale and small-scale aspects of asymmetry. Left: The same quadratic spline, sectioned at twice the scale. Right: The same after three bilaterally paired landmarks in the orbital region are deleted. The local asymmetry (shift of the midline) has now vanished, leaving only the familiar isotropic left–right size difference.

key in Figure 6). The midline shift is now obliterated, leaving only the global quadratic trend for mean left–right asymmetry in every direction of this frontal presentation.

Other Coordinate Systems with Their Own Symmetries

The same `Edgewarp` display mode that underlay the dynamic grids of Figure 5 can also be used to show the actual content of solid biomedical images. One of these images, available without charge from the National Library of Medicine of the United States National Institutes of Health, is in my judgment the most magnificent single medical image yet published: the complete solid Visible Female, "Eve," a fully registered true-color data set of some 7 billion bytes. `Edgewarp` has built-in utilities that access Eve directly over the Internet, and its manual (Bookstein and Green, 2002) explains how to construct dynamic displays of arbitrary sections of Eve and compare them to analogous sections of your own CT or MRI image or any other three-dimensional anatomical resource to which you have access. For publication these continuously moving displays must be reduced, like the grids in Figure 5, to a discrete series of static diagrams in a common coordinate system. Figure 8 (see color insert) shows Eve's corpus callosum as displayed in this manner. The plane facing the viewer is one conventional "midsagittal plane" (plane of approximate symmetry) of this structure. The other planes shown are a selection from a full circle of normal sections of the callosum centered precisely on its midcurve (which is not a plane structure). For an explanation of this construction, along with an example of its medical importance, see Bookstein et al., 2002; the figure here is adapted from the cover of the journal issue in which that article appeared. Figure 9 (see color insert) shows a structure of approximately the same net size, a virtual "preparation" of the diploë of Eve's calva following a curve beginning near the vertex and proceeding down the surface of the parietal bone. The upper half of the figure is a sample of a continuous surface, such as might be possible from an elegant anatomical preparatory; but the lower figure represents an impossible preparation in which this surface is combined with its own normal sections, so as to place it in local anatomical context as well as any medical artist could do.

These two examples have in common the liberation of the *report* of anatomy, or of its comparisons, from any dependence on the coordinate systems characterizing the imaging physics by which the raw data were gathered. When morphometric data were limited to landmark points, these issues did not arise, as landmarks do not specify any particular frame of reference, and all the Procrustes methods for landmark analysis are coordinate-free. But the *biology* of the extensions of the landmark methods to curves and surfaces is not coordinate-free—these structures represent the real processes by which the coordinates have been (literally) *coordinated*. In this way, Figures 4 through 9 hint at a future toolkit of ways in which reports can align with the obvious morphogenetic alignments pertinent to how the data were generated in the first place.

SUGGESTION 3: THE EXTENSION OF CARTESIAN INFORMATION TO DERIVATIVES

Semilandmarks are *lacking* a Cartesian coordinate or two: the information they convey is limited to one direction in the plane (a curve), one direction in space (a surface), or two directions in space (a curve). The combination of the thin-plate spline with Procrustes methods for analysis of this data (cf. Mitteroecker et al., this volume) is straightforward enough that we can ignore the difference between landmarks and semilandmarks (i.e., between data that comes two coordinates at a time vs one coordinate at a time in the plane, or three coordinates vs fewer than three in space) for most morphometric purposes.

Some years ago Bill Green and I published an extension of the landmark toolkit that modifies the formalism in the opposite way. This was the method of *edgels* (edge information at landmarks) of Bookstein and Green, 1993. An edgel is made from a landmark by *adding* additional coordinates—augmenting the actual Cartesian coordinates of location by additional information regarding derivatives of the deformation (presumably based on knowledge coming from edges or textures of the original data). The mathematics of the edgel will not be reviewed here—it is set out in full in the 1993 paper—but the general colonization of three-dimesional data by the old two-dimesional toolkit suggests that now might be a good time to resurrect that possibility as well.

The easiest way to introduce the formalism is by reanalysis of what was actually the first semilandmark data analysis in anthropology: the study of the medial frontal bone published by Bookstein et al. (1999). That study

Figure 8. Corpus callosum of the Visible Female, "Eve," as represented by the pencil command in Edgewarp. The image is in true anatomical color, except for the bright blue, which is some latex injected to keep nasal cavities from collapsing during postmortem sectioning. All planes shown are computed sections through a volume of some 2.7 billion $(300 \,\mu m)^3$ voxels. Large square: approximate midsagittal section. Perpendicular planes: normals to the exactly observed symmetry curve of the callosum at a sampling of positions around the midline. From Bookstein et al., 2002, cover figure.

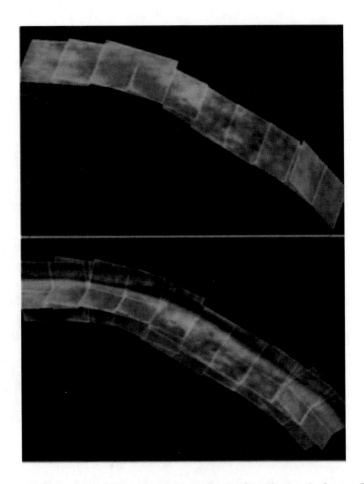

Figure 9. Another view of Eve at much smaller scale. Above: A sheet of sections tangent to the diploë from the vertex laterally along the parietal bone. Below: The same sheet, accompanied by a series of normal planes indicating the cross-section of the calva used to locate the diploë. The lower image evidently does not correspond to any possible real anatomical preparation.

was of a total of 24 anthropoids represented by 15 landmark points and 21 semilandmarks derived from simulated midsagittal sections of actual CT images of 16 *Homo sapiens,* six fossil *Homo,* and two chimpanzees. The principal finding of the paper combined an observation about invariability of the inner border of the frontal bone with an observation about localization vis-a-vis the inner and outer borders in comparisons of modern *Homo* with fossils. The interesting part of the data analysis involved representation of these two borders of the sectioned frontal bone by a pair of semilandmark structures. The corresponding grid (Figure 10, top panel) clearly indicates the concentration of the shape difference at the frontal sinus.

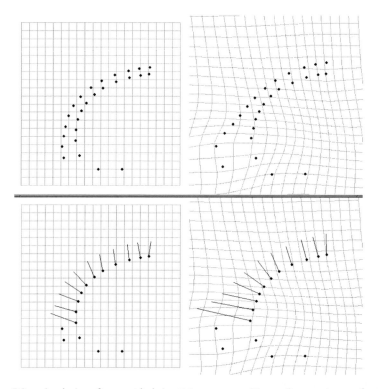

Figure 10. Analysis of an archel in Edgewarp. Top: Comparison of average form of the midsagittal frontal bone and vicinity, 16 *Homo sapiens* vs six fossils, using 26 landmarks and semilandmarks as published in Bookstein et al., 1999. Bottom: A nearly identical grid representing the thickness of the frontal bone by derivative information (edgels) instead of the locations of the second curve of semilandmarks.

The idea of the edgel is that the effect of a grid on pairs of nearby landmarks, or landmarks paired with nearby semilandmarks, can be modeled as an explicit datum pertaining to the *derivative* of the mapping. Below in Figure 10 is the edgel representation of the same deformation scheme. Now there is only one curve of semilandmarks, rather than two, and the separation between inner and outer tables of the frontal bone is represented instead by ratios of change of length (and, to a limited extent, direction) of the little vectors drawn. (In reality, each of these vectors is simply double the separation of the corresponding pair of semilandmarks from the upper panel, projected in the direction normal to the inner curve.)

Evidently this alternative formalism results in very nearly the same description of deformation as the extended two-curve semilandmark representation above. It is, however, a good deal more convenient than the upper version, in that the natural descriptor (the thickness of the frontal bone) appears here as an actual data term, not a derived variable emerging from manipulation of semilandmark coordinates acquired in some sense "separately" (compare the analysis of this thickness in Bookstein et al., 1999, and also in Bookstein, 2002).

Combinations of these two formalisms, semilandmarks and edgels, make possible a great increase in the descriptive power of morphometrics for actual structural questions at multiple scales of geometric observation simultaneously. The original edgel formalism can specify any number of derivatives at a landmark (or, as here, at a semilandmark), and can specify direction and magnitude of directional derivatives separately. The example of Figure 10, dealing as it does with an arc of tissue, might as well be called an "archel"—a curve of semilandmarks with a field of derivatives in the perpendicular direction. Extended to three-dimensional surfaces, these specify thick shells, and so one is tempted to call them "shelels." For semilandmarks on space curves with a full specification of the derivative structure in normal section, we have "tubels" or (the pun is irresistible) "vessels." Bookstein and Green (1994) show, in principle, how to specify curvature changes (second derivatives) by combinations of edgels at close spacing. Extended to three dimensions, these become "ridgels," a notion that will, at long last, objectify the quantitative description of evolutionary changes in ridge curve form (cf. Mitteroecker et al., 2003), upon which so many of our actual evolutionary inferences about skulls rely.

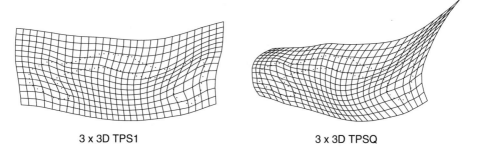

3 x 3D TPS1   3 x 3D TPSQ

Figure 11. Sexual dimorphism of the senile human corpus callosum, exaggerated threefold. Left: In three dimensions, the kernel of the linear-trend thin-plate spline insulates the space surrounding an arch or sheet from reorientations of that structure. Right: The quadratic-trend spline propagates such changes into the surrounding grid. Data are from Davatzikos et al., 1996, as previously reanalyzed in Bookstein, 2003.

There is an interesting interaction between the archel and the algebra of the thin-plate spline as I have been exploring it in this chapter. The kernel function r of the ordinary three-dimensional thin-plate spline is nondifferentiable at landmarks or semilandmarks. As a consequence, transformations applying the old (linear-drift) TPS to sheets of semilandmarks in three dimensions do not constrain the grid to bridge the deformation on one side of the pair of sheets smoothly to the deformation on the other side of the pair. The result (cf. Figure 11, left) is to insulate changes within arches from interaction with the structures around them—not a helpful property. The alternative thin-plate spline introduced in this chapter, with kernel r^3 for data from landmarks or semilandmarks in three dimensions, does *not* have this problem (Figure 11, right). In allowing changes within a sheet of tissue to extrapolate into the tissue normal to the sheet for some distance, it presents an alternative report of local features that may be a better guide to actual morphogenetic processes in several applications.

Toward the Next Revolution

All this has presumed "named location data," but of course there is a tremendous amount of additional information in biomedical images. The companion field of medical image analysis, in particular, has chosen (owing to its historical

origin in more classical applications of signal-processing) to emphasize a
different formalism in which the image contents are represented as functions
of Cartesian coordinates rather than coordinate values *per se*.

In a celebrated editorial some years ago, Rohlf and Marcus (1993)
announced a "revolution" in morphometrics. They may have been more
prophetic than they intended, in that it is typical of revolutions in the
quantitative sciences (cf. Kuhn, 1959) that parts of a data tradition are left
behind in the rush to establish new formalisms that are more powerful in
particular applications. The domain in which any such revolution has taken
place must then spend the next few years in a "complex mopping-up operation" (Kuhn, *op. cit.*) in which practitioners have to go back for the scientific
signals that were left behind in the rush to the new formalism. That is
certainly the case for the new geometric morphometrics, which is overdue
to go back and pick up pictorial information, textures, and all the rest of
the information that was available in earlier data representations (such as
photographs) beyond the landmarks and semilandmarks reviewed here. In
an editorial for a medical image journal (Bookstein, 2001), I tried to set
a context for this "next revolution" by setting out a mathematical formalism in which analysis by image statistics (gray-scale gradients) and analysis
by geometric morphometrics (landmarks and semilandmarks) were equivalent. The trick is to expand the variation of image contents in the vicinity of
a mean image in a Taylor series for shifts of arguments taken at the locations
of actual landmarks; then (at least for small ranges of image variation) the
least-squares functional analysis of the Taylor series expansions approximates
the geometric morphometrics of the same landmark locations. That no simple
worked examples of this next round of techniques are available for inclusion
in this chapter owes solely to the obduracy of certain standing study sections
at NIH.

Realistic expansions of this simple tautology will result in tools that mix
schemes of labeled coordinates with representations of the "remaining"
information over a fairly broad range of statistical styles. The interchangeability
will make possible a radical extension indeed of our current morphometrics
toolkit, an extension combining the explicit coordinates of points, curves,
or surfaces with the "implicit" landmarking carried by the parameters of
those function spaces. In other words, the combinations of semilandmark schemes with derivative information need not be discretized. Instead,
composite data resources could be constructed that in effect balance this

formalism of deformation grid against the present formalism of generalized landmarks (points, curves, surfaces) itself. These extensions, along with others I haven't thought up yet, represent the thrust of a new, much more automated morphometrics, aimed at winnowing large numbers of very large data resources (solid images, or their time-series) for sparse or fugitive signals that have great empirical import should they happen to occur. Such a merger between morphometrics and biomedical image analysis would surely result in a better methodology than either field can presently offer on its own.

ACKNOWLEDGMENT

Preparation of this chapter was supported in part by USPHS grant GM–37251 and contract LM–03511 to the University of Michigan. I am grateful to the morphometrics team in the Institute of Anthropology, University of Vienna (Horst Seidler, Katrin Schäfer, Hermann Prossinger, Philipp Gunz, Philipp Mitteroecker) for comments on an earlier version, and to another group member, volume editor Dennis Slice, for inviting me to contribute even though I was unable to attend the actual symposium session. Edgewarp software, developed by Bill Green of the University of Michigan, is available at no charge for linux operating systems from our website ftp://brainmap.med.umich.edu/pub/edgewarp3.

REFERENCES

Adams, D., Rohlf, F. J., and Slice, D. E., 2004, Geometric morphometrics: Ten years of progress following the "revolution," *Ital. J. Zool.* 71:1–11.

Bookstein, F. L., 1991, *Morphometric Tools for Landmark Data: Geometry and Biology*, Cambridge University Press, New York.

Bookstein, F. L., 1996, Combining the tools of geometric morphometrics, in: L. F. Marcus, M. Corti, A. Loy, G. J. P. Naylor, and D. E. Slice, eds., *Advances in Morphometrics*, NATO ASI Series A: Life Sciences, volume 284. Plenum.

Bookstein, F. L., 2001, "Voxel-based morphometry" should never be used with imperfectly registered images, *Neuroimage* NIMG.2001.0770, 14: 1454–1462.

Bookstein, F. L., 2002, Creases as morphometric characters, in: N. Macleod, ed., *Morphometrics, Shape, and Phylogeny: Proceedings of a Symposium*, London: Taylor and Francis, pp. 139–174.

Bookstein, F. L., 2003, Morphometrics for callosal shape studies, in: E. Zaidel and M. Iacoboni, eds., *The Parallel Brain: Cognitive Neuroscience of the Corpus Callosum*, MIT Press, pp. 75–91.

Bookstein, F. L. and Green, W. D. K., 1993, A feature space for edgels in images with landmarks, *J. Math. Imag. and Vis.* 3:231–261.

Bookstein, F. L. and Green, W. D. K., 1994, Hinting about causes of deformation: Visual explorations of a novel problem in medical imaging, in: F. Bookstein, J. Duncan, N. Lange, and D. C. Wilson, eds., *Mathematical Methods in Medical Imaging III. SPIE Proceedings*, volume 2299, pp. 1–15.

Bookstein, F. L., Schäfer, K., Prossinger, H., Seidler, H., Fieder, M., Stringer, C. et al., 1999, Comparing frontal cranial profiles in archaic and modern *Homo* by morphometric analysis, *Anat. Rec.–The New Anatomist* 257:217–224.

Bookstein, F. L., Athey, B. D., Wetzel, A. W., and Green, W. D. K., 2000, Navigating solid medical images by pencils of sectioning planes, in: *Mathematical Modeling, Estimation, and Imaging*, D. Wilson, H. Tagare, F. Bookstein, F. Préteaux, and E. Dougherty, eds., *Proc. SPIE*, vol. 4121, pp. 63–76.

Bookstein, F. L. and Green, W. D. K., 2002, *User's Manual, EWSH3.19*, posted to the Internet, ftp://brainmap.med.umich.edu/pub/edgewarp3/manual.

Bookstein, F. L., Sampson, P. D., Connor, P. D., and Streissguth, A. P., 2002, Midline corpus callosum is a neuroanatomical focus of fetal alcohol damage, *Anat. Rec.–The New Anatomist* 269:162–174.

Dean, D., Hans, M., Bookstein, F. L., and Subramanyan, K., 2000, Three-dimensional Bolton-Brush growth study landmark data: ontogeny and sexual dimorphism of the Bolton Standards cohort, *Cleft. Palate. Craniofac. J.* 37: 145–156.

Huxley, J., 1932, *Problems of Relative Growth*, Methuen.

Kuhn, T. P., 1959, The function of measurement in modern physical science, in: *Quantification*. H. Woolf, ed., Bobbs-Merrill, Indianapolis, IN, pp. 33–61.

Mardia, K.V., Bookstein, F. L., and Moreton, I. J., 2000, Statistical assessment of bilateral symmetry of shapes, *Biometrika* 87:285–300.

Miller, M. I., Trouvé, A., and Younes, L., 2002, On the metrics and Euler-Lagrange equations of computational anatomy, *Ann. Rev. Biomed. Eng.* 4: 375–405.

Rohlf, F. J. and Bookstein, F. L., 2003, Computing the uniform component of shape variation, *Syst. Biol.* 52:66–69.

Rohlf, F. J. and Marcus, L. F., 1993, A revolution in morphometrics, *Trends Ecol. Evol.* 8:129–132.

Sneath, P. H. A., 1967, Trend-surface analysis of transformation grids, *J. Zool.* 151:65–122.

Wahba, G., 1990, *Spline Models for Observational Data*, Society for Industrial and Applied Mathematics.

CHAPTER THREE

Semilandmarks in Three Dimensions

Philipp Gunz, Philipp Mitteroecker, and Fred L. Bookstein

Today there is a fully developed statistical toolkit for data that come as coordinates of named point locations or landmarks. Because all the statistical methods require these landmarks to be homologous among the specimens under investigation it is challenging to include information about the curves and surfaces in-between the landmarks in the analysis. The problem is that these correspond biologically as extended structures rather than lists of distinct points. This chapter is devoted to the method of semilandmarks (Bookstein, 1997), which allows these homologous curves and surfaces to be studied with the existing statistical toolkit. Information from the *interior* of homogeneous tissue blocks is not accessible by these methods.

An earlier morphometric practice uses some nonlandmark points from curves or surfaces as if they were landmarks: the *extremal points* (*Type* III of Bookstein, 1991) that have definitions like "most anterior" or "widest point." These locations, however useful for traditional distance measurements, are ambiguous

Philipp Gunz, Philipp Mitteroecker, and Fred L. Bookstein • Institute for Anthropology, University of Vienna, Althanstrasse 14, A-1091 Vienna, Austria. **Fred L. Bookstein** • Michigan Center for Biological Information, University of Michigan, 3600 Green Court, Ann Arbor, MI 46103.

Modern Morphometrics in Physical Anthropology, edited by Dennis E. Slice.
Kluwer Academic/Plenum Publishers, New York, 2005.

regarding the one or two coordinates "perpendicular to the ruler." We will call those coordinates *deficient*, and the points to which these coordinates belong, *semilandmarks*. The methodology of semilandmarks this chapter reviews eliminates the confounding influence of the deficient coordinates by computing them solely using the part of the data that is *not* deficient. To be specific, they are treated as missing data and estimated, all at once, in order to minimize the net bending energy (see below) of the data set as a whole around its own Procrustes average. This concept of semilandmarks appeared first in an appendix to the Orange Book (Bookstein, 1991) and was first applied to two-dimensional outline data in Bookstein (1997). Here we explicitly extend the algebra to curves and surfaces in three dimensions and give practical advice on how to collect and interpret this kind of data.

HOMOLOGY

All approaches to landmark-driven morphometrics make one fundamental assumption: that the landmark points are homologous across specimens. The notion of homology invoked in this assumption is not the classic biological notion of that name, which entails similarity of structure, physiology, or development owing to common descent (Ax, 1984; Cain, 1982; Mayr, 1963, 1975; Remane, 1952). In this classic diction, only explicit entities of selection or development can be considered homologous.

Since points *per se* are not likely to be explicit targets of selection, this criterion is too strict—it would rule out almost any use of point coordinates in the course of evo-devo research. Hence for some 30 years morphometrics has used a distinct but related notion of homology, traceable perhaps to an article by Jardine (1969), that centers on variation in the relationships among locations of structures across samples. This notion of homology, often called *geometrical homology*, is embedded in arguments that draw inferences from the appearance of mapping functions, by which we mean the (Cartesian) transformation grid diagrams invented by Albrecht Dürer and rediscovered by D'Arcy Thompson early in the 20th century. The landmarks and semilandmarks that serve as data for the methods of this chapter both arise as careful spatial samples of this underlying mapping function.

For two-dimensional data, landmark locations from photographs or drawings are often sufficient in number to sustain powerful statistical analysis. In three dimensions, however, the number of truly homologous point locations is often very limited. On the skull, true landmarks are typically located on bony processes, at the intersections of sutures, or at foramina (Richtsmeier et al., 1995). But many curving structures lack punctate landmarks of this sort, and on others candidate points cannot be declared with any assurance to correspond across realistic ranges of variation. The method of semilandmarks begins with structures that are known to correspond as parts (the classic biological notion of homology), and then represents them by geometric curves or surfaces that, in turn, generate reasonable mapping functions. In this way the biological notion of homology has most of its power and sweep restored, as the notion of point-landmark has proved too stringent for effective biometrics in most three-dimensional anthropological applications.

OTHER APPROACHES

There have been earlier attempts to include information from regions lacking landmarks in biometric analysis. Moyers and Bookstein (1979) placed *constructed landmarks* using geometric combinations of defined landmarks along lines erected at specific angles to define new landmarks, but the authors later discarded the method because the prerequisite of homology could not be fulfilled by these new points. Extensions of the thin-plate spline to include curvature information can be found in Bookstein and Green (1993, see also: Bookstein this volume) and Little and Mardia (1996). *Smooth surface analysis* introduced by Court Cutting, David Dean, and Fred Bookstein in 1995 (Cutting et al., 1995) combines the idea of constructed landmarks with previous work on parametric averaging of surfaces (Cutting et al., 1993) for analysis of skull shape in a congenital syndrome, Crouzon Disease. After a thin-plate-spline unwarping to the Procrustes average landmark configuration, equally spaced points are declared homologous along ridge curves and geodesics, and then evenly spaced points are declared homologous on the surface patches lofted above triangles or quadrilaterals woven out of those curves. A statistical analysis separates the total geometric signal into one part from the true landmark points, together with the residual. Andresen et al. (2000) automatically capture semilandmarks using shape features by an algorithm called *geometry constrained diffusion*. Ridge lines, characterized by a minimax property of directional surface curvature, are extracted and matched in order to establish object correspondence. The semilandmarks are mapped into Procrustes space and analyzed using principal coordinates.

Each of these approaches is *ad-hoc* or algebraically inconsistent in one or another important way. There are some Procrustes steps, some Euclidean projection steps, some unwarping steps, and some operations of equal spacing, under the control of no particular governing equation. It would be preferable to have an approach that is matrix-driven at all its steps, so that in studies of modest variation, such as characterizes most quantitative evo-devo work in vertebrate zoology, the variation and covariation of all parameters, whether interpreted, modelled, or discarded as nuisance or noise, can be treated together. To build such a protocol, we exploit the very convenient fact that to the thin-plate spline interpolant, the familiar graphical warping/unwarping operator, there is associated a scalar quantity, the *bending energy*, that is a quadratic form in the locations of the "target" landmark structure. Just as a Procrustes analysis minimizes the sum of squares of a set of forms in one feature space (isometric or affine shape coordinates), so the bending energy can be used to minimize an analogous sum of squares in the complementary feature space of bending, a sum of squares that corresponds surprisingly well to the signals by which features of a geometric homology map are interpreted over a wide range of applications. The combination of these two steps results in an essentially unique set of shape coordinates for the semilandmarks describing most realistic assemblages of landmarks, curves, and surfaces on three-dimensional forms.

WHAT IS WRONG WITH EQUIDISTANT SAMPLES?

To justify a method more complicated than equally spaced points on curves or even triangulations of surfaces, it is necessary to show what goes wrong with those temptingly simple alternatives.

Figure 1a shows a rectangle with one landmark in the lower left corner along with 27 other points spaced equally around the outline. Figures 1b and 1c show a slightly different rectangle with two different sets of semilandmarks. In 1b the points are spaced equally along the outline whereas 1c represents the positions that optimize the bending energy (namely, at zero, for affine transformations). The left thin-plate spline grid in Figure 2 shows a remarkably suggestive pattern of gradients and twists. But since they can *all* be made to disappear by respacing of the semilandmarks on the outline, none of this apparent bending is credible (in the absence of corroborative information, for instance from histology, that some tissue sheet did indeed "turn the corner"). The comparisons we publish, and the statistics that support them, need to apply

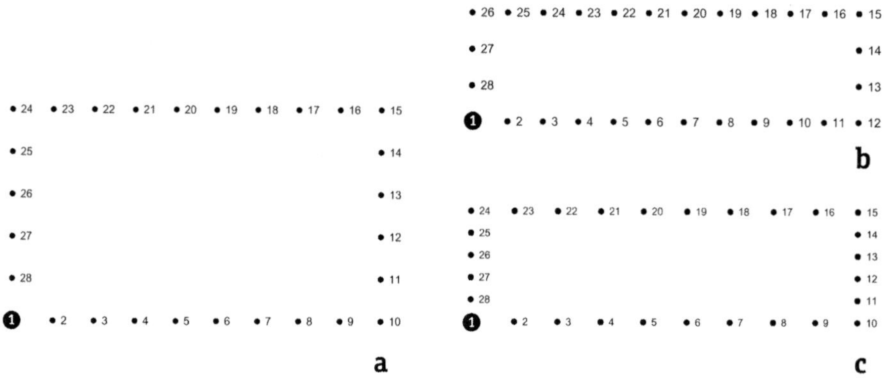

b

• 26 • 25 • 24 • 23 • 22 • 21 • 20 • 19 • 18 • 17 • 16 • 15
• 27   • 14
• 28   • 13
❶ • 2 • 3 • 4 • 5 • 6 • 7 • 8 • 9 • 10 • 11 • 12

• 24 • 23 • 22 • 21 • 20 • 19 • 18 • 17 • 16 • 15
• 25   • 14
• 26   • 13
• 27   • 12
• 28   • 11
❶ • 2 • 3 • 4 • 5 • 6 • 7 • 8 • 9 • 10

a

• 24 • 23 • 22 • 21 • 20 • 19 • 18 • 17 • 16 • 15
• 25   • 14
• 26   • 13
• 27   • 12
• 28   • 11
❶ • 2 • 3 • 4 • 5 • 6 • 7 • 8 • 9 • 10

c

Figure 1. (a) Rectangle with one landmark in the lower left corner along with 27 other points equally spaced around the outline. (b) A more elongated rectangle with the semilandmarks still equally spaced while in (c) the positions are chosen to optimize bending energy (see text).

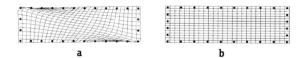

a   **b**

Figure 2. Thin-plate splines corresponding to Figure 1. (a) Deformation grid from the rectangle in 1a to 1b, (b) Deformation grid from the rectangle in 1a to 1c.

in the presence of this ignorance about actual spacing. The only way we can think of to achieve this invariance is to *produce* the spacing as a by-product of the statistical analysis itself.

Figure 3 shows a similar problem for outline structures that bend at large scale. When the points are distributed on the bent form under the criterion of equidistancy (3b), their positions relative to the corners do not correspond to the points in (3a). A better solution is presented in 3c. Figure 4 shows that the TPS grid from 3a to 3c is much smoother (and thus, in this application, less misleading) than the one from 3a to 3b.

While the two generic examples of elongating or bending rectangles might have been resolved in part by placing true landmarks at the corners and *Type III* landmarks at the midpoints of the sides, in many applications Nature is less generous with sharp corners or other shape features that could serve as landmarks. This is the case for the midline of the *corpus callosum*, the structure that connects the two hemispheres of the brain. Figure 5 shows a dataset composed of *corpus callosum* outlines taken from midsagittal sections of MRI scans representing

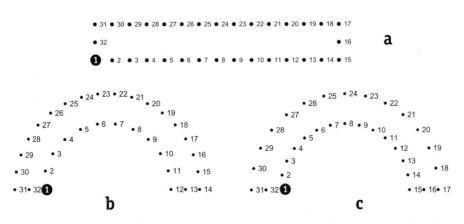

Figure 3. (a) Form with one true landmark in the lower left corner and 31 other points equally spaced along the outline. (b) Bent form with one true landmark (1) and 31 other points in equal spacing. (c) The position of the points now optimizes bending energy.

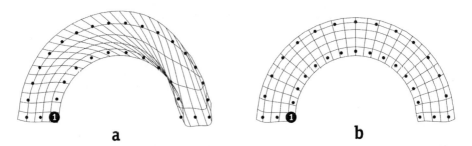

Figure 4. Splines corresponding to Figure 3. (a) Deformation grid from the form in Figures 3a and 3b. (b) Deformation grid from the form in Figures 3a and 3c.

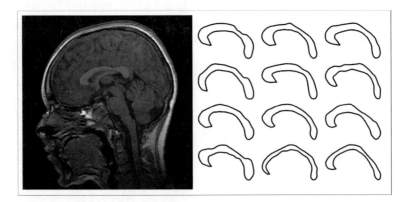

Figure 5. Midsagittal section of an MRI scan and some *corpus callosum* outlines.

normal variation of adults and children. These curves elongate and bend but have only one landmark (*rostrum*).

Figure 6 shows deformation grids between the average (consensus) form and a form with equidistant points compared with the same form captured by semilandmarks. When the consensus is compared to the specimen with the equidistant points, the thin-plate spline deformation grid shows strong local shape differences. Again, there is no reason to consider these changes to be in any way real, as they are very sensitive functions of the arbitrary spacing assumption. By comparison, the points of the form in the lower right corner

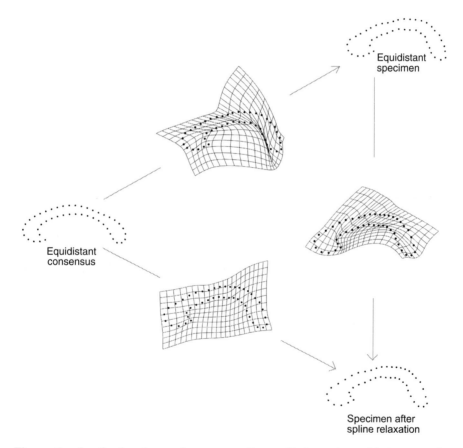

Figure 6. Semilandmarks on the *corpus callosum*. Deformation grids between the consensus form (left side) and a form with equidistant points (upper right) compared with the same form captured by semilandmarks (lower right). Note that the strong local shape effects suggested by the left upper thin-plate spline are an artifact of the equidistancy; the lower left spline, reflecting the real shape difference, is much smoother.

have been placed so as to minimize the bending energy of the interpolation being drawn. Semilandmarks like these can then be treated as homologous, without artifact, in many multivariate analyses, including those that attend to local features of the spline. All those shears along the callosal outline in the upper two grids are meaningless scientifically, regardless of their stark visual effect. These examples typify the ways in which minimizing bending energy serves to protect the scientist from interpreting misleading aspects of a transformation grid in the class of applications concerning us here.

The bending energy that we are minimizing in the course of our analyses is, of course, not itself a biological quantity. It is instead a convenient numeraire for cutting through true ambiguity of empirical representations, rather as the least-squares principle cuts through what would otherwise be the difficult problem of choosing a single line to represent a data scatter. In either case, the aim is to sequester that about which we are truly ignorant (in the linear case, the true errors about predicted values; in the morphometric case, the true spacing of geometric homologues along biologically homologous curves or surfaces). The information that remains stems from the shapes to be studied; arbitrary choices required for digitization have been cancelled out by algorithm. The reason for choosing bending energy instead of, say, Procrustes distance or some other elementary quantity is that in studies where biological interpret-ation will proceed via features of the grid (rather than, for instance, in terms of phenetic distance or some other narrowly systematic quantity), the bending energy corresponds to the visual signal actually detected by the scientist. It is the local contribution to the variation of second derivatives of the interpolated mapping (see Bookstein, 1991), the rate of change of size and shape of those little grid cells in the deformation diagram, and so is very close to a quantific-ation of the actual information purported to demonstrate any finding claimed. Conversely, bending energy is *invariant* under the operations of a Procrustes superposition—it doesn't change under rescaling, translation, or rotation of landmark sets—and so computing with it won't interfere with the established Procrustes part of the current geometric morphometrics toolkit.

ALGORITHM

Algebraic Preliminaries

The first two sections following assemble previously published formulas at the core of the method here. This section presents the algebraic setup for the thin-plate spline on landmarks and for the extension to minimizing bending energy over points sliding on lines, as originally set out by Bookstein (1997). The section on Spline Relaxation on Surfaces shows the notation for the extension to surfaces and section on Flow of Computations sets out the algorithmic cycle we actually follow, which combines these algebraic steps with Procrustes averaging and with projection of semilandmarks from tangent structures back down to actual curving data sets.

In 3D, let U be the function $U(\vec{r}) = |r|$, and consider a reference shape (in practice, a sample Procrustes average) with landmarks $P_i = (x_i, y_i, z_i)$, $i = 1, \ldots, k$. For data in three dimensions, let U be the function $U_{ij} = U(P_i - P_j)$, and build up matrices

$$
\mathbf{K} = \begin{pmatrix} 0 & U_{12} & \cdots & U_{1k} \\ U_{21} & 0 & \cdots & U_{2k} \\ \vdots & \vdots & \ddots & \vdots \\ U_{k1} & U_{k2} & \cdots & 0 \end{pmatrix}, \quad \mathbf{Q} = \begin{pmatrix} 1 & x_1 & y_1 & z_1 \\ 1 & x_2 & y_2 & z_2 \\ \vdots & \vdots & \vdots & \vdots \\ 1 & x_k & y_k & z_k \end{pmatrix}, \quad (1)
$$

$$
\mathbf{L} = \begin{pmatrix} \mathbf{K} & \mathbf{Q} \\ \mathbf{Q}^t & \mathbf{O} \end{pmatrix},
$$

where $\mathbf{O}$ is a 4×4 matrix of zeros. The thin-plate spline $f(P)$ having heights (values) h_i at points $P_i = (x_i, y_i, z_i)$, $i = 1, \ldots, k$, is the function $f(P) = \sum_{i=1}^{k} w_i U(P - P_i) + a_0 + a_x x + a_y y + a_z z$ where $W = (w_1, \ldots, w_k, a_0, a_x, a_y, a_z)^t = \mathbf{L}^{-1} H$ with $H = (h_1, h_2, \ldots, h_k, 0, 0, 0, 0)^t$. Then we have $f(P_i) = h_i$, all i: f interpolates the heights h_i at the landmarks P_i. Moreover, the function f has minimum *bending energy* of all functions that interpolate the heights h_i in that way: the minimum of $\iiint_{\mathbf{R}^3} \sum \sum_{i,j=1,2,3} \left(\frac{\partial^2 f}{\partial x_i \partial x_j} \right)^2$. This integral is proportional to $-W^t H = -H_k^t L_k^{-1} H_k$, where L_k^{-1}, the *bending energy matrix*, is the $k \times k$ upper left submatrix of $\mathbf{L}^{-1}$, and H_k is the corresponding k-vector of "heights" $(h_1, h_2, \ldots, h_k)$. For morphometric applications, this procedure is applied separately to each Cartesian coordinate: $H = (x_1' \cdots x_k' \, 0 \, 0 \, 0 \, 0)$, then $H = (y_1' \cdots y_k' \, 0 \, 0 \, 0 \, 0)$, then $H = (z_1' \cdots z_k' \, 0 \, 0 \, 0 \, 0)$ of a 'target' form.

In the application to real landmarks, the bending energy of the thin plate spline is the global minimum of the integral squared second derivatives. In the case of semilandmarks this same property can be used as a criterion for optimization: The semilandmarks are allowed to slide along tangents to the curve or surface until the bending energy between a template and a target form is minimal. For curves, we seek the spline of one set of landmarks

$X_1 \ldots X_k$ (the template) onto another set of landmarks $\Upsilon_1 \ldots \Upsilon_k$ of which a subset of m elements are semilandmarks. In the following notation, $i_1 \ldots i_m$ is the list of landmarks that actually slide—this is a sublist of the complete list of landmarks/semilandmarks numbered from 1 through k—so that we use a double notation: Υ_i for the ith landmark/semilandmark, but Υ_{i_j} for the jth sliding landmark. Write Υ^0 for the "starting positions" of all these landmarks. The semilandmarks, Υ_{i_1} through Υ_{i_m}, are free to slide away from their starting positions $\Upsilon^0_{i_j}$ along tangent directions $v_{i_j} = (v^x_{i_j}, v^y_{i_j}, v^z_{i_j})$ to the curve, while the remaining (nonsliding) landmarks cannot move from their starting locations Υ^0_i. To simplify the following equations, we rearrange the coordinates of all the Υ^0s, sliding or nonsliding, in a vector of the x-coordinates, then the y-coordinates, then the z-coordinates: $\Upsilon^0 = (\Upsilon^x_1, \Upsilon^x_2, \ldots, \Upsilon^x_k, \Upsilon^y_1, \ldots, \Upsilon^y_k, \Upsilon^z_1, \ldots, \Upsilon^z_k)$. To describe the new positions of the m sliding landmarks Υ_{i_1} through Υ_{i_m}, we set out m parameters $T_1 \ldots T_m$ (T for "tangent"), so that the positions after sliding are $\Upsilon_{i_j} = \Upsilon^0_{i_j} + T_j(v^x_{i_j}, v^y_{i_j}, v^z_{i_j}), j = 1, \ldots, m$. In the ordering of the vector Υ^0, build up a matrix of all these directional constraints together:

$$\mathbf{U}(3k \times m): \quad \begin{aligned} U_{i_j,j} &= v^x_{i_j} \\ U_{k+i_j,j} &= v^y_{i_j} \\ U_{2k+i_j,j} &= v^z_{i_j}, \end{aligned} \tag{2}$$

where $j = 1, \ldots, m$, all other elements zero.

The sliding now proceeds all at once, all the Υ_{i_j} moving from $\Upsilon^0_{i_j}$ to $\Upsilon^0_{i_j} + T_j(v^x_{i_j}, v^y_{i_j}, v^z_{i_j})$, in order to minimize the bending energy of the resulting thin-plate spline transformation as a whole. This bending energy turns out to be

$$-\Upsilon^t \begin{pmatrix} L_k^{-1} & 0 & 0 \\ 0 & L_k^{-1} & 0 \\ 0 & 0 & L_k^{-1} \end{pmatrix} \Upsilon \equiv -\Upsilon^t L_k^{-1} \Upsilon \tag{3}$$

in the notation introduced earlier in this section. It has to be minimized over the hyperplane $\Upsilon = \Upsilon^0 + \mathbf{U}T$ and the solution to this weighted least squares problem is

$$T = -(\mathbf{U}^t L_k^{-1} \mathbf{U})^{-1} \mathbf{U}^t L_k^{-1} \Upsilon^0. \tag{4}$$

Anatomical landmarks affect the sliding of semilandmarks, in that they appear in the matrix $\mathbf{L}$ and thus determine the amount of bending energy associated with translations along the tangent vectors T_j semilandmark by semilandmark.

But if you have sufficiently many semilandmarks in general position (at least six points on curves in 2D or 3D, or at least twelve on surfaces in 3D), the semilandmarks can be made to slide "all by themselves," without any need for landmarks to anchor them. For a great deal more explanation of all these matters, the reader is referred to the original journal publication of Bookstein (1997).

Spline Relaxation on Surfaces

The extension of the formalism for surfaces is straightforward: Instead of tangent vectors the semilandmarks are allowed to slide on tangent planes. We seek the spline of one set of landmarks $X_1 \ldots X_k$ (the template) onto another set of landmarks $Y_1 \ldots Y_k$ of which a sublist $Y_{i_1} \ldots Y_{i_m}$ are free to slide away from their positions along the tangent plane to the surface spanned by two tangent vectors v_{i_j} and w_{i_j} at the original position of the semilandmark. For sliding on tangent planes $Y_{i_j} = Y_{i_j}^0 + T_j^1 v_{i_j} + T_j^2 w_{i_j}$, where v_{i_j} and w_{i_j} are unit vectors spanning the tangent plane. Corresponding to the two directions of sliding per semilandmark, the matrix U of directional information now has two columns per semilandmark: it becomes

$$U(3k \times 2m): \quad \begin{aligned} U_{i_j,j} &= v_{i_j}^x \\ U_{k+i_j,j} &= v_{i_j}^y \\ U_{2k+i_j,j} &= v_{i_j}^z \\ U_{i_j,j+m} &= w_{i_j}^x \\ U_{k+i_j,j+m} &= w_{i_j}^y \\ U_{2k+i_j,j+m} &= w_{i_j}^z, \end{aligned} \quad (5)$$

where $j = 1, \ldots, m$, all other elements zero.

With this matrix U, equation (4) still supplies the m by 2 matrix of parameters T for which the corresponding semilandmark locations Y_{i_j} minimize the bending energy (equation 3). Our actual formalism concatenates these two matrices U, one for the curves and one for the surfaces; all the semilandmarks, on curves or on surfaces, slide at once.

Flow of Computations

Our splined semilandmark analysis begins with any convenient selection of semilandmarks on all the curves or surfaces of a data set. The semilandmarks representing any curve should be equal in number across the sample and should begin in rough geometrical correspondence (e.g., equally spaced); those representing a surface should be reasonably evenly and similarly spaced. Clearly observable curves on surfaces, such as ridges, should be treated as curves instead of surface points; clear local extremes of curvature on curves should be treated as *Type II* landmarks rather than semilandmarks.

The tangents for curves can be calculated as the standardized residual vector of the two neighboring (semi)landmarks. For surfaces the first two principal components of the surrounding landmarks can serve as the two tangent vectors spanning the tangent plane. If the curve or surface is strongly bent in some regions this way of calculating the tangents may become to imprecise. Then either the spacing of semilandmarks should be reduced which results in a larger number of landmarks, or the calculation of tangents should be based on additional information like a denser sampling of curve or surface points or a parametric representation of the curvature (see also section on Data Acquisition).

The *basic algorithm* we propose is then a simple alternation of a Procrustes superimposition with a splined optimization step, each minimizing its own specific sum of squares:

(1) Calculate tangents for each semilandmark.
(2) Relax all specimens against the first specimen.[1]
(3) Compute the Procrustes average configuration.
(4) Calculate new tangents.
(5) Relax all specimens against Procrustes average of step (3).
(6) *Iterate (3)–(5) until convergence.*

During spline relaxation the semilandmarks do not slide exactly on the curves or surfaces but along the curves' or surfaces' tangent structures. Although that reduces the computational effort because the minimization problem is now linear, the sliding along tangents lets the semilandmarks slip off the data. After the relaxation step these points can be placed back on the outline (Figure 7), resulting in a better *extended algorithm*:

(1) Calculate tangents for each semilandmark.
(2) Relax all specimens against the first specimen.

[1] There is no initial Procrustes superimposition step necessary because bending energy is invariant to translation, scaling and rotation.

(3) Replace each slid semilandmark by its nearest point on the (curving) surface.
(4) Compute the Procrustes average configuration.
(5) Calculate new tangents.
(6) Relax against Procrustes consensus of step (4).
(7) Replace each slid semilandmark by its nearest point on the surface.
(8) *Iterate steps (4) to (7) until convergence.*

This extended algorithm should be used when sharp curvatures are present in the data set (e.g., the splenium of the *corpus callosum* data set in Figures 5

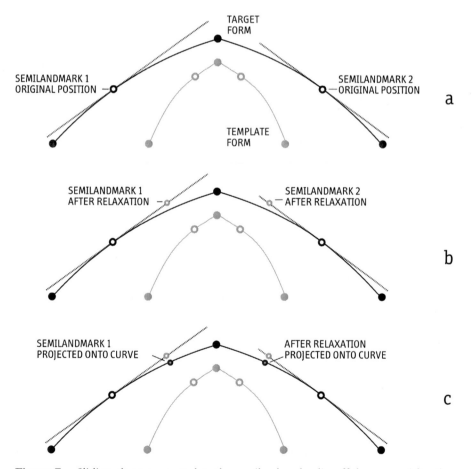

Figure 7. Sliding along tangents lets the semilandmarks slip off the curve. After the relaxation step these points can be placed back on the outline.

and 6). When applying semilandmarks solely to rather smooth curves or surfaces (e.g., human cranial vault) the basic algorithm usually is sufficient.

STATISTICAL ANALYSIS OF SEMILANDMARK DATA

The more semilandmarks, the better as far as the representation of a geometric form is concerned. In general, the sampling of semilandmarks depend on the complexity of curves or surfaces and the detail of curvature that is of interest. Sampling experiments can help finding an "optimal" number of semilandmarks in the sense of how much information additional landmarks would contribute. For the human neurocranial vault we found 150–200 semilandmarks to be a good representation.

In detailed morphometric data sets, there are far more semilandmarks than specimens (e.g., Bookstein et al., 1999, 2003). This would ordinarily cause a problem for parametric statistical inference, and in the case of semiland-marks there seem to be no actual statistical models available (For instance, Gaussian models for individual semilandmark variation, such as the familiar Mardia-Dryden (1998), do not apply to landmarks bound to lines; notions of independent variation at the multiple semilandmarks of a single curve or sur-face do not apply; etc.). The conventional approach to variable-rich problems, which is to project according to the Procrustes or similarly convenient geometry onto a lower-dimensional empirical eigenspace, will often suffice for such classic comparative themes as allometry or sexual dimorphism. But for more general investigations, it is better to abandon classic statistical models altogether for the more modern alternative that presumes nothing about data distributions at all. Hence excess of variables over cases ends up causing no problems. To pursue this issue (the so-called "high-P low-n issue") would take us far outside the limits of this chapter.

In this model-free context, surveys of empirical data sets proceed by principal coordinates of some distance function (the familiar *relative warps*, for instance, are principal coordinates for Procrustes distance). We don't need to review these methods here (but see e.g., Slice, this volume), as they are the backbone of most of the Procrustes empirical findings ever published; indeed, one principal justification for the semilandmark methods here is that they require no changes whatever in that part of the Procrustes toolkit. Statistical *inference*, on the other hand, requires a somewhat more nuanced adjustment. In our practice, most testing goes via the *randomization methods* first sketched by R. A. Fisher and now, with the ubiquity of personal computers, perfectly practical for most morphometric studies (Good, 2000).

In general, a permutation test deals with two sets of data vectors for the same specimens. In anthropological applications, one vector will likely be a set of Procrustes shape coordinates for some landmark/semilandmark configuration, and the other vector might be a group i.d. code, another set of shape coordinates, or a collection of non-morphometric measurements. Some statistic relating these two data blocks (such as a group mean difference, or a multiple correlation) is claimed to be interesting and informative, and we want to test this claim against a null hypothesis of no relationship, without making any assumptions whatever about theoretical distributions (Gaussian noise, etc.). We carry out this challenge by considering, or sampling, all the different ways that the rows of the first data matrix could be paired with the rows of the second (i.e., all the permutations of one case order with respect to the other: hence the name of the technique). For each such permutation, compute the same statistic that was claimed interesting in the first place, and collect all the values of that pseudo-statistic (in general there will be $N!$ of them, where N is the total sample size; for a two-group comparison there will be $N!/k!(N-k)!$ nonredundant permutations) in one big histogram. Under the null hypothesis of no meaningful association between the data blocks, the statistic you actually computed should have been drawn randomly from this distribution. So the P-value (technically, the α-level) of the association you actually observed is, exactly, the fraction of this permutation distribution that equals or exceeds the statistic observed. (The word "exact" in the preceding sentence is the same as in the "Fisher exact test" and other familiar contexts. These methods are exact in the sense in which all F-tests and other multivariate Gaussian-assumption approaches are merely approximate under the same conditions.)

For a very small sample with two groups, 3 cases against 3, there are 20 possible rearrangements of the subgroups; thus the best possible P-value you could get is $1/20$, or 0.05. For 8 cases against 8, this minimum P-value is $1/12870$; that is the largest data set for which we have ever computed the exact permutation distribution. For larger samples, the universal custom is to sample from the permutation distribution using a suitable random-number generator (e.g., this is the alternative offered in Rohlf's and Slice's packages available for free at http://life.bio.sunysb.edu/morph/). The observed data set (i.e., the "permutation" with the actual case order preserved between blocks) is to be taken as the first permutation "sampled." For this Monte-Carlo version, one reports an approximate P-value of m/n with s.e. of $\sqrt{m}/n$, where n is the number of permutations generated and m is the total number of permutations sampled for which the test statistic equals or exceeds the value actually observed. The larger the value of n, the more accurate this approximation.

The power of the test varies by the choice of the test statistic. The authors of this chapter prefer Procrustes distance; others use t-tests, F-ratios, or lower-dimensional multivariate summaries such as T^2. While there is nothing special about a randomization test that is applied to semilandmarks, nevertheless there is something special about the way semilandmarks are used for these statistics. The coordinates that would have been considered "deficient" if these points had been used as landmarks are explicitly *omitted* from statistical manipulations of the resulting Procrustes coordinates. This means, in practice, that the variables consist of distances of the semilandmarks normal to the average curve or surface, or their sums of squares in Procrustes superposition.

WHICH LANDMARKS SHOULD SLIDE?

Bookstein (1991) defined three classes of landmarks, based upon the amount and quality of shape information they represent. Landmarks of TYPE I (juxta-positions of tissues) or II (maxima of curvature) are defined in all coordinates and should as a general rule be taken as real landmarks. TYPE III landmarks, defined by phrases like "the most anterior" or "the farthest from," would better be treated, along with neighboring points, as semilandmarks. Their defini-tions stem from distance measurements and are therefore informative just in one direction. The other coordinates are deficient and should be estimated by the sliding algorithm whenever using landmark based statistics. Occasionally, however, biological questions warrant the sliding of TYPE II and even TYPE I land-marks that lie on curves and surfaces also captured by semilandmarks. These exceptional landmarks include points on sutures (such as *frontomalare orbitale* on the orbital ridge), particularly crossing points of sutures (such as *lambda* or *bregma* on the neurocranial vault). When the *functional* shape of the neuro-cranium is of principal interest, landmarks like these should be allowed to slide. Taken as anatomical landmarks, they yield information mainly about develop-ment instead (i.e., how the particular neurocranium manages to realize its shape ontogenetically or phylogenetically).

The locations of true landmark points interact with the shape of curves and surfaces in producing the final locations of semilandmarks. Omitting landmarks

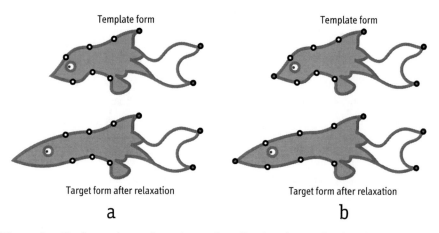

Template form   Template form

Target form after relaxation   Target form after relaxation

a   b

Figure 8. To force a better homology of semilandmarks, use landmarks. (a) For the limited landmark set shown here, minimizing the bending energy slides the semiland-marks (hollow circles) to inappropriate positions. (b) A better set of semilandmarks arises when an additional anatomical landmark (filled circle) is placed at the tip of the "jaw."

when they are easily available, or spacing semilandmarks too sparsely with respect to reliable features of curve or surface form, can produce obviously incorrect results. Figure 8 demonstrates one of these predictable pathologies, as semilandmarks can depart from true landmarks they should accompany or can ignore obvious features of curving form that happen not to have been referred to. We do not set down rules here, as in practice these problems are obvious, once inspected, and the solutions intuitive.

DATA ACQUISITION

The algorithms described above require two kinds of data for each specimen: coordinates of named point locations/landmarks and coordinates of a discrete representation of curving form in-between. In principle there are three types of data sources: discrete landmark point data, discretely sampled curve or sur-face data, and volume image data. When data begin with image volumes, the first step is usually the explicit location of the curves or surfaces along which semilandmarks will be spaced. This operation is computationally demanding and hardly possible in an algorithmic way, in spite of many experiments in the medical imaging literature. For instance, standard methods for mesh genera-tion fail when patches fold anyway. For a typically *ad-hoc* response to this, see the surface remeshing step in Andresen et al. (2000).

In our experience with primate crania it is faster, more accurate and less expensive to locate samples of semilandmarks explicitly by a device like a Microscribe or Polhemus digitizer that directly yields coordinate data. In particular, one powerful source of information about surfaces is the *ridge curve*, the locus of points with an extreme of surface sectional curvature in the direction perpendicular to the curve. Tracing ridge curves on a virtual specimen is quite tedious, whereas tracing them on a physical specimen is relatively easy. When the physical specimen is not available or one wants to measure internal structures and is hence obliged to use a virtual specimen, we recommend using a software package like Edgewarp3D (Bookstein and Green, 2002) that allows the explicit visualization of sectional curvatures. For surface-semilandmarks one needs to extract a dense cloud of points from the volumetric information—this surface extraction is available in many medical imaging software packages.

How to Measure?

All reasonable approaches to this praxis are constrained by the prerequisites of the semilandmark-algorithm. Semilandmarks have to have the same counts on every curve or surface of the assembly and have to be in the same relative order with respect to each other and to any true landmark points that may be present.

Curves in three dimensions: Procedures for three-dimensional curves are a straightforward extension of those for two-dimensional curves. Although the algebraic formalism does not require the endpoints of curves to be point landmarks, we strongly advise that they be delimited in this way, or else semilandmarks might slip off the available curving data in the course of sliding along tangent lines. To get the same number of semilandmarks in the same order on each specimen, it is convenient to begin with points equidistantly spaced along outline arcs, perhaps through automatic resampling of a polygonal approximation to the curve. In the case of volume image data, one can begin with points spaced inversely to radius of curvature on a typical form, then warp them into the vicinity of every other specimen using only true landmarks, and finally, project them down onto the apparent curve in the image. This is how the curves of Bookstein et al. (2002) were located.

Surfaces: Techniques for surfaces differ substantially from those for curves in that except for planes and cylinders there is no straightforward analogue to the notion of "equal spacing." Along with Andresen et al. (2000), we recommend beginning with a hugely redundant sample of points on each surface. These can be produced just by "scribbling" around the surface using a device such as a Microscribe digitizer set to stream mode. Alternatively, one can use point clouds generated by a surface scanner or extracted surfaces from volumetric data.

On one single reference specimen, we then carefully produce a mesh of far fewer points, relatively evenly spaced, by thinning the redundant point cloud (Figure 9). (Points should be more dense near ridges of the surface even if those are not to be treated as curves.) The reference specimen is then warped to the landmark configuration of another specimen. On the surface representation of this target specimen the points nearest to the warped mesh are taken as starting positions of the semilandmarks. This procedure is repeated for every specimen in the data set until every specimen possesses a starting configuration for the subsequent relaxation step.

At the same time, the other surface points of each specimen, the ones not used as semilandmarks, continue to supply information for the sliding algorithm; the two dominant eigenvectors of their variation in small neighborhoods around the

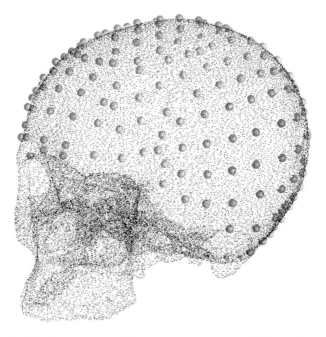

Figure 9. By thinning a dense, discrete representation of the surface we produce a mesh of relatively evenly spaced points, which are then used as starting positions for the semilandmarks (Gunz et al., 2002; Mitteroecker et al., 2004).

semilandmarks are used to specify the vectors v_{i_j}, w_{i_j} of the tangent planes along which they slide. The slid semilandmark can be projected down to the original surface according to the quadric approximation of the surface perpendicular to this best-fitting plane or any other parametric representation like a thin-plate spline.

EXAMPLE

We illustrate the method of semilandmarks using a sample of 52 human crania to study sexual dimorphism (this sample is part of the larger data set of Bernhard, 2003). On each of the 20 adult males, 20 adult females, and 12 subadults we placed 435 landmarks: 37 anatomical landmarks, 162 semilandmarks on three-dimensional-ridge curves, and 236 semilandmarks on surfaces. Most of the anatomical landmarks are in the face and cranial base, with only a few on the neurocranium. The semilandmarks are distributed on seven curves and on the surface of the neurocranial vault. Landmarks and semilandmarks were captured by a Microscribe G2X, and the surfaces resampled as explained in the previous section. All data handling and statistical analysis was done using *mathematica*-routines programmed by the authors.[2] The data set was treated by the basic algorithm described in the section on Flow on Computations. The Procrustes coordinates of the resulting semilandmark locations are shown in Figure 10.

A plot of the first pair of relative warp (RW) scores (Figure 11) shows that the first RW represents ontogenetic development with the children at one extreme and the male adults at the other. Figure 12 visualizes RW1 as a three-dimensional TPS grid computed using all 435 points but drawn as if restricted to the midsagittal plane only. There is general enlargement of the face relative to the neurocranium, marked prognathism, and maxillary extension. Figure 13 visualizes RW2 by the effect of the corresponding TPS on the triangulated surface from one single typical specimen. The effect of RW2 is mostly on relative cranial width.

We performed a Monte Carlo permutation test to assess the statistical significance of the shape difference between adult males and adult females. Using Procrustes distance as test statistic, in 116 out of 3,000 cases the distance between

---

[2] Two-dimensional semilandmarks can conveniently be handled by existing software packages: Bookstein & Green's Edgewarp2D and James Rohlf's "TPS"-programs. Three-dimensional semilandmarks are available in Edgewarp3D (Bookstein and Green, 2002).

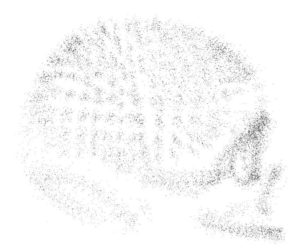

Figure 10.   Procrustes coordinates of 52 *H. sapiens* crania.

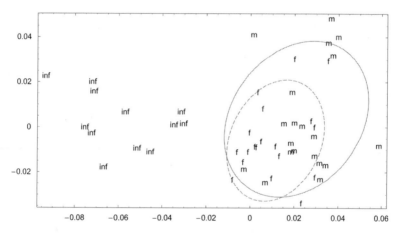

Figure 11.   Scores of the first relative warp against the second for the full data set. Individuals labelled with "inf" are children, "m" are male adults and "f" are female adults. Note that the 75% confidence ellipse for females (dashed) lies within the male variation (solid ellipse).

randomly relabeled groups was equal to or larger than the actual distance; hence the significance level of the dimorphism is $P \sim 116/3,000 \approx 0.04$. Figure 14 exaggerates this mean difference by a series of factors in both directions. Females have higher orbits and males wider ones; females have a smaller alveolar process, males a broader and more prognathic upper jaw; females have a somewhat globular neurocranial shape, smaller zygomatic arches, and a less pronounced supraorbital torus.

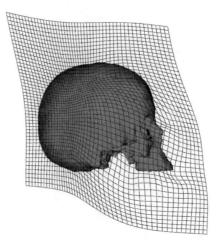

Figure 12. Visualization of the first relative warp (the abscissa of Figure 11) as a midsagittal thin-plate spline. Note the relative enlargement of the face during postnatal development. The specimen shown is the template specimen; the effects of the grid are exaggerated by a factor 4.5.

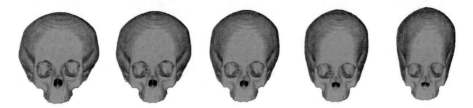

Figure 13. Visualization of the second relative warp as a series of unwarped specimens.

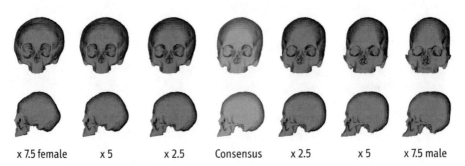

x 7.5 female   x 5   x 2.5   Consensus   x 2.5   x 5   x 7.5 male

Figure 14. Visualization of sexual dimorphism of *H. sapiens* in a sample of 40 adult specimens. The consensus form in the middle is unwarped to the female mean (left side) and to the male mean (right side). The shape differences are exaggerated to ease interpretation.

P-values like the 0.04 reported just now won't vary much by spacing of evenly spaced semilandmarks, as the Procrustes distances to which they contribute are so redundant. But changes in the coverage of a curving form (addition or deletion of parts, or analysis first by curves and then by surfaces) can alter the strength of statistical findings to an arbitrary extent. There is no general solution to this problem, because a *P*-value is not the answer to any sort of scientific question. As shown, however, the Procrustes methods recommended here result in visualizations of form change in every region of an extended structure. Statistical inferences can go forward quite well in terms of the parts separately even when sliding is in terms of an overall bending energy formalism such as that used here. For an example, see Marcus et al. (1999).

Figure 15 divides this empirical mean difference into a component for static allometry (i.e., the regression of each shape coordinate upon Centroid Size) and a remainder. The upper row shows the relocation of each of the 435 landmarks or semilandmarks that is predicted by sexual size dimorphism; the lower row, the remainder of the actual mean landmark or semilandmark shift between the sexes. The difference between allometry and residual is clearest in the parietal bone,

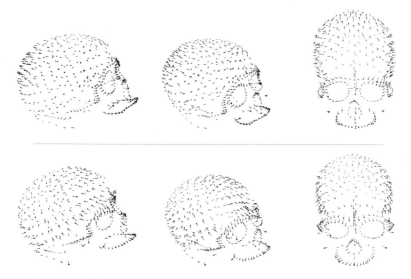

Figure 15. Sexual dimorphism within the adult subsample, separated into allometric and non-allometric components (see text). The differences between allometry (upper row) and non-allometry (lower row) are most visible in the parietal bone, the zygomatic region, the piriform aperture, and the orbits.

the zygomatic region, the piriform aperture, and the orbits. The multivariate shape vectors for allometry and sexual dimorphism have an angle of 76.3°.

Figure 16 is a different decomposition of the same total sexual dimorphism signal. The left column shows the total mean shift as a little vector at each landmark or semilandmark. When the female consensus configuration is warped to the male using only the true anatomical landmarks, the true landmarks are exactly on the average male position but the semilandmarks' positions are just estimated by the true ones. The middle column of Figure 16 shows this true landmark-driven warping as little vectors—this is the technique of Ponce de Leon and Zollikofer (2001). The picture comes close to the left one because a lot of the information about sexual dimorphism is captured by the traditional landmarks already. The right column shows the residuals from the mean female configuration to the estimated male configuration from the middle column. Notice that many regionally specific aspects of the dimorphism—especially the parietal bosses, the lower temporal bone, the orbits, and the alveolar process—are not accounted for by shifts of landmarks alone. Although a major part of shape change of curves and surfaces in this sample can be reconstructed from landmark positions only, other important local features can be accounted for only by exploiting the additional information in semilandmarks.

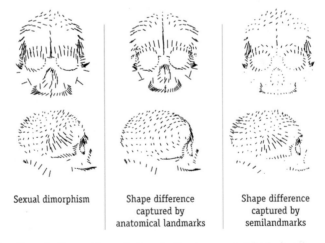

| Sexual dimorphism | Shape difference captured by anatomical landmarks | Shape difference captured by semilandmarks |

Figure 16. Sexual dimorphism shown as Procrustes residuals between adult male and female average forms (left). Shape differences that are captured by the anatomical landmarks (middle), and shape differences captured by semilandmarks (right) after a landmark-driven warping. All shifts are exaggerated by a factor 5.

ACKNOWLEDGMENTS

We want to thank Markus Bernhard, Peter Brugger and Daniela Prayer for access to their data, Horst Seidler, Katrin Schaefer, Dennis Slice, and Gerhard Weber for their advice and support. Our research is supported by the Austrian Ministry of Culture, Science and Education, and the Austrian Council for Science and Technology P200.049/3-VI/I/2001, GZ 200.093/3-VI/I/04 and by the Austrian Science Foundation P14738.

REFERENCES

Andresen, P. R., Bookstein, F. L., Conradsen, K., Ersboll, B. K., Marsh, J. L., and Kreiborg, S., 2000, Surface-bounded growth modeling applied to human mandibles, *IEEE Trans. Med. Imaging* 19:1053–1063.

Ax, P., 1984, *Das phylogenetische System. Systematisierung der lebenden Natur aufgrund ihrer Phylogenese.* G. Fischer, Stuttgart.

Bernhard, M., 2003, Sexual dimorphism in the craniofacial morphology of extant hominoids. Ph.D. Thesis, University of Vienna, Austria.

Bookstein, F. L., 1991, *Morphometric Tools for Landmark Data: Geometry and Biology,* Cambridge University Press, Cambridge (UK), New York.

Bookstein, F. L., 1997, Landmark methods for forms without landmarks: Morphometrics of group differences in outline shape. *Med. Image. Anal.* 1:225–243.

Bookstein, F. L. and Green, W. D. K., 2002, *Users Manual, EWSH3.19,* ftp://brainmap.med.umich.edu/pub/ewsh.3.19.manual.

Bookstein, F. L., Sampson, P. D., Connor, P. D., and Streissguth, A. P., 2002, Midline corpus callosum is a neuroanatomical focus of fetal alcohol damage, *Anat. Rec.* 269:162–174.

Bookstein, F. L., Gunz, P., Mitteroecker, P., Prossinger, H., Schaefer, K., and Seidler, H., 2003, Cranial integration in Homo: Morphometrics of the mid-sagittal plane in ontogeny and evolution, *J. Hum. Evol.* 44:167–187.

Bookstein, F. L., Sampson, P. D., Streissguth, A. P., and Connor, P. D., 2001, Geometric morphometrics of corpus callosum and subcortical structures in the fetal-alcohol-affected brain, *Teratology* 64:4–32.

Bookstein, F. L., Schaefer, K., Prossinger, H., Seidler, H., Fieder, M., Stringer, C., et al. 1999, Comparing frontal cranial profiles in archaic and modern homo by morphometric analysis, *Anat. Rec.* 257:217–224.

Cain, A., 1982, On Homolgy and Convergence, in: *Problems of Phylogenetic Studies,* K. Joysey and A. Friday, eds., Academic Press, London.

Cutting, C., Bookstein, F. L., Haddad, B., Dean, D., and Kim, D., 1998, A spline-based approach for averaging three-dimensional curves and surfaces, in: *Mathematical Methods in Medical Imaging II*, J. Wilson and D. Wilson, eds., SPIE Proceedings.

Cutting, C., Dean, D., Bookstein, F. L., Haddad, B., Khorramabadi, D., Zonneveld, F. W., and McCarthy, J. G., 1995, A three-dimensional smooth surface analysis of untreated Crouzon's syndrome in the adult, *J. Craniofac. Surg.* 6:444–453.

Dryden, I. L. and Mardia, K. V., 1998, *Statistical Shape Analysis*, John Wiley & Sons, New York.

Gunz, P., Mitteroecker, P., Teschler-Nicola, M., and Seidler, H., 2002, Using Semi-landmarks on surfaces to analyze a Neolithic hydrocephalus, *Am. J. Phys. Anthropol. Suppl.* 33:79.

Mitteroecker, P., Gunz, P., Teschler-Nicola, M., and Weber, G. W., 2004, New morphometric methods in Paleopathology: Shape analysis of a Neolithic Hydroceph-alus, *Computer Applications and Quantitative Methods in Archaeology.*

Good, P., 2000, *Permutation Tests: A Practical Guide to Resampling Methods for Testing Hypotheses*, Springer, New York.

Jardine, N., 1969, The observational and theoretical components of homology: A study based on the morphology of the dermal skull-roofs of rhipistidian fishes, *Biol. J. Linn. Soc.* 1:327–349.

Marcus, L. F., Frost, S. R., Bookstein, F. L., Reddy, D. P., and Delson, E., 1999, Comparison of landmarks among living and fossil Papio and Theropithecus skulls, with extension of Procrustes methods to ridge curves. Web publication, http://research.amnh.org/nycep/aapa99/aapa6.html.

Mayr, E., 1963, *Animal Species and Evolution*. Harvard University Press, Cambridge, MA.

Mayr, E., 1975, *Grundlagen der zoologischen Systematik*: P. Parey.

Moyers, R. E. and Bookstein, F. L., 1979, The inappropriateness of conventional cephalometrics, *Am. J. Orthod.* 75:599–617.

Ponce de Leon, M. S. and Zollikofer, C. P., 2001, Neanderthal cranial ontogeny and its implications for late hominid diversity, *Nature* 412(6,846):534–538.

Remane, A., 1952, *Die Grundlage des natürlichen Systems, der vergleichenden Anatomie und der Phylogenetik*. Akademische Verlagsgesellschaft Geest und Portig, Leipzig.

Rohlf, F. J., 1993, Relative warp analysis and an example of its application to mosquito wings, in: *Contributions to Morphometrics*, L. F. Marcus, E. Bello, and A. García-Valdecasas, eds., Monografias, Museo Nacional de Ciencies Naturales, Madrid.

CHAPTER FOUR

An Alternative Approach to Space Curve Analysis Using the Example of the Neanderthal Occipital Bun

David Paul Reddy, Katerina Harvati, and Johann Kim

INTRODUCTION

In the course of several morphometric analyses that used outlines or curves (also simply called lines, or often "space curves" in three-dimensional), which had been resampled to produce semilandmarks, it became clear to the authors that an alternative to conventional techniques for reducing unwanted variance caused by the placement of semilandmarks at regular intervals along the curves (this variance being considered an artifact of resampling, and therefore neither of biological nor statistical significance) could be devised to take better advantage of the higher density of information in the original unresampled curves.

David Paul Reddy • Vertebrate Paleontology, American Museum of Natural History, 79th St. at Central Park West, New York, NY 10024. **Katerina Harvati** • Department of Anthropology, New York University, 25 Waverly Place, New York, NY 10003. **Johann Kim** • Vertebrate Paleontology, American Museum of Natural History, 79th St. at Central Park West, New York, NY 10024.

Modern Morphometrics in Physical Anthropology, edited by Dennis E. Slice.
Kluwer Academic/Plenum Publishers, New York, 2005.

A remark by F. James Rohlf that such a technique could have advantages, but that it had not yet been implemented, proved to be the catalyst for development of the algorithm presented here and for its application to the re-analysis of data collected by Harvati, which has been the subject of other papers and abstracts (Harvati, 2001; Harvati et al., 2002; Harvati and Reddy, in prep.).

Various techniques to minimize extraneous variation along curves as an artifact of resampling have been and are employed in geometric morphometric applications in physical anthropology. Dean et al. made a study of a series of space curves (glabellar and lateral brow ridge, temporal line, coronal suture, and superior nuchal line) considered individually in *Homo erectus* and modern *Homo* using a chi-square test to classify transitional specimens (Dean et al., 1996). A two-step process was used to construct first an average curve by averaging the points obtained by resampling individual splined curves, which had been aligned to best fit at equal intervals of arc length, and then the average curve was splined and resampled at equal intervals of arc length, planes projected orthogonal to the tangents of points resampled at equal arc length on the splined average curve, and the intersections of the original splined curves with these orthogonal planes taken to produce a set of resampled curves with minimized variance. The points on these resampled curves were then averaged within group (*H. erectus* and modern) to produce an average curve for each group. Curves from specimens labeled as transitional were then fit endpoint-to-endpoint with each of the group average curves (though in our opinion this would seem to have the effect of reintroducing some variance along the curve, assuming the curves are not again resampled) and rotated around the chord between the joint endpoints for minimum variance. Summed distances between each transitional curve and the fit and rotated group average curves were calculated and compared to the same statistics calculated within-group, and chi-squared and empirical probabilities used to make assignments of transitional specimens to one of the two groups.

Bookstein et al. undertook a study very similar in aim to the present study, comparing inner and outer mid-sagittal frontal cranial profiles in archaic and modern *Homo* (Bookstein et al., 1999) using Procrustes analysis, sliding, and permutation tests. Bookstein's classic "sliding" technique (Bookstein, 1997) minimizes the mean-squared variance between uniformly sub-sampled or resampled curves which have already been fitted by Generalized Procrustes Analysis (GPA), by allowing the semilandmarks, those resampled points along the curves that are not Type I, II, or III landmarks (Bookstein, 1991), to slide along the curve. The sliding technique minimizes or "relaxes" the bending energy of the resampled points, using the model known as the "thin-plate spline," which is based on the physics of the deformations of an infinite and infinitely thin plate of metal (Bookstein, 1989). It should be noted that in these studies the curves were deliberately sampled with a uniform number of semilandmarks.

A completed manuscript by Marcus and coauthors applies Bookstein's sliding to ridge curves on fossil and living Papio and Theropithecus (Marcus et al., 2000). In that study landmarks and curves were collected in three-dimensions using a jointed three-dimensional digitizing arm, and the curves were densely and nonuniformly sampled. The curves were subsequently uniformly resampled, and were then projected into the Procrustes space defined by the landmarks alone and slid by Bookstein. The goal of the study was to see what additional information could be gleaned from the statistical analysis of the curves.

What all these studies have in common is that the sliding or relaxation is accomplished on uniformly sub-sampled or resampled curves. The technique proposed in this chapter attempts to provide an alternative to conventional techniques for reducing extraneous variation along resampled curves, accomplishing the same functional goals, while preserving as much fidelity to the original space curve information as possible. While our technique does not require any "typed" landmarks to exist in the input curves, it is advisable to have important features, such as the endpoints of the curve segments, fixed by biologically meaningful landmarks.

MATERIALS

The occipital "bun," or chignon, is one of the most frequently discussed Neanderthal characteristics and often considered a derived Neanderthal trait. The presence of a weak occipital bun, or "hemibun," in many Late Paleolithic European specimens has been seen by some as evidence of continuity or interbreeding between Neanderthals and early modern humans in Europe, particularly in the Central European fossil record and the Mladec crania (see Harvati, in prep. for further citations). The occipital bun is described variably as a posterior projection of the occipital squama or a great convexity of the occipital plane, and is often associated with the presence of a depression of the area around lambda on the occipital and parietal bones. This trait has been applied to a range of morphological patterns and is difficult to assess using either traditional caliper measurements (Dean et al., 1998; Ducros, 1967) or landmark-based geometric morphometrics methods (Harvati, 2001; Yaroch, 1996). It is therefore usually described qualitatively.

Along with the previous work (Harvati, 2001; Harvati et al., 2002; Harvati and Reddy, in prep.), this study evaluates the chignon morphology quant-itatively and assesses its usefulness in separating Neanderthals from modern humans, as well as the degree of similarity of the Late Paleolithic "hemibuns" to the Neanderthal occipital buns. This approach uses geometric morphomet-ric analysis of space curves of the midline plane of the posterior part of the skull, which outlines the occipital bun in lateral view. The sample consists of nine recent modern human populations and a fossil sample comprising sev-eral Middle and Late Pleistocene hominid specimens. The modern human populations consist of 20–30 individuals each, comprising a total of 255 spe-cimens. Only adult crania were included, as determined by a fully erupted permanent dentition. Sex was unknown in most cases and was assessed by inspection during data collection and from the literature. When possible, equal numbers of male and female specimens were measured. For further details on the recent human samples see Harvati (2001); also Harvati and Reddy (in prep.).

The fossil human sample included nine Neanderthal specimens from Europe and the Near East (Amud 1, Circeo 1, La Chapelle, La Ferrassie 1, La Quina 5, Saccopastore 1, Shanidar 1, Spy 1, Tabun C1); two pre-Neanderthal specimens (Biache, Reilingen); and eight Late Paleolithic anatomically modern humans from Europe (Cro Magnon 1 and 2, Mladec 1, 4, 5, and 6, Predmosti 3 and 4). Where the original fossils were unavailable, high-quality casts from the Anthropology Department of the American Museum of Natural History were measured. As most fossil specimens did not preserve a complete nuchal plane, the analysis was limited to the posterior cranial midline plane from bregma to inion, rather than the complete outline from bregma to opisthion.

While previous studies showed clear distinctions between the mean morphology of modern human, Middle and Late Paleolithic, and Neanderthal populations, some Late Paleolithic specimens were misclassified as Neander-thal when subjected to a discriminant analysis. The present study aims to improve upon the previous results by attempting to increase the fidelity of the data, while maintaining the same degree of sub-sampling and while removing extraneous variation "along the curve" in a manner similar to the Bookstein sliding technique.

THE METHOD

The data for the present study are the same original data as the previous studies (Harvati, 2001; Harvati et al., 2002; Harvati and Reddy, in prep.). They were collected as three-dimensional outline curves composed of landmarks and semilandmarks. All data were collected by Harvati, using the Microscribe 3DX digitizer, with specimens placed in the Frankfurt horizontal plane. Two curve segments were collected in three dimensions between standard osteometric landmarks, from bregma to lambda and from lambda to inion (Figure 3). The two curve segments were resampled so that each specimen comprised the same number of equivalent semilandmarks for each curve segment. To produce the semilandmarks, resampling was done using linear interpolation between original curve points at equal distances along the integrated curve lengths using a custom C program written by Reddy. The two segments were then concatenated, creating a single curve with a total of 25 points, the two endpoints and 23 semilandmarks. This number was chosen by trial-and-error as this was the minimum number that seemed to preserve the visual impression of the original curves. A copy of this data was fit by Generalized Procrustes Analysis (GPA) using Morpheus (Slice, see http://life.bio.sunysb.edu/morph/). This procedure aligns the specimens by translating, rotating, and scaling them for size, so that remaining differences are due to "shape."

Next, the transformation matrix between each set of original, unaligned landmarks and its aligned counterpart was calculated. The affine transformation matrix x of A onto b can be calculated using the *"normal equations"* of linear algebra (Strang, 1976)

$$x = (A^{\mathrm{T}}A)^{-1}A^{\mathrm{T}}b.$$

where A is the original, unaligned specimen with an additional dimension whose values are all 1, to convert the resulting matrix from a projection matrix to a transformation matrix, A^{T} is the transpose of the matrix A, and b is the aligned specimen. While the formula is for an affine projection, the constraints imposed by the rigid transformation provided by Morpheus restrict this operation to translation, rotation, and scaling, with no shearing component. The projection or transformation of A onto b is then

$$p = Ax = A(A^{\mathrm{T}}A)^{-1}A^{\mathrm{T}}b.$$

As this procedure was coded in MATLAB, it was both faster computationally and potentially more stable numerically to calculate the transformation matrix using the "mldivide" or backslash operator, which uses QR Factorization to accomplish the same operation, avoiding the matrix inversion step:

$$x = A \backslash b; p = A * x;$$

The unresampled curve for each specimen was then projected by its transformation matrix into the Procrustes space of its resampled counterpart (see Figures 1 and 3). Mean positions for each resampled semilandmark point were calculated from the resampled, GPA'ed curves (see Figures 2 and 4), and mean tangents to the curve were calculated from normalized tangents at points along the unresampled curves closest to the resampled semilandmarks. The endpoints of the curves, anchored at Type I landmarks, were considered fixed and were excluded from these calculations. These mean positions and mean tangents were then used to define perpendicular planes through the data, with the mean positions defining points of rotation for the planes, and the tangents defining normals for the planes.

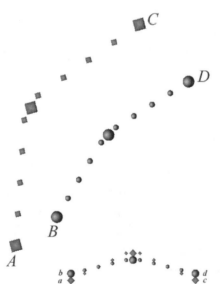

Figure 1. Cartoon of the unresampled points being carried along in a rigid rotation and scaling into Procrustes space with the resampled points. The points on the original curves (small spheres and rhomboids) are carried along with the resampled points (large spheres and rhomboids). The curves AC and BD are projected into the Procrustes space as ac and bd.

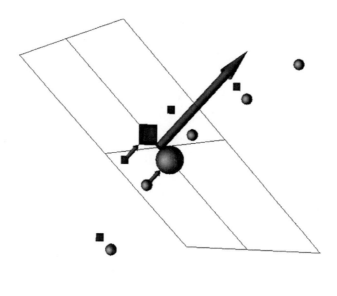

Figure 2. Cartoon of the construction of the mean line points and tangents. The base of the large rocket is placed at the mean of the resampled points, and the point of the large rocket illustrates the tangent. This becomes the normal to the plane, and the points on the unresampled line are relaxed onto the plane, illustrated by the small rockets.

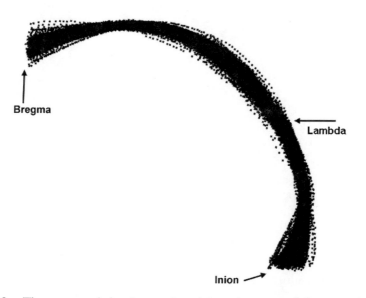

Figure 3. The unresampled points projected into the space of the resampled and Procrustes aligned points.

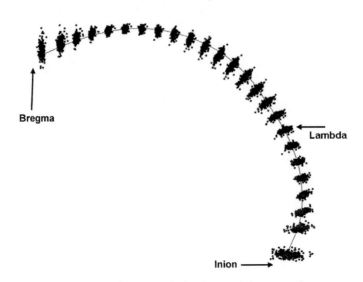

Figure 4.   The resampled points and the mean line.

Next, the projection λ of each point in each of the unresampled curves on the perpendicular planes was calculated as

$$\lambda = n^{\mathrm{T}}(c - d)$$

where c is the pivot point of the plane, n is the normal, and d is successively each point in the curve. The point closest to the plane, namely the one having the minimum magnitude of λ, was then projected onto the plane, the projection z being

$$z = d_{\min} + \lambda_{\min} n$$

These steps applied to the actual data and the final result can be seen in Figures 3, 4, and 5.

It should be noted here that for curves whose trajectories are very convoluted, this definition of "closest" can fail, as the curves may cross the infinite perpendicular planes more than once, hence a more sophisticated test incorporating proximity would have to be devised. Also, we took the projection of the closest point onto the plane as an approximation of the intersection of the curve segment crossing each plane with the plane itself. In cases where the original, unresampled curves are perhaps less densely sampled than in the present study, this procedure could introduce errors, and it would then be advisable to correctly calculate the intersection, a more expensive calculation.

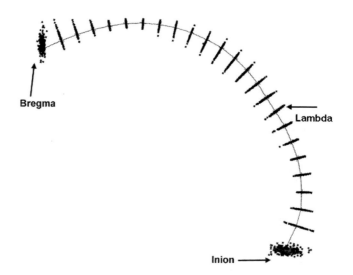

Figure 5.   The nonuniform curve relaxation points and the mean line.

We here coin a term for this entire procedure, which we will call "nonuniform curve relaxation," to indicate that the relaxation in the bending energy (the reduction in extraneous variance) comes from choosing points at which the unresampled (nonuniformly sampled) curves intersect planes that are orthogonal to the tangent of the mean curve, rather than from sliding points along the tangents or splines of uniformly sampled curves.

In order to more directly compare results with our previous study of this material, which used two-dimensional curves limited to the mean sagittal plane, the three-dimensional curve segments were then reduced to two dimensions, using a singular value decomposition. The dimension with the lowest variance, the third or Z dimension in this case, was dropped, leaving only XY coordinates projected into the approximate mid-sagittal plane of each specimen.

RESULTS

A detailed presentation of the statistical analysis of the sliding data set, including discussion of each of the Neanderthal, pre-Neanderthal, and Late Paleolithic specimens appears elsewhere (Harvati and Reddy, in prep.) and should be consulted. The results presented here focus only on the salient differences between the two methods.

Principal Components Analysis

The two PCAs were similar in their results (Figures 6 and 7). In both analyses PC 1 (53.2% and 60.8% of total variance respectively for the *sliding* and *nonuniform curve relaxation data sets*) did not separate Neanderthals completely from modern humans. Along this component, Neanderthals were significantly different in their PC scores from all modern human populations except Late Paleolithic in the *sliding data set*; and from all modern humans except Late Paleolithic Europeans and Inugsuk Eskimos in the *nonuniform curve relaxation data set* (Bonferroni t-test). PC 2 (25.1% and 22.5% respectively) did separate Neanderthals from modern humans more completely in both analyses. Neanderthals were significantly different in their PC 2 scores from all modern human populations in both cases.

In both PCAs two of the Near Eastern Neanderthals fell well within the modern human range along PC 2: Shanidar 1 and Tabun C1. Furthermore,

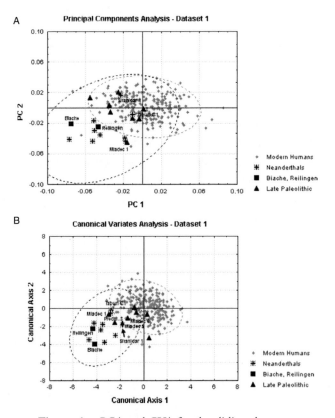

Figure 6. PCA and CVA for the sliding data set.

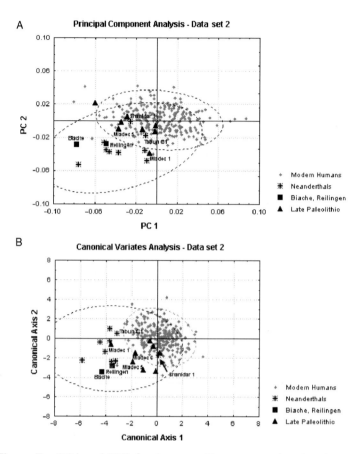

Figure 7. PCA and CVA for the nonuniform curve relaxation data set.

in both analyses one Upper Paleolithic specimen, Mladec 1, fell at the extreme of the modern human range and with Neanderthals on PC 2. Both pre-Neanderthal specimens fell with Neanderthals.

Canonical Variates Analysis

For both the sliding and the nonuniform curve relaxation data sets a CVA was performed on the first 15 principal components (99.5% and 99.3% of the total variance respectively, Figures 6 and 7), and using population rather than species as grouping variable. In both cases the first canonical axis (30.6% and 19.7% of the total variance respectively) separated Neanderthals from modern humans. Neanderthals were significantly different in their scores from all modern human populations along this axis.

The CVAs differ in that: in the *sliding CVA*, Shanidar 1 and Tabun C1 fell very close to each other at the end of the Neanderthal range and close to modern humans along Can 1 (Figure 6). Furthermore, in the sliding analysis, three Upper Paleolithic specimens fell in the area of overlap between Neanderthals and modern humans along Can 1 (Mladec 1 and 5 and Predmost 3). Mladec 6 was also close to the two Near Eastern Neanderthal specimens. The two pre-Neanderthal individuals fell with the Neanderthals. In the *nonuniform curve relaxation analysis*, Shanidar 1 fell well within the modern human range, whereas Tabun C1 did not. All the Late Paleolithic specimens fell in the area of overlap between Neanderthals and modern humans, due to the position of Shanidar 1. However, it has been suggested that this specimen may be artificially deformed (Trinkaus, 1982). Among the Late Paleolithics, the only specimen that fell within the rest of the Neanderthals along the first canonical axis 1 was Mladec 1.

Mahalanobis Squared Distances

The Mahalanobis squared distances were calculated using a correction for unequal sample sizes (Marcus, 1993), and are presented in Tables 1 and 2. The two matrices are very similar. The general dichotomy between Neanderthals and modern humans was evident in both. In both analyses the pre-Neanderthal specimens were very close to each other, with Biache also being closest to Neanderthals while Reilingen was about equidistant from Neanderthals and the modern human populations. The *nonuniform curve relaxation data set* Mahalanobis squared distance matrix differed in that the distance between Neanderthals and the Late Paleolithic specimens was somewhat greater than that found when the sliding data set was used. This Neanderthal-to-Late Paleolithic distance was still the smallest distance between Neanderthals and modern humans, although it was now almost equal to the Neanderthal–Khoisan distance.

Discriminant Analysis

When treated as unknown specimens to be classified by posterior probability in a discriminant analysis, in the *sliding data set*: Mladec 1 was the only Late Paleolithic specimen classified as Neanderthal when asked to classify to population. In this analysis, Shanidar 1 was classified as modern human (Inugsuk Eskimo).

Table 1. Bias-corrected Mahalanobis squared distances matrix for the sliding data set

	Nea	And	Aus	Brg	Dgn	Epi	Esk	Eur	LP	San	Tol	Bch	Reil
Nea	0.00												
And	20.95	0.00											
Aus	18.43	4.63	0.00										
Brg	21.32	4.04	6.83	0.00									
Dgn	16.90	4.38	5.66	10.35	0.00								
Epi	20.93	1.33	3.32	3.70	6.74	0.00							
Esk	13.62	4.25	4.90	3.18	7.20	4.70	0.00						
Eur	18.42	2.29	6.84	NS0.39	7.86	3.10	2.77	0.00					
LP	10.06	8.64	4.87	7.24	9.81	8.63	7.73	6.97	0.00				
San	12.97	2.53	4.39	5.84	2.33	4.23	4.28	4.70	7.17	0.00			
Tol	20.72	2.79	5.71	3.43	6.31	4.43	2.97	2.99	10.04	4.89	0.00		
Bch	NS2.57	38.58	32.13	34.68	33.03	34.91	24.43	32.35	19.20	31.30	34.41	0.00	
Reil	37.65	57.98	44.58	54.17	48.32	49.8	52.26	55.30	37.65	52.33	60.17	NS11.17	0.00

Note: Nea: Neanderthal; And: Andaman Islanders; Aus: Australian; Brg: Austrian Berg; Dgn: W. African Dogon; Epi: Afalou/Taforalt; Esk: Inugsuk Eskimo; Eur: W. Eurasian; LP: Late Paleolithic; San: Khoisan; Tol: Melanesian Tolai; Bch: Biache; Reil: Reilingen.

Table 2. Bias-corrected Mahalanobis squared distances matrix for the nonuniform curve relaxation data set (group labels as in Table 1)

	Nea	And	Aus	Brg	Dgn	Epi	Esk	Eur	LP	San	Tol	Bch	Reil
Nea	0.00												
And	19.51	0.00											
Aus	14.15	2.99	0.00										
Brg	18.53	3.84	3.07	0.00									
Dgn	14.45	3.45	5.08	7.49	0.00								
Epi	20.31	3.02	2.39	1.75	6.11	0.00							
Esk	15.06	4.27	3.34	1.94	5.59	5.09	0.00						
Eur	17.67	3.10	3.65	NS0.57	5.57	3.06	2.62	0.00					
LP	11.03	10.48	4.60	6.63	11.10	8.40	7.21	4.94	0.00				
San	11.63	4.55	3.05	5.72	1.72	4.52	4.08	4.61	7.56	0.00			
Tol	18.34	2.00	3.60	2.29	5.05	4.27	1.76	2.81	11.37	5.25	0.00		
Bch	NS9.67	40.44	37.57	34.79	37.67	41.55	32.29	35.17	23.89	35.69	38.02	0.00	
Reil	46.58	58.28	56.12	53.59	57.01	58.59	61.19	59.43	45.98	58.07	62.04	NS19.33	0.00

When asked to classify to species, Mladec 1, 5, and 6, as well as Predmost 3, were classified as Neanderthals; whereas Shanidar 1 was classified as Neanderthal. Both pre-Neanderthals were classified as Neanderthal in both instances. A cross-validation classification performed on the entire data set succeeded in classifying eight out of the nine (88.9%) Neanderthal specimens and 249 of the 255 (97.6%) modern human specimens to species correctly.

With the *nonuniform curve relaxation data set*, when asked to classify to population, Mladec 1 was classified as Neanderthal and Shanidar 1 was classified as modern human (Austrian Berg). When asked to classify to species, Mladec 1 and 6, but not 5, were also classified as Neanderthals. Unlike with the sliding data set, Shanidar 1 was now classified incorrectly as modern human when asked to classify to species. In both discriminant analyses the pre-Neanderthal specimens were classified as Neanderthal. The cross-validation classification succeeded in classifying seven out of the nine (77.8%) Neanderthal specimens and 253 of the 255 (99.2%) modern human specimens to species correctly.

CONCLUSIONS

Sub-sampling curves to achieve uniform sampling is inherently a low pass filtering operation, reducing the local, high-frequency information inherent in the original, nonuniformly sampled curve data. Nonuniform curve relaxation, by taking this local, high-frequency information into consideration when constructing relaxed, uniformly sampled curves, can help to preserve subtle features of the curves through resampling. It avoids the degradation of signal quality that occurs when sub-sampling is performed first and the curves are subsequently relaxed based on their reduced information content. The true strength of the technique lies in combining the uniform resampling and the relaxation, or sliding, process into one optimized algorithm and in treating the set of curves as a whole. In contrast, standard techniques apply sub-sampling to the individual curves, reducing their information content, and then apply relaxation as a statistical correction.

Nonuniform curve relaxation may be an attractive alternative to sliding for better preserving subtle shape information, which may in turn reduce misclassification of individual specimens. Although in this case the results of the two analyses did not differ dramatically, we feel that the increased accuracy of the data does result in fewer inconsistencies in the analysis of the occipital bun shape. The nonuniform curve relaxation analysis did result in somewhat greater discrimination between modern humans with posterior cranial profile shapes similar to, yet subtly different from, Neanderthals and the true Neanderthal specimens. On the other hand, some of the Late Paleolithic specimens were consistent in showing similarities to the Neanderthal sample. Finally, the Near Eastern Neanderthal Shanidar 1 was consistently found in the nonuniform curve relaxation analysis to fall with modern humans in all statistical analyses, underscoring the problematic nature of the posterior cranial profile of this individual. These results and their implications for the relationship between Neanderthals and early modern humans are explored further in Harvati and Reddy (in prep.).

REFERENCES

Bookstein, F. L., 1989, Principal warps: Thin-plate splines and the decomposition of deformations. *IEEE Trans. Patt. Anal. Mach. Intell.* 11:567–585.

Bookstein, F. L., *Morphometric Tools for Landmark Data*. Cambridge Press, Cambridge, 435 pp.

Bookstein, F. L., 1997, Landmark methods for forms without landmarks: Localizing group differences in outline shape, *Med. Image. Anal.* 1:225–243.

Bookstein, F. L., Schaefer, K., Prossinger, H., Seidler, H., Fieder, M., Stringer, M., Weber, G. et al., 1999, Comparing frontal cranial profiles in archaic and modern Homo by morphometric analysis, *Anat. Rec. (New Anat.)* 257:217–224.

Dean, D., 1993, The middle Pleistocene *Homo erectus/Homo sapiens* transition: New evidence from space curve statistics, Ph.D. dissertation, City University of New York, New York.

Dean, D., Hublin, J. J., Holloway, R., and Ziegler, R., 1998, On the phylogenetic position of the pre-Neanderthal specimen from Reilingen, Germany, *J. Hum. Evol.* 34:485–508.

Ducros, A., 1967, Le chignon occipital, mesure sur le squelette, *L'Anthropologie* 71:75–96.

Harvati, K., 2001, The Neanderthal problem: 3-D geometric morphometric models of cranial shape variation within and among species, Ph.D. dissertation, City University of New York, New York.

Harvati, K., Reddy, D. P., and Marcus L. F., 2002, Analysis of the posterior cranial profile morphology in Neanderthals and modern humans using geometric morphometrics, *Am. J. Phys. Anthropol.* S34:83.

Harvati, K. and Reddy, D. P., in prep., Quantitative assessment of the Neanderthal occipital "bun" using geometric morphometric ridge-curve analysis.

Marcus, L. F., 1993, Some aspects of multivariate statistics for morphometrics, in: L. Marcus, F. Bello, and A. García-Valdecasas, eds., Contributions to Morphometrics, Madrid: Monografias Museo Nacional de Ciencias Naturales, pp. 99–130.

Marcus, L. F., Frost, S. R., Bookstein, F. L., Reddy, D. P., and Delson, E., 2000, *Comparison of Landmarks among Living and Fossil Papio and Theropithecus skulls, with Extension of Procrustes Methods to Ridge Curves*, completed manuscript, http://research.amnh.org/nycep/aapa99/aapa6.html, 2000.

Strang G., 1976, *Linear Algebra and its Applications*, Academic Press, New York, pp. 106–107.

CHAPTER FIVE

Correcting for the Effect of Orientation in Geometric Morphometric Studies of Side-View Images of Human Heads

Waleed Gharaibeh

INTRODUCTION

For all of the promise of laser scanners and three-dimensional (3D) digitizers, digitizing two-dimensional (2D) landmarks on photographic images is still the most convenient way of sampling the shapes of biological structures in geometric morphometric studies. Geometric morphometric methods extract shape variables from landmark configurations such that they are invariant to the configuration's location, orientation and scale. Biological 3D structures can be variously rendered into 2D representations depending on, among other things, the choice of a point of view (i.e., object orientation with respect to the "camera") that is employed in the process of imaging. In the case of side view images

Waleed Gharaibeh • Department of Ecology and Evolution, State University of New York at Stony Brook, Stony Brook, NY 11794-5245.

Modern Morphometrics in Physical Anthropology, edited by Dennis E. Slice.
Kluwer Academic/Plenum Publishers, New York, 2005.

of human heads, a generalized Procrustes superimposition (GPA; Rohlf and Slice, 1990) of the 2D digitized landmarks removes the variability due to the size of the head in the image, its x and y position and its orientation around the imaginary z-axis (emanating orthogonally from the surface of the image). This means that geometric morphometrics methods control for variation among individuals in "nodding" (what would be called "pitch" in the aeronautical convention), but not for deviation from *norma lateralis* by way of head turning (rotation around the y-axis or "yaw") and head tilting (x-rotation or "roll").

In general, variation in the orientation of 3D objects, other than a rotation around the z-axis, results in variation in aligned 2D landmark configurations that cannot be distinguished from "true" shape differences in the original 3D structures. Morphometricians have considered the problem of object orientation (e.g., Dean, 1996; Roth, 1993), but they have generally assumed that the solution requires the acquisition of 3D coordinates, either directly by use of 3D hand digitizers and 3D scanners, or by way of photogrammetric reconstruction of 3D coordinates from multiple 2D images (e.g., Fadda et al., 1997; Spencer and Spencer, 1995; Stevens, 1997). The 3D configuration can then be rigidly rotated to the target orientation (e.g., Weiss et al., 2003) and the 2D coordinates are obtained.

This approach, however, requires the researcher to have access to the studied 3D object, and, as such, cannot address the orientation problem in photographic and other kinds of 2D imaging records which are often unique and irreplaceable. This is particularly true of the considerable archives of side and front view photographs of human heads that were collected by physical anthropologists in the last century. The American anthropologist Henry Field and his collaborators, for example, photographed thousands of individuals from Iraq and neighboring countries, and recorded their complementary ethnographic and anthropometric data (Field, 1935, 1952). The craniofaciometric variability in this record, among other things, reflects the unique dietary, healthcare, and other environmental circumstances of particular populations at a particular time that cannot be easily reproduced. Sampling contemporary descendant populations would not replace that record, but, by way of comparison, would allow us to study the biological effect of the radical shift in living conditions that has since ensued.

An example from that record can be used to illustrate the serious practical significance of the problem at hand. The craniofaciometric diversity in a pooled sample of 219 Iraqi Army soldiers (IA), mostly Arabs from southern Iraq (Field, 1935), and 93 Assyrian Levies recruits (Field, 1952) was sampled in

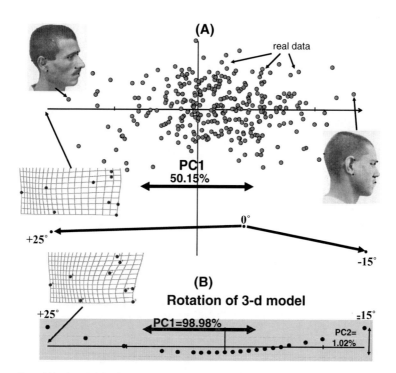

Figure 1. (A) A principal component analysis (PCA) of shape data extracted from nine-landmark configurations digitized on the side-view images of 219 Iraqi Army (IA) soldiers (Field, 1935) and 93 Assyrians (Field, 1952). The two IA soldiers at the extremes of PC1 show a high degree of head-turning deviation from true lateral in either direction. (B) A PCA of shape change due to *y*-rotation of a 3D nine-landmark human head configuration in the +25° to −15° range. Throughout this chapter the standard position of the head is *norma lateralis* facing left; rotation toward and away from the viewer is given in positive and negative degrees, respectively.

configurations of nine landmarks registered on side view images of the subjects. A principal component analysis (PCA) of the shape variables revealed that the first principal component (PC1) is strikingly associated with the angle of head turning (Figures 1A and B), while the "nuisance" effect of head tilting is not as readily identifiable. Visual inspection of similar samples showed the subjects of the IA-Assyrian sample to have been comparatively well disciplined in their posture, thus suggesting that deviation from *norma lateralis* might contribute even more nuisance variability to more typical samples. As a matter of expedience, this study focuses on removing the effect of head turning (yawing) in studies of side view images of human heads, but the same approach suggested here can also be applied to removing head tilting (rolling) from side view images; to removing both head turning and "nodding" (pitching) from front view images; to the problem of 3D object orientation in 2D images in general; and, in essence, to removing other optical distortions and confounding effects.

THE METHOD OF ORTHOGONAL PROJECTION

Burnaby (1966) suggested a simple approach to correct for undesirable effects in multivariate data. The undesirable variability is characterized by a set of vectors and is then removed from the space of total multivariate variability by projecting the data onto a subspace orthogonal to those vectors. In Burnaby's formulation, a data matrix $\mathbf{X}$ of n specimens and p variables can be adjusted for the effect of a p by 1 nuisance vector $\mathbf{f_1}$ as

$$\mathbf{X'} = \mathbf{XL}$$

where $\mathbf{X'}$ is the adjusted n by p data matrix and the p by p correction matrix $\mathbf{L}$ is computed as

$$\mathbf{L} = \mathbf{I_p} - \mathbf{f_1}(\mathbf{f_1^t f_1})^{-1}\mathbf{f_1^t}$$

where $\mathbf{I_p}$ is a p by p identity matrix. A p by q matrix $\mathbf{F}$ can be substituted for the vector $\mathbf{f_1}$ to remove q variables from the data at once, as was done by Rohlf and Bookstein (2003) for the computation of the uniform shape component for 3D data.

SIMULATING SHAPE CHANGE DUE TO ROTATION

In Burnaby (1966), orthogonal projection was proposed as a method of multivariate size correction by considering PC1 of within-group data as a size-related vector defining the subspace onto which the data are projected (Rohlf and Bookstein, 1987, discuss this method and compare it to other size correction techniques). Even though PC1 of the IA-Assyrian data is associated with head turning (whether the data are pooled or examined within each group), one should not be tempted to use it as a vector characterizing head turning in a Burnaby-style correction (which can be achieved simply by dropping PC1 from the analysis and using the remaining nonzero principal components as the corrected data set). As revealed by Figure 2A, a marked difference between the Assyrians and IA is localized at the back of the head (sampled by landmark 8). The deformation grid illustrating this difference in Figure 2A shows a striking

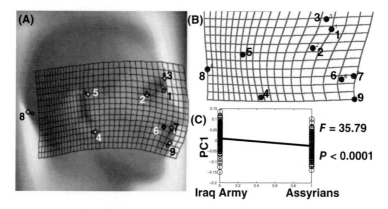

Figure 2. (A) A thin-plate spline showing the deformation required to change the consensus landmark configuration of the IA group (the average unwarped image of which is shown in the background) into that of the Assyrian group. (B) A thin-plate spline representing shape change due to a simulated turning of the head towards the camera, reproduced from Figure 1B. (C) An ANOVA showing a highly significant difference between the PC1 scores of the IA and Assyrian groups.

resemblance to the effect of head turning toward the camera in Figure 1B (reproduced in Figure 2B) and this relationship is consistent with a significant difference between the IA and Assyrian PC1 scores as shown by the ANOVA in Figure 2C. However, this group difference is also consistent with directly measured anthropometric data and visual inspection of the sample at hand (Field, 1952). Removing PC1 from of the data would result in the loss of not only this aspect of shape difference between the groups, but also perhaps other important biological differences that are not geometrically correlated with head turning. More problematic still is the situation in which the head-turning effect may not be so strongly associated with PC1, but instead is apportioned among a number of subordinate PCs. In such an event, the nuisance variability would be practically invisible and impossible to characterize by way of a PC analysis, but nonetheless may still be statistically important enough to suggest a spurious effect or obscure a real one.

A more direct estimate of the nuisance effect can be obtained by simulation experiments in which the nuisance variables are made to vary along a practically relevant range and the resultant shape change is tracked in multivariate shape space. Slice's (1999) geometric motion analysis follows a similar idea for characterizing shape change associated with biomechanical movement in Kendall's tangent space. More in the vein of correcting for nuisance variables, specifically to control for variability in the angle between two articulated structures, Adams (1997) set jointed 2D landmark configurations to two specified angles, computed the shape variables for all specimens (set to both angles) using GPA and thin-plate splines (TPS), and calculated the "articulation vector" for each specimen as the difference of shape variables over the two settings. He then used the articulation vector averaged over all specimens as the $\mathbf{f_1}$ vector in the orthogonal projection method.

Unfortunately, head turning and tilting cannot be simulated using 2D coordinates alone and the z-coordinates cannot be readily estimated from the images. Instead, simulations can be conducted using specimens that are not part of the sample, as long as it can be shown that the results are likely to hold for the sample specimens. One approach to simulation involves recording the coordinates of 3D landmarks on human heads, or 3D models of human heads, rotating the landmark configurations to different angles to represent deviations from *norma lateralis* by head turning and tilting and then orthogonally projecting the coordinates along the z-axis to obtain x- and y-coordinates. Alternatively, a series of images can be taken of human subjects with their heads turned and tilted to different angles and 2D landmark locations are subsequently recorded on the photographs. In both approaches, the 2D realizations are superimposed via GPA and are projected onto tangent space to yield shape variables. Any variability in the shape of the rotated realizations of any one model (head) has to be the consequence of simulated or actual change in that model's orientation. A PCA is performed on the shape variables of each model's set of rotated realizations separately to summarize this variability using as few dimensions as possible. This is repeated over all models used, across which the estimates of the nuisance variables can be averaged.

The first approach allows complete flexibility in the choice of angles, while the second approach offers a more realistic simulation. For example, the series of images might trace a particular combination of turning and tilting angles that the biomechanics of neck articulation determine. Furthermore, the researcher's perception of landmark position can be affected by the angle of head turning or tilting, especially at extreme angles where the true position of the landmarks is obscured behind the solid volume of the head. In such a situation, a human observer is more likely to change the operational definition of the hidden landmark, often forcing the point to lie on the visible outline, than to attempt to estimate the real position of the landmark. As such, there is more to the nuisance effect of rotation than pure geometry, and the second approach is inherently a more realistic simulation. For practical reasons, this study explores the head-turning effects using the first, simpler and more formal, approach; the need to follow up with a study using the second approach is then assessed in the light of the results.

SIMULATION BY ROTATION OF 3D STRUCTURES

The effect of head tilting and turning on 3D landmark configurations can be simulated by using the rotation equation

$$\mathbf{C_r} = \mathbf{C R_x R_y}$$

where $\mathbf{C_r}$ is an l (number of landmarks) by 3 matrix of rotated 3D coordinates; $\mathbf{C}$ is the l by 3 matrix of pre-rotation coordinates; $\mathbf{R_x}$ is a

$$\begin{bmatrix} 1 & 0 & 0 \\ 0 & \cos \alpha & -\sin \alpha \\ 0 & \sin \alpha & \cos \alpha \end{bmatrix}$$

matrix of rotation around the x-axis by the angle α, that is, head tilting; and $\mathbf{R_y}$ is a

$$\begin{bmatrix} \cos \beta & 0 & \sin \beta \\ 0 & 1 & 0 \\ -\sin \beta & 0 & \cos \beta \end{bmatrix}$$

matrix of rotation around the y-axis by the angle β, that is, head turning. The angles of rotation are varied over a relevant interval of α or β values and the x- and y-coordinates are retained as if projecting orthogonally along the z-axis. The rotated 2D coordinates can then be GPA aligned and projected orthogonally onto tangent space to yield Kendall tangent space coordinates (Rohlf, 1999), and the shape change due to rotation can be characterized by performing a PCA on the resultant shape variables.

Desirable Properties of the Projection Vectors

Considering one nuisance variable at a time, two conditions are required of the simulated shape variability for the suggested application of the orthogonal projection method to be effective: (i) the shape variability due to the nuisance variable should be nearly linear so that it can be adequately approximated by

PC1 and (ii) the estimated PC1 vectors should be homogeneous (parallel) over the range of the target sample's shape diversity and relevant angle interval.

It is desirable for PC1 to dominate nuisance variability because even though the orthogonal projection equation allows for any number of vectors representing the effects of numerous nuisance variables, the more dimensions that are dropped from the data the greater the chance of losing interesting biological variability. It would be especially ill-advised to eliminate a dimension in shape space that accounts for only a small proportion of nuisance variability since such a procedure would purge little nuisance variability from the data, while possibly discarding important biological shape variability.

The homogeneity of PC1s is requisite because nuisance vectors extracted from rotating a number of extraneous 3D configurations will be averaged to define the subspace onto which all sample specimens will be projected. The success of this procedure in eliminating nuisance variability, and nothing else, from the total variability of a given sample will be inversely proportional to the difference in angle between the actual nuisance PC1 vectors, which would have been obtained from the sample's member specimens (had they been available to us in 3D), on one hand, and the PC1 vectors estimated from the model 3D configurations on the other. If the model 3D configurations are assumed to collectively cover the sample's shape diversity, and if the PC1 nuisance vectors extracted from rotating them are all homogeneous, then it is warranted to assume that these PC1 vectors are also homogeneous with those that would have been obtained from rotating the sample specimens. If this condition of homogeneity is satisfactorily met, a single vector—for example, the mean of the PC1s extracted from 3D model simulations—can be considered for use as the basis for correction by orthogonal projection for the entire sample.

The interval of angle deviations from true lateral over which the sample specimens vary affects both conditions of linearity and homogeneity; however, this interval might be difficult to estimate reliably from the sample. Therefore, it is important to ascertain that linearity is maintained over a wide angular interval and that the direction of estimated nuisance PC1 vectors is not overly sensitive to errors in interval choice.

Covering the Relevant Portion of Shape Space

The desirable properties of the projection vectors need to hold over the entire shape diversity of the target human sample. Once the rough boundaries of that relevant part of shape space are delineated by the sample, 3D landmark configurations that are "representative" of that region can then be employed in the rotation simulations. Instead of attempting to match each specimen in the target sample with a rotating 3D configuration, a sufficient number of configurations can be generated such that they enclose the observed variability of the sample. These artificial configurations, or pseudo-specimens, can be generated by adding noise—sampled from a random uniform distribution of a given interval and centered around zero—to the x-, y-, and z-coordinates of landmark configurations registered on different representations of human heads. This model of error that is equally isotropic around each core landmark position, but independent among landmarks is similar to the simplest case of Goodall's perturbation model (Goodall, 1991), except in that the variation is uniformly, rather than normally, distributed around each landmark. The consequences of using this model with regard to statistical testing are not relevant for this side of the study; it is merely a method of "sampling" configurations from the relevant portion of the shape space.

This study employed configurations of 17 3D landmarks registered on seven models of human heads, three of which represented males of European ancestry; one represented a female of European ancestry; one represented a Middle Eastern male; one represented a Southeast Asian female and one was the vaguely anthropomorphic Mr. Potato Head toy (Hasbro, Inc., Pawtucket, RI). The 3D configurations were rotated to what was adjudged to be left-facing *norma lateralis* by matching corresponding landmarks across the plane of lateral symmetry, and the nine landmarks on the midsagittal and the right side of the face corresponding to those registered on the IA-Assyrian sample were retained. Five hundred pseudo-specimens were generated from five of the 3D configurations (100 each) using the perturbation model. The x- and y-coordinates of the pseudo-specimens were pooled with the IA-Assyrian sample, and all 2D configurations were GPA superimposed to yield shape variables, which were then rotated to their principal components. The interval parameter of the uniform error distribution was iteratively set so that the most extreme scores on all, but the first, eigenvectors of a PCA of the shape variables always belonged to pseudo-specimens (not shown). The difficulty in enclosing PC1 is not surprising considering that, at both extremes, PC1, at least partly, reflects head-turning variability. To make the pseudo-specimens enclose PC1 (without setting the random error to extremely high levels), the landmarks of one of the human models, a bust of George Washington, were adjusted so as to generate two configurations with extreme positive and negative PC1 scores (this mostly involved landmark no. 8, corresponding to the occiput, the position of which on the statue was not clear to begin with).

The two new Washington configurations were then used along with the five other human configurations to generate 700 pseudo-specimens that easily occupied the most extreme positions on the PCs (Figure 3). The pseudo-specimens occupied the most extreme positions on all principal components (the largest four are shown in Figure 3), but they were clustered with gaps appearing in between; this study assumes that the properties of projection vectors within these gaps are not very different from those attained for pseudo-specimens at their perimeter. Figure 3B demonstrates that having the pseudo-specimens occupy the most extreme positions on all PCs does not guarantee that they will enclose the shape variability along every direction of the shape space (consider the combination of low PC3 and PC4 scores), but they do come close, and—considering the vast portion of the shape space they occupy beyond the boundaries of the sample—they should provide a conservative appraisal of the properties of estimated shape change due to head-turning vectors. Perhaps it would have been conceptually purer to project the pseudo-specimens onto PC axes computed from the IA-Assyrian sample alone; however, doing so does not change the basic pattern: pseudo-specimens are at, or very near to, the most extreme values along all PC directions, but not along all possible directions in the space (not shown).

Even with the extreme variability of the pseudo-specimens, the relationship between the Procrustes distances in the Kendall shape space and the tangent distances is strongly linear with an uncentered correlation coefficient that is effectively unity (0.99999). This strong relationship suggests that the portion of the total shape space that is occupied is small enough for statistical analyses in the tangent space to be valid and that these analyses should be fairly robust to small changes in the point of tangency within this portion. Indeed, some experimentation showed that excluding the target sample from the GPA superimposition does not affect the results while it greatly increases the computational speed. In this study, all of the generated pseudo-specimens, in their variously rotated realizations, will be GPA superimposed and projected onto a single tangent space using their overall mean as the point of tangency. Subsequently, the rotated realizations of each single pseudo-specimen will be PCA-decomposed separately in order to investigate the properties of shape variability due to head turning and, ultimately, the appropriateness of the orthogonal projection correction.

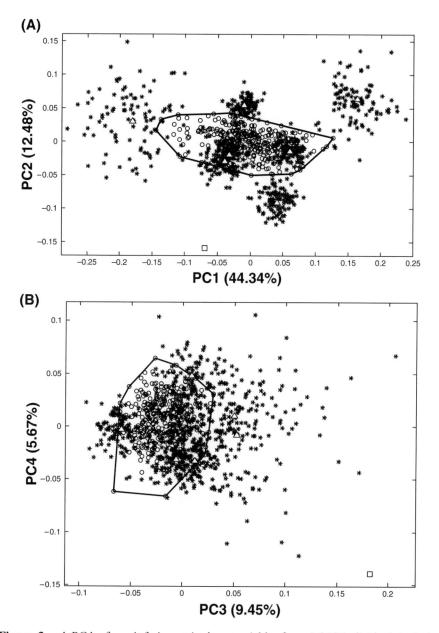

Figure 3. A PCA of craniofaciometric shape variables from 1,019 individuals including 312 human subjects belonging to the IA-Assyrian sample (circles; circumscribed by a hull) and 700 simulated pseudo-specimens (asterisks) generated using the perturbation model from 3D landmark configurations of 7 human-like models (triangles). The position of the Mr. Potato Head toy is indicated by a square. (A) PCs 1 and 2. (B) PCs 3 and 4.

Estimating the Relevant Range of Rotation

The interval of head-turning angle, β, operating in the IA-Assyrian sample can be estimated using the rotated realizations of the 3D configurations as a yardstick (or a protractor, Figure 1A). However, PC1 is unlikely to be due entirely to head rotation, and so one is required to make assumptions about the sample's "true" shape variability (i.e., its shape variability excluding that due to head turning and other nuisance factors). An assumption about the probability distribution of different angles within a range is also required. I assume that the probability distribution of head-turning angles within a specified range is uniform, because it is simple to simulate and express in terms of fixed interval limits (e.g., $-30°$ to $+30°$). The estimation will proceed by generating n pseudo-specimens (where n is the size of the target sample, 312 in the case of the IA-Assyrians) from a seed 3D configuration using the model for independent uniform isotropic error. The interval parameter of the model specifies the magnitude of the target sample's "true" shape variability excluding that due to head turning. The pseudo-specimens are then rotated randomly to different angles within a specified head-turning range and GPA superimposed, along with the target sample. PC axes are computed from the shape variables of the pseudo-specimens *only* to obtain the rough rotation blackboard onto which the original sample is projected. This process is repeated iteratively until the head-turning range is found that just succeeds in enclosing PC1.

At one extreme, assuming that almost all of the variability in the PC1 of the IA-Assyrian sample was due to head turning, I repeatedly estimated the range enclosing the sample to be around $40°$, regardless of the choice of the seed 3D configuration. At the other extreme, assuming that only the most extreme 10 outliers on the sample's PC1 were due to head turning, I estimated the total range to be less than $10°$. The range of rotation, regardless of its estimated magnitude, seems to be fully exhibited within the IA group (both individuals shown with the highest and lowest PC1 scores in Figure 1A belong to it), with the range of the Assyrian group being only slightly narrower (Figure 2C).

The angular range of head turning in the IA-Assyrian sample is likely to be somewhere between those two extremes ($10°$ and $40°$). However, in order to allow for samples with less-disciplined subjects, for intervals of rotation that are not centered exactly on true lateral ($0°$), and for the minor underestimation of the rotation range that results from sampling coarsely from the uniform distribution, the properties of the nuisance variability are investigated thoroughly in the −40° to +40° interval (80° range, twice the upper bound of the estimate). Certain aspects of the variability are investigated even beyond this range.

CHARACTERIZING SHAPE CHANGE DUE TO ROTATION

To appreciate the general trends in shape change due to head turning, one of the 3D configurations was rotated in 1° increments around the y-axis to varying ranges of β. Figure 4 aids in visualizing the shape change in the GPA-superimposed landmark coordinates as the configuration is rotated between −40° and +40°. To characterize the nuisance variability within shape space, the different rotated realizations within a given range were GPA superimposed and orthogonally projected onto tangent space. The shape variables from the aligned configurations for each range of rotation were PCA-decomposed separately, and the first and second principal components scores for all of the ranges were graphed on the same ordination plot (Figure 5).

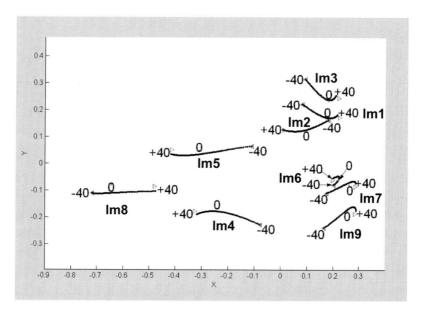

Figure 4. GPA superimposed realizations of a single nine-landmark 3D human head configuration as it is y-rotated between −40° (away from the viewer) and +40° (toward the viewer). Among other shape changes, the figure shows the proportional shortening of the area between the ear (landmarks 4 and 5) and the occiput (landmark 8) as a result of the head turning toward the camera.

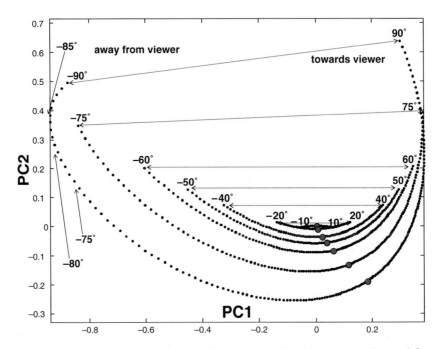

Figure 5. The "trajectories" of shape change due to head turning as obtained from *y*-rotating a single 3D nine-landmark configuration registered on a human head model to varying angular intervals. The first two PCs accounted for over 96% of total shape variability even for the widest interval, with the share of PC2 increasing as the interval widened (see Table 1). The large circles indicate *norma lateralis* (0° head turning).

Even for the widest range, the first two principal components accounted almost entirely for all of the shape variation (Table 1). Evidently, the linearity and symmetry of the "trajectory" describing the shape change due to rotation increase as the range of the rotation is narrowed around the true lateral. The points seem to get gradually more dispersed toward the ends of the trajectories suggesting an accelerated rate of shape change per degree of rotation as the head is turned farther away from true lateral. This acceleration is more pronounced for turning away from the eye than toward it. The asymmetry of the trajectories is not surprising given that four of the landmarks are on the right side of the face, and indeed it is greatly reduced when the corresponding landmarks on the left side of the face are included in the analysis (not shown).

The degree of linearity is also asymmetric around the true lateral: the segment of the trajectory between 0° and −60° is slightly more linear than the corresponding 0° to +60° segment (as was confirmed by performing separate PC

Table 1. PCA decomposition of shape variability due to rotating one 3D nine-landmark configuration around the *y*-axis to different angular intervals, as shown in Figure 5

Head-turning interval	Proportion of shape variability				
	PC1	PC2	PC3	PC4	PC5
5° to −5°	0.9993	0.0007	0	0	0
10° to −10°	0.9974	0.0026	0	0	0
20° to −20°	0.9901	0.0099	0	0	0
30° to −30	0.9774	0.0225	0.0001	0	0
40° to −40°	0.9590	0.0408	0.0002	0	0
50° to −50°	0.9337	0.0656	0.0007	0	0
60° to −60°	0.9006	0.0975	0.0020	0	0
75° to −75°	0.8327	0.1591	0.0081	0.0001	0
90° to −90°	0.7214	0.2390	0.0385	0.0008	0.0003

analyses on those intervals). This means that although the rate of shape change is higher when turning away rather than toward the eye within the 0° to −60° interval, more of that change is accounted for by PC1, and, as such, turning away from the eye might be slightly more amenable for correction using the orthogonal projection method (ignoring for now the problem of landmarks disappearing behind the outline of the face). The curvature of the "turning away" segment dramatically increases at about −75° angle at which point this pattern no longer holds.

The asymmetry of the trajectory and differences in the rate of shape change per degree means that narrowing or expanding the range around a fixed point would, in principle, have some effect on the direction of the PC1 vector (in addition to the obvious effect on its proportion of variance explained). However, the figure suggests that this effect may be negligible within the practically relevant head-turning interval.

At the more relevant −40° to +40° interval, PC1 nearly constitutes 96% of the total variance and dominates further as the range is narrowed. However, the curvature of the trajectory is substantial enough that a PC1 extracted from a rotation interval centered on a point near the −20° end might be different in direction from that of an interval centered near the +20° end. This suggests that the consequences of misestimating the midpoint of the angular interval when carrying out an orthogonal projection correction are worth investigating.

To show that the properties of the trajectory within the −40° to +40° interval are not unique to the particular 3D configuration used, the shape

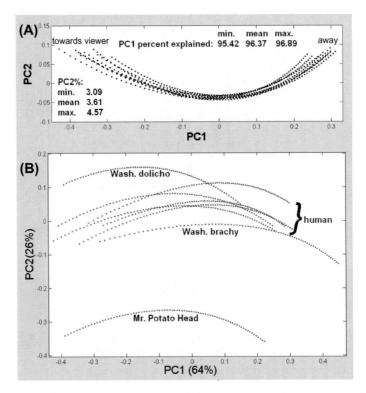

Figure 6. (A) The trajectories of shape change due to head turning over the −40° to +40° interval as exemplified by eight diverse 9-landmark 3D human head configurations. This is in fact eight PC analyses performed on the rotated realizations of each configuration separately. (B) A PCA performed on all of the realizations from the eight configurations pooled together as one sample.

change trajectories resulting from simulating head turning using eight 3D configurations (including the two configurations modified from the Washington bust and the Mr. Potato Head toy) are presented in Figure 6A. These were obtained by performing a PCA on the (globally) aligned rotated realizations of each configuration separately, but then plotting the PC1 and PC2 scores for all configurations on the same graph. PC1 always accounted for more than 95% of the total shape variance and even the trajectory for Mr. Potato Head shows the same properties as those of the more realistic human head models. A PCA of the aligned realizations from all configurations pooled together as one sample (Figure 6B), suggests that despite the considerable diversity of the configurations, the trajectories are reasonably parallel. This is better established

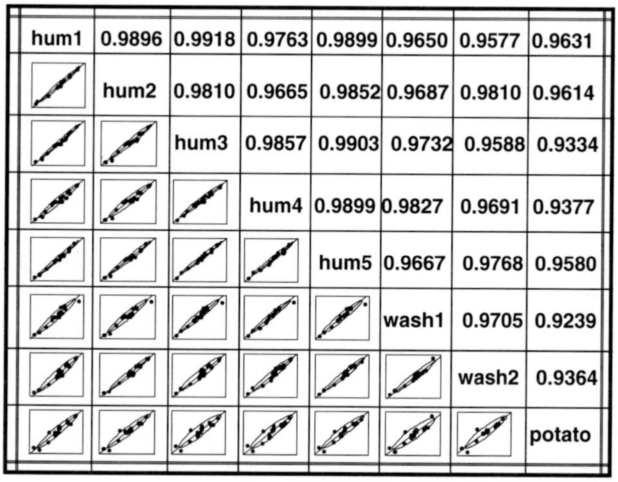

hum1	0.9896	0.9918	0.9763	0.9899	0.9650	0.9577	0.9631
	hum2	0.9810	0.9665	0.9852	0.9687	0.9810	0.9614
		hum3	0.9857	0.9903	0.9732	0.9588	0.9334
			hum4	0.9899	0.9827	0.9691	0.9377
				hum5	0.9667	0.9768	0.9580
					wash1	0.9705	0.9239
						wash2	0.9364
							potato

Figure 7. The correlation coefficients and bivariate plots for pairs of PC1s extracted from the rotated configurations in Figure 6A. High correlations correspond to a tight dispersal around the major axis, as marked by the confidence ellipses.

by plotting the PC1s extracted from the different configurations against each other and by computing Pearson's product moment correlations among them (Figure 7). The correlations, which correspond to the cosines of the angle between the PC1 vectors in an 18-dimensional dual space, are always over 0.965 (angle less than 15°) for the five realistic human models, and are never lower than 0.92 (23°) for Mr. Potato Head.

Sampling the curves representing shape change due to rotation using 1° increments, as was done in Figure 6, is computationally intensive. It is then of practical relevance to study the effect of the increment size in which head turning is simulated on the estimates of eigenvectors and eigenvalues of the resultant shape change. Figure 8 shows that for uniform angle increments the estimate of PC1 variance explained asymptotes quickly as the increment is reduced (sampling is refined). In the presented example, any sampling scheme that involves more than the endpoints of the interval (−40° and +40°) puts the estimate almost within 1% of the asymptotic variance explained value. In this same example the estimates of the eigenvectors seem hardly affected at all by varying angle increments; on average, pairs of PC1s estimated using varying increments had a correlation coefficient of 0.9998 with a maximum of unity and a minimum of 0.9992. Thus, by increasing the size of the rotation increments, simulations involving a large number

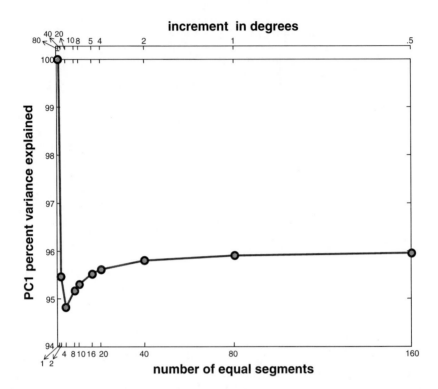

Figure 8. The effect of the resolution with which head turning is simulated, using one 3D configuration, on estimating the percent of the resultant shape variance explained by PC1. The x-axis is given in the degrees by which the angle of head turning is incremented (top) and the number of equal segments into which the total interval, −40° to +40°, is subsequently broken (bottom). For example, when head turning is sampled in 2° increments, the 80° range is broken into 40 equal segments (i.e., it is sampled at 41 angles including the endpoints) and PC1 is estimated to explain 95.8% of the total shape variance. When PC1 is estimated using the endpoints of the interval (i.e., the head is rotated to −40° and +40° only; equivalent to an 80° increment) the shape change is by necessity a straight line and PC1 is estimated to explain 100% of the variance.

of pseudo-specimens can be undertaken without sacrificing the validity of the results.

For a thorough investigation of the properties of head-turning shape variability over the relevant portion of shape space, sets of 700 pseudo-specimens, generated using the same human head models and parameters as those presented earlier in Figure 3, were rotated in increments of 10° over the −40°

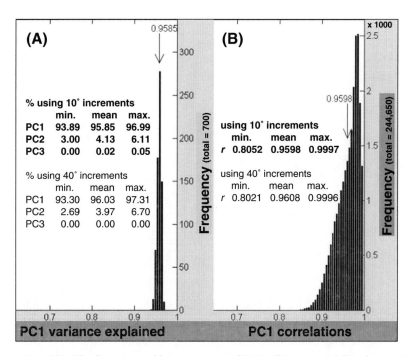

Figure 9. (A) The linearity and homogeneity of PC1 of shape variability due to simulated head turning in the range of −40° to +40° using 700 pseudo-specimens generated with the same parameters as those in Figure 3A. A histogram showing the distribution of percent variance explained by the 700 PC1s. (B) The distribution of the correlation coefficients of all 244,650 possible pairs of PC1s.

to +40° interval. The x- and y-coordinates of the rotated pseudo-specimens (i.e., 2D configurations) were GPA superimposed to yield shape variables. PC axes and the variance they explained were then computed from the globally aligned rotation set of each pseudo-specimen separately and the correlations among the PC1s of the different rotation sets were calculated (Figure 9). The variance explained by PC1 fell tightly between 93% and 97% with PC2 accounting for nearly all of the remaining variability (Figure 9A). The PC1s were fairly homogeneous with an average correlation of 0.96 and a minimum of 0.80 (Figure 9B).

By varying the range of rotation of the 700 pseudo-specimens around a fixed midpoint, some of the earlier conjectures were affirmed. Namely, as the range of rotation is narrowed, the proportion of variance accounted for by PC1 increases while its vector direction is hardly affected. Figure 10 shows the results for

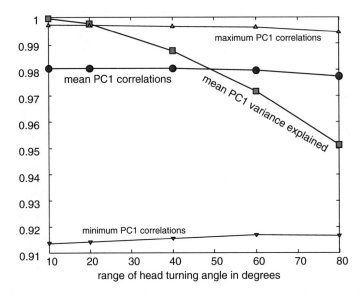

Figure 10. The effect of expanding the angular range of rotation around a given midpoint on the direction of the PC1 vector of shape change due to head turning. Seven hundred pseudo-specimens similar to those in Figure 3 were *y*-rotated around a fixed midpoint of 0°, but to varying ranges (e.g., between −40° and +40° for a range of 80°). For each range, the correlations between the 700 PC1 vectors and the mean PC1 for the 10° range (used as a standard) were calculated and summarized. The mean variance explained was also computed for each set of generated PC1s.

the 0° midpoint; spot-checking using other midpoints confirmed the pattern. In other words, within the relevant range of rotation, knowing the midpoint of an interval of rotation angles is nearly sufficient to determine the direction of the major axis of the resultant shape variability.

To investigate the affect of misestimating the midpoint of the rotation interval on the estimates of the PC1 vector, eight midpoints were designated between −35° and +35° in regular 10° increments (i.e., −35°, −25°, −15°, −5°, +5°, +15°, +25° and +35°) and the head turning of 700 pseudo-specimens (similar to those described earlier) was simulated using a 20° rotation range centered around these midpoints. The correlation coefficients between the PC1s extracted from rotating around each midpoint and the average PC1 vector for the 0° midpoint were computed (Figure 11). Figure 11 suggests that the mean correlation between the PC1 vectors generally decays exponentially as a function of the difference in angle between the midpoints of their rotation intervals.

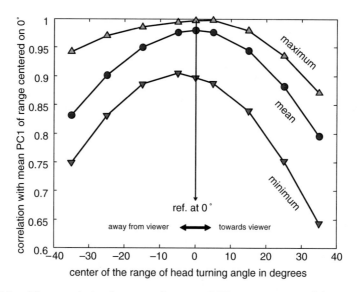

Figure 11. The correlation between the mean PC1 vector extracted from a rotation angle range centered on 0° and PC1 vectors extracted from ranges with midpoints varying between −35° and +35°. Minimum (triangles pointing down), mean (circles), and maximum values (triangles pointing up) of the correlation coefficients are shown. The PC1 vectors were computed by rotating the 700 pseudo-specimens described earlier in 20° ranges centered on the specified midpoints.

The decay in correlation seems faster when the midpoint is moved, so to speak, from the reference vector toward the eye than when it is moved away from it. This is compatible with the earlier observation that more of the shape change is linear (explained by PC1) when turning away rather than toward the eye within the −60° to +60° interval. The same patterns were observed when using the PC1 vectors of midpoints −25° and +25° as references (not shown). The three sets of simulations suggest that correlations higher than 0.9 are expected for PC1s of interval midpoints that are within 20° from each other. Together with the previous results, the high degree of PC1 vector homogeneity over considerable differences between the midpoints of simulated intervals suggests that, unless the midpoint of the sample rotation interval is grossly misestimated, PC1 vectors extracted from rotation simulations reasonably characterize the shape change due to rotation in a sample. This is especially true if care is taken when sampling 3D pseudo-specimens (for the purpose of estimating f_1) so as to closely imitate the variability exhibited by the sample.

POWER SURFACE COMPARISONS

For many, the usefulness of a method like the one proposed here is ultimately assessed by its success in correcting for the confounding effects of the nuisance variables on tests of statistical significance. A separate study considers the effect of head-turning variability on the size and power of tests for between-group shape differences, and the extent to which the suggested orthogonal projection method remedies this problem (Gharaibeh, unpublished data); following is a summary of its most pertinent results.

First, power surface analyses show that Goodall's test for group differences is so adversely affected by nuisance orientation variability that it cannot be salvaged by the orthogonal projection correction, or by permutation tests, and this test is not recommended whenever such variability is suspected.

When MANOVA test procedures are used, the confounding effects of the nuisance orientation variability, and subsequently the efficacy of the orthogonal projection correction, depend on the association between the true shape differences and the nuisance variability. If true shape differences are orthogonal to the nuisance shape variability, the orthogonal projection method is effective in correcting the inflation in rejection rates brought about by mean differences in head-turning angle between the groups, with a minimal loss of power—as long as the between-group differences in head orientation are not unrealistically large. Furthermore, within-group nuisance variability does not obscure true group shape differences that are orthogonal to it.

On the other hand, the stronger the association between nuisance and true shape variability, the more conservative the orthogonal projection correction becomes.

APPLICATIONS

To illustrate the effect of orthogonal projection correction, the method was applied to the eight rotated model configurations depicted in Figure 6B. The PC1 vectors were computed from each rotated set separately, and their average was designated as f_1. After the shape variables of the eight aligned configurations were projected onto the subspace orthogonal to f_1, a principal component analysis of the residual shape variability was computed and the first two PCs plotted (Figure 12). Shape variability due to head turning seems to have been largely eliminated, leaving mostly the small component of the variability that

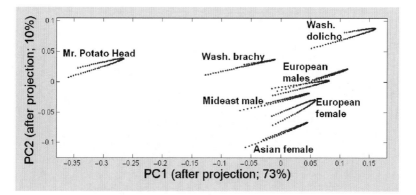

Figure 12. The effect of the orthogonal projection correction on the pooled rotated realizations depicted in Figure 6B. This PCA was performed on the residual variability of the sample after applying the correction.

spilled over to PC2 before the correction (less than 5% of the total variability, represented by the slight curvature of the trajectories in Figure 6B). In place of the nuisance effect, the difference between Mr. Potato Head and the rest of the models now dominates the residual shape variability.

The effect of applying the orthogonal projection correction to the IA-Assyrian sample is illustrated in Figure 13 by projecting that sample's residual shape variability (after the correction) onto its original PC axes (before the correction). The figure clearly shows that the variability in PC1 was compressed into a narrow band that hopefully represents, for the most part, the component of PC1 variability that was not due to (nor coincidentally correlated with) head turning. The nuisance vector, f_1, used in this correction was estimated from 500 pseudo-specimens, generated by the independent isotropic error model from five 3D configurations, and rotated (the pseudo-specimens) over the interval $-10°$ to $+10°$. However, similar results were obtained by using a reasonable variety of other 3D seed configurations, isotropic error model parameters, and head-turning angle intervals; even using a single rotated configuration seems to result in an effective correction.

The large difference between the IA group and the Assyrians (Wilks' lambda $= 0.5088$; $F = 15.7173$; $p < 0.0001$, as low as the number of permutations would allow) is slightly increased after the correction (Wilks' lambda $= 0.4994$; $F = 16.3147$).

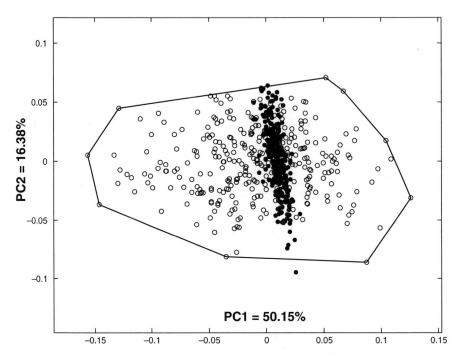

Figure 13. The effect of the orthogonal projection correction on the IA-Assyrian sample. The GPA aligned IA-Assyrian sample was plotted onto its first two principal components (open circles circumscribed by a convex hull). Using y-rotated 3D configurations, shape variability due to head turning was characterized by a single vector $\mathbf{f}_1$, which was then eliminated from the IA-Assyrian sample using the method of orthogonal projection. The residual variation of the corrected sample was then projected onto the original PC axes (black dots). The dramatic collapse of PC1 as a result of this correction occurred under a variety of choices for $\mathbf{f}_1$.

SUMMARY AND CONCLUSIONS

This study investigated the properties of nuisance shape variability due to head turning (or yawing—the deviation of the midsagittal plane from parallelism with the receptive surface of the imaging device) in samples of side-view images of human heads. Head turning was implicated with the first principal component of shape variability in one such sample and is expected to be of even greater importance in other samples. Simulations based on the rotation of 3D human head models—selected such that their diversity of shape surpasses that of the observed population—showed that nuisance shape variability due to head turning is highly linear. For example, about 96% of the simulated nuisance shape variability, on average, was accounted for by the first principal component alone when allowing a $-40°$ to $+40°$ interval of rotation (at least twice the range that was observed in the examined population). The same simulations showed the PC1 vectors of head-turning variability, extracted from highly diverse 3D models, to be very close to homogeneous (parallel) with an average correlation coefficient of 0.96. Furthermore, the PC1 vectors were found to be fairly robust to errors in midpoint interval estimation. Together, these results suggest that the nuisance variability due to head turning can be easily and reliably characterized in a sample using a single vector, and subsequently removed using the orthogonal projection method. The maintenance of linearity and homogeneity over relatively extensive shape diversity and wide rotation ranges is promising for the application of the orthogonal projection method to correct for disorientation variability in other biological structures, although that prospect should be explored on a case-by-case basis.

A preliminary inspection of shape change due to head tilting (rolling) in side-view images and head turning (yawing) in front-view images showed that both are also predominantly linear. Furthermore, the PC1s extracted from head turning and head tilting in side-view images seem to be largely uncorrelated, which implies that the two effects can be characterized separately using 3D models and corrected additively, without having to worry about their interaction. Characterizing nuisance shape variability due to different kinds of head motion using staged photograph series, instead of 3D rotations, might still be worthwhile for estimating the relevant intervals of rotation by direct visual comparison, and because it incorporates the effect the rotation has on the observer's perception of landmark positions.

This study assessed the efficacy of the suggested correction in application using the exploratory statistical technique of principal component analysis. A related study that utilized power analysis to investigate the effect of head turning on testing hypotheses of mean shape differences among groups and the success of the orthogonal projection method in countering this effect will be presented elsewhere.

The generally favorable results of this study suggest that the same approach of linearly characterizing undesirable shape variability and subsequently eliminating it by way of orthogonal projection might be useful in correcting for other kinds of optical distortions (e.g., parallax due to varying the distance between the object and the camera; Mullin and Taylor, 2002).

ACKNOWLEDGMENTS

Much of the theoretical basis for this study came about from discussions with F. James Rohlf. Most of the analyses, simulations, and graphs in this study were run in Matlab (Mathworks, Inc., Natick, MA) using specially written routines, many of which called upon routines from Richard E. Strauss' library (http://www.biol.ttu.edu/Strauss/Matlab/Matlab.htm). F. James Rohlf's tps series and Dennis Slice's Morpheus et al. (both available from http://life.bio.sunysb.edu/morph) were also used in generating some of the graphs and analyses. William Jungers kindly allowed me the use of his MicroScribe-G2 digitizer (Immersion Corporation, San Jose, CA). Revisions made to this manuscript by Dennis Slice, Windsor Aguirre, and an anonymous reviewer were much appreciated.

REFERENCES

Adams, D. C., 1999, Methods for shape analysis of landmark data from articulated structures. *Evolutionary Ecology Research* 1:959–970.

Burnaby, T. P., 1966, Growth-invariant discriminant functions and generalized distances. *Biometrics* 22:96–110.

Dean, D., 1996, Three-dimensional data capture and visualization, In: *Advances in Morphometrics*, L. F. Marcus, M. Corti, A. Loy, G. Naylor, and D. Slice, eds., Plenum Press, New York, pp. 53–69.

Fadda, C., Faggiani, F., and Corti, M., 1997, A portable device for the three dimensional landmark collection of skeletal elements of small mammals, *Mammalia* 61:622–627.

Field, H., 1935, Arabs of central Iraq, their history, ethnology, and physical characters, *Memoirs of the Field Museum of Natural History*, vol. 4.

Field, H., 1952, The anthropology of Iraq, *Papers of the Peabody Museum of American Archaeology and Ethnology* vol. 46 no. 2.

Goodall, C. R., 1991, Procrustes methods in the statistical analysis of shape, *Journal of the Royal Statistical Society, Series B* 53:285–339.

Mullin, S. K. and Taylor, P. J., 2002, The effects of parallax on geometric morphometric data, *Comput. Biol. Med.* 32:455–464.

Rohlf, F. J., 1999, Shape statistics: Procrustes superimpositions and tangent spaces, *Journal of Classification* 16:197–223.

Rohlf, F. J. and Bookstein, F. L., 1987, A comment on shearing as a method for "size correction," *Syst. Zool.* 36:356–367.

Rohlf, F. J. and Bookstein, F. L., 2003, Computing the uniform component of shape variation, *Syst. Biol.* 52:66–69.

Rohlf, F. J. and Slice, D. 1990, Extensions of the Procrustes method for the optimal superimposition of landmarks, *Syst. Zool.* 39:40–59.

Roth, V. L., 1993, On three-dimensional morphometrics, and on the identification of landmark points, In: *Contributions to Morphometrics*, L. F. Marcus, E. Bello, and A. Garcia-Valdecasas, eds., Museo Nacional de Ciencias Naturales (CSIC), vol. 8. Madrid, Spain.

Slice, D. E., 1999, Geometric motion analysis, *Am. J. Phys. Anthropol.* 28:253–254.

Spencer, M. A. and G. S. Spencer, 1995, Technical note: video-based three-dimensional morphometrics, *Am. J. Phys. Anthropol.* 96:443–453.

Stevens, W. P., 1997, Reconstruction of three-dimensional anatomical landmark coordinates using video-based stereophotogrammetry, *J. Anat.* 191:277–284.

Weiss K. L., Pan, H., Storrs, J., Strub, W., Weiss, J. L., Jia, L. et al., 2003, Clinical brain MR imaging prescriptions in Talairach space: technologist and computer-driven methods, *Am. J. Neuroradiol.* 24:922–929.

CHAPTER SIX

Fourier Descriptors, Procrustes Superimposition, and Data Dimensionality: An Example of Cranial Shape Analysis in Modern Human Populations

Michel Baylac and Martin Frieß

INTRODUCTION

The analysis of cranial shape variation in human populations, present or past, is an area of anthropological research in which geometric morphometric techniques are frequently and successfully applied, as can be seen throughout the present volume. A quick survey of the applied techniques emphasizes the obvious, that is, landmarks, rather than outlines, are predominantly used,

Michel Baylac • Muséum National d'Histoire Naturelle, CNRS FRE 2695 and Department "Systématique et Evolution," Service Morphométrie, Paris. **Martin Frieß** • New York Consortium of Evolutionary Primatology, American Museum of Natural History, New York.

Modern Morphometrics in Physical Anthropology, edited by Dennis E. Slice.
Kluwer Academic/Plenum Publishers, New York, 2005.

though the extension of GPA to outlines through the introduction of sliding landmarks (Bookstein, 1997; Bookstein and Mardia, 1998) has somewhat filled the gap between both approaches. With respect to Generalized Procrustes Analysis (GPA, Rohlf, 2000), the choice of specific landmarks and, to some degree, their number, play a crucial role for the assessment of the biological shapes under study. The human skull can be used to illustrate this point: Although it has homologous type I landmarks (Bookstein, 1991) that can be and have been used to analyze shape differences (Penin and Baylac, 1995; Frieß, 1999), they are somewhat spread out over the entire skull, leaving large areas uncovered for subsequent analyses. In the case of human midsagittal profiles, for instance, most of the major vault bones (frontal, parietals, occipital) carry at each end but one suitable type I landmark *sensu* Bookstein (1991), and consequently their actual contours are barely taken into consideration. This simple issue underlines the need to quantify variation between landmarks by using methods such as landmark relaxation (Bookstein, 1997) or Fourier analysis (Kuhl and Giardina, 1982; Lestrel, 1997; Rohlf, 1986; Rohlf, 1990). The purpose of this article is to further investigate the advantages of combining outline- and landmark-based approaches for the quantification of cranial shape. We use a common framework mixing outline and landmark superimposition already used in a similar context (Frieß and Baylac, 2003) together with a direct visualization of the outlines differences. In this framework, homologous landmarks (also called control-points in the present context) entirely control the orientation of the outlines, which are superimposed following the rotation parameters of the Procrustes superimposition of the landmarks. One of the advantages of multiple control points is to reduce the dependence upon a particular orientation. An additional benefit is that landmark- and outline-based results become comparable since both share a common reference system for the translation and rotation parameters. Previous studies comparing outlines and landmarks used data sets that were not directly comparable in this respect. They nevertheless recurrently showed the better statistical results achieved by outlines (see e.g., McLellan and Endler, 1998). Therefore, our purpose is not to confirm this result again. It is merely to argue that combining outlines and landmarks using the same objects provide a more detailed insight into the shape differences. Since outlines carry supplemental shape information located between landmarks, comparing the results of both analyses may allow for a preliminary subsetting of the shapes and therefore could provide useful information about the pertinent geometric scale of the shape differences. We address also the important point of data dimensionality reduction. Fourier descriptors, as well as most alternative contour descriptors, result in an inflation of the dimensions of the data space. Most solutions to the ensuing statistical dilemma—more variables imply much more objects proportionally—have been solved by substituting principal component scores for the original data. The problem comes from the necessity to determine how many components should be kept for the statistical analyses. We compare some traditional selection procedures such as the broken-stick and the Jolliffe rules to a selection based on the maximization of the correct classification percentages using cross-validation.

Artificial Cranial Deformation

This study deals with cranial shape analysis in the case of artificial deformation, a cultural practice especially known from South America (Dingwall, 1931). However, the main goal of this contribution is to concentrate more on methodological issues when dealing with shape quantification and less on the anthropological issues involved in artificial deformation, which are the main focus of another contribution (Frieß and Baylac, 2003). The following section briefly summarizes some of the general aspects of artificial deformation, providing thereby the background for the present study. The subsequent sections discuss more of the methodological details.

Artificial cranial deformation is known as a quasi-universal cultural practice from numerous historic and possibly prehistoric populations (Antón, 1989; Antón and Weinstein, 1999; Brothwell, 1975; Dingwall, 1931; Trinkaus, 1983). Among the numerous attempts to group and distinguish deformation types, the work of Dembo and Imbelloni (1938) and Falkenburger (1938) is most commonly cited, and their terminology was also used for the present study. Using criteria that take into account the means of deformation as well as the resulting shapes, two basic types of artificial cranial deformation are recognized, namely anteroposterior (AP) and circumferential (C). Deformation of the AP type leads to a flattening of the frontal and/or occipital bone that is typically compensated by a lateral expansion of the parietals. Conversely, a long, narrow, and conical vault shape characterizes circumferentially deformed skulls (Antón, 1989; Dembo and Imbelloni, 1938; Falkenburger, 1938) (Figure 1).

Besides the characteristic vault modifications, secondary effects on the face and the base have been reported for both deformation types, but these reports are inconsistent. It is generally assumed that both deformation types lead to

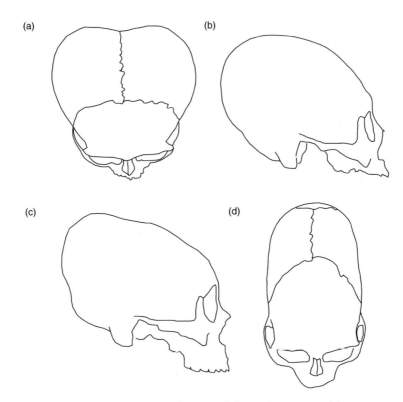

Figure 1. Schematic illustration of major deformation types: (a) anteroposterior, superior view (b) anteroposterior, lateral view (c) circumferential, lateral view (d) circumferential, superior view.

a basicranial and basioccipital flattening (Antón, 1989; Cheverud et al., 1992; Frieß and Baylac, 2003; Kohn et al., 1993), although increased flexion has been reported for the AP type (Moss, 1958). The two deformation types affect the face differently: While circumferential deformation leads to long, narrow, and protruding faces, AP deformed skulls have been described as exhibiting short and broad faces. This very brief summary should be sufficient to stress that whatever the associated modifications are, their assessment and understanding requires quantitative approaches that are not easily provided by conventional point-to-point measures.

MATERIALS AND METHODS

Raw data consisted of two-dimensional landmarks and contour drawings of four cranial samples in left lateral view (Figure 2; Table 1). The drawings were

Table 1. Cranial samples used for the cranium and the vault analyses

Sample and origin	N cranium	N vault
Peruvian AP	35	51
Peruvian C[a,b]	13	36
Peruvian PND[a]	36	37
Japanese JAPA[a]	46	46
Total	130	170

Notes: Origins of the samples.
[a] Musée de l'Homme, Paris.
[b] University of Pennsylvania, Philadelphia. The majority of the Peruvians, whether deformed or not, come from Pachacamac and Ancon. The Japanese sample comes from Hiogo-Kobe (Set-Tsu province), the Inuit are equally of Alaskan and Greenland origin.

made by parallel projection at natural scale (see Frieß, 1999; Frieß and Baylac, 2003) and represent two different configurations: A relatively complete contour of the cranium including portions of the face and the occipital clivus, and a second outline that is reduced to the vault portion of the first configuration (nasion to basion).

In addition to the outlines, we recorded a series of landmarks for superimposing the craniofacial outlines. A total of nine landmarks (cranium) or seven (vault) were used. With nasion as the starting point in the left lateral view, the points are in clockwise order: Glabella, bregma, lambda, inion, opisthion, basion, maxillary tuberosity, and prosthion, the latter two being the points omitted in the analyses of the vault. Landmarks and outlines were digitized in TPSDig (Rohlf, 1996). The landmarks were then used for a GPA (Rohlf, 2000). Similarly, outlines were used as input for Elliptic Fourier Analysis (EFA) (Kuhl and Giardina, 1982; Lestrel, 1997; Rohlf, 1990).

Extending Elliptic Fourier Analysis

Given that Fourier parameters are sensitive to starting point, location, size, and orientation of objects (Rohlf, 1990), outlines must be superimposed using a common reference system before being analyzed. Kuhl and Giardina (1982) used a normalization based on outline information, which is geometrically but rarely biologically justified (Bookstein, 1991; Rohlf, 1990). For instance, normalization for orientation uses the major axes of the first ellipse, which is sensitive to contour irregularities and for which homology from outline to outline hardly applies. Rohlf (1990) presented an orientation along a common axis defined by two homologous landmarks, an approach that has been used in subsequent studies and yielded reliable results (Monti et al., 2001; Tangchaitrong et al., 2000; Sakamaki and Akimoto, 1997). However, the common use in geometric morphometrics of the Procrustes superimposition method has clearly shown that alignments based on multiple points provide better results (Dryden and Mardia, 1998; Rohlf, 1990) compared to two-points registration methods, as well as to the orientation planes traditionally used in anthropology (Frieß, 1999; Penin and Baylac, 1995; Penin and Baylac, 1999). In the present study we applied a multiple point reorientation, which uses homologous landmarks as control points for the alignment (Figures 2a and 2b). This procedure uses a Generalized Procrustes Analysis (GPA, Rohlf, 2000) that consists of preliminary centering and size normalization by centroid size, that is, the square root of the sum of the squared distances between the centroid location and all control points of an object. Objects are iteratively rotated until the sum of the squared distances to the mean or consensus configuration is minimized. Outlines are first centered and size-normalized by dividing, specimen by specimen, the coordinates by the square root of the outline surface. Rotation parameters calculated for the control points are applied to their corresponding outlines at each iteration. Alternatively, one could normalize the outlines by the centroid size parameter of the corresponding landmark

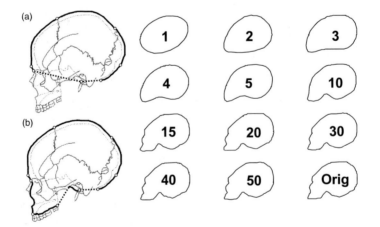

Figure 2. Landmarks and outlines recorded: (a) cranium (b) vault. The right panel illustrates the reconstruction of a skull outline using increasing numbers (in bold) of harmonics.

configurations. We nevertheless preferred to normalize the outlines by their own size parameter that is more closely fit to the form of the contours. Both size normalizations applied to landmarks and outlines are isometric procedures that do not modify the proportions of the objects. Therefore, allometries are not eliminated by the size normalization, but may be specifically visualized or extracted by multivariate regression of harmonic coefficients on size (Frieß and Baylac, 2003; Monti et al., 2001).

We used the elliptic Fourier approximation described by Kuhl and Giardina (1982). It consists of the decomposition of a curve into a sum of harmonically related ellipses. Each harmonic yields four coefficients that are used as input variables for standard multivariate statistics, such as discriminant analysis. We used 30 harmonics (plus the fundamental) for the cranium summing up to 122 Fourier coefficients (31×4, minus 2 for translations). For the vault we used only 15 harmonics resulting in 62 Fourier coefficients. In both cases these harmonic numbers were selected because they visually provided a good accuracy when outlines were reconstructed (Figures 2a and 2b).

Statistical Analyses

Outlines, described by numerous coordinates and/or harmonic coefficients, have the side effect of inflating the dimensionality of the shape space. In many instances, the number of variates become higher than the (within-) group sample sizes, leading to intractable statistical results. In the case of homologous landmarks, Rusakov (1996) proposed a Monte Carlo approach based on PCA allowing for the selection of the significant landmarks. Such a solution does not apply to outlines, where coordinates do not describe homologous locations. One alternative solution could be to reduce the number of harmonics. This would be at the expense of both a poor outline recovery and a reduced statistical power. The only real alternative seems to be to use as many harmonics as needed to fully represent the shapes, and then to reduce the dimensionality of the Fourier space. The most common approach in dimensionality reduction uses progressive subsets of principal components (Jolliffe, 1986; Krzanowski, 1987, 1988). This procedure has the additional statistical benefit of transforming correlated variates into uncorrelated new ones, a property particularly useful in the case of the highly redundant sets of harmonic coefficients. The problem now becomes one of selecting the dimension of the subspace. Most rules of selection are heuristic or *ad hoc* methods (Jolliffe, 1986; Krzanowski, 1988) largely leading to subjective and arbitrary decisions. Since our purpose was to discriminate between *a priori* defined groups, rather than searching the best subspace representation, we chose to select the number of retained components in each analysis so as to minimize the total cross-validated misclassification percentages. This procedure has been already exemplified in Dobigny et al. (2002), and Baylac et al. (2003). We used also two standard rules, the random broken stick (Jolliffe, 1986) and the Jolliffe's rule that retain only the components greater than 0.7 times the mean of the eigenvalues (Jolliffe, 1986). The latter is a less stringent version of the classical mean of eigenvalues (Jolliffe, 1986; Krzanowski, 1988). Canonical variate and discriminant analyses used the same selected reduced sets of data (see Table 2).

Misclassification percentages were estimated using concurrently linear (LDA) and K-nearest neighbors discriminations (KNN). We used KNN discriminations because preliminary results showed their results to be less dependent upon the number of principal components in the model than LDA. The number K of nearest-neighbors was selected between 1 and 30, again in order to minimize

Table 2. Cross-validated misclassification percentages (in bold) for the cranium and the vault (calculated using multiple linear discriminant analyses (LDA) and K-nearest neighbors (KNN) discriminations)

	Cranium		Vault	
	Outlines	Landmarks	Outlines	Landmarks
LDA AP deformed	**2.86**	**2.86**	**11.76**	**15.69**
LDA C deformed	**15.38**	**23.08**	**5.56**	**13.89**
LDA PND	**13.89**	**25.00**	**8.11**	**37.84**
LDA JAPA	**2.17**	**10.90**	**0.00**	**8.70**
LDA total	**5.39**	**13.85**	**6.47**	**18.24**
Ncomp (% variance)	17 (98.13%)	10 (96.90%)	6 (94.81%)	10 (100.00%)
LDA total broken stick	**18.46**	**27.69**	**28.24**	**35.88**
Ncomp (% variance)	4 (88.80%)	4 (73.62%)	3 (87.12%)	2 (52.20%)
LDA total Jolliffe's rule	**9.23**	**25.38**	**6.47**	**28.82**
Ncomp (% variance)	10 (96.03%)	5 (81.35%)	6 (94.81%)	5 (89.25%)
Total KNN	**15.38**	**20.00**	**7.65**	**21.18**
Ncomp (% variance)	15 (97.73%)	12 (98.98%)	6 (94.81%)	10 (100.00%)
KNN	6	20	7	7

Note: The first four lines detail the results obtained using LDA. The last four lines compare the total cross-validated misclassification percentages using different rules for dimensionality reduction, and provide the numbers of retained principal components (Ncomp), the corresponding percentages of explained variance (% variance), plus the number of KNN used in the computations of KNN discriminations (see text for further explanations).

the overall cross-validated misclassification percentages. In total, four canonical variates analyses (CVA) and multiple discriminant analyses were computed and compared: each of the two configurations (cranium, vault) was tested once with each of the two types of input variables (outlines, landmarks).

Visualizations of the outline deformations along the canonical axes were made using the procedure presented by Monti et al. (2001). In a first step, the predicted Fourier coefficients were calculated by multivariate regression (Krzanowski, 1988) over a variate representing the direction of interest (projections onto the canonical axes). These coefficients were then used in an inverse Fourier transformation in order to reconstruct the deformed outlines. GPA, Elliptic Fourier Coefficients, and graphical outputs were calculated using MATLAB functions devised by one of us (MB). Statistical analyses used the R statistical language (Ihaka and Gentleman, 1996) for Windows v. 1.81 (http://cran.r-project.org/).

RESULTS

Since inclusion of size in the discriminations did not improve the misclassification percentages, we present the results only for the shape components.

Cranium

Outlines: The MANOVA was highly significant: Wilks' Lambda$=0.01135$, $F=26.0545$, $df=45/334$, $p<10^{-4}$. Plots onto the first canonical plane (91.17% of variance), as well as the deformations along the axes (Figure 3), show that the circumferentially deformed skulls are the most distant to the other three groups. Their cranial outline exhibits a marked conical vault that protrudes posteriorly and superiorly and is flattened frontally as well as occipitally. The facial projection is very marked anteriorly and inferiorly in this group, contributing to the effect of an overall elongation of the cranium. But the relative length of the upper face (nasion to prosthion) is not significantly modified compared to the other samples. In circumferentially deformed skulls, nasion protrudes almost to the same level as glabella. The occipital and basioccipital portions also show significant modifications associated with circumferential deformations. The general orientation of the clivus, *foramen magnum* axis, and the nuchal plane is very different in C deformed skulls. The overall orientation of

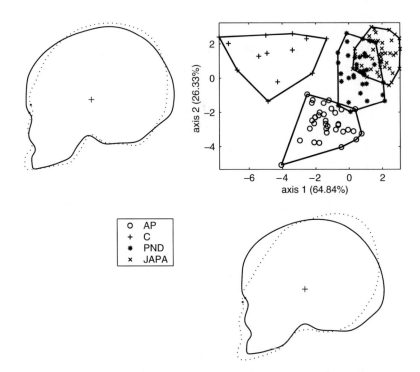

Figure 3. Canonical variate analysis of the four groups using elliptic Fourier parameters derived for the crania. Abbreviations: AP = anteroposterior deformations, C = circumferential deformations, PND = undeformed Peruvians, and JAPA = undeformed Japanese. Dotted reconstructed outlines illustrate the extreme shapes for the negative sides of the axes, solid lines are for the positive ones. In both cases the deformations illustrate the observed range of variation without magnification. Group limits are illustrated by their convex hulls.

the *foramen magnum* is much more in line with either clivus or nuchal plane, making this entire portion less angled than it is in the three other populations.

The projection onto the second canonical axis (Figure 3) basically reflects differences between the AP group and the three other samples (C-type and undeformed). Anterior–posterior deformation, besides frontal and occipital flattening, shows different modifications in the basioccipital and face. Note that the angle between the *foramen magnum* axis and nuchal plane is much more acute in AP-deformed skulls than it is in the other groups, and that the clivo-foraminal angle appears more obtuse. The axis of the *foramen magnum* slopes down toward opisthion when compared to the other populations. Also, in comparison to the C-group, effects of AP deformation appear to produce a longer and more anteriorly projected alveolar process rather than a combined anterior–inferior protrusion of the entire face.

Cross-validated misclassification percentages (Table 2, first column) are particularly low for the AP deformed (only one misclassified into PND) and the undeformed Japanese samples. Figure 4 illustrates the evolution of both raw and cross-validated total misclassification percentages as a function of the number of retained components for the cranial outlines. In this case, we retained the first 17 principal components (98.13% of the total variance), which resulted in a misclassification percentage of 5.4%. As seen on the Figure 4, a perhaps more parsimonious minimum could be achieved by using only the first nine components (Figure 4, 95.39% of total variance) but this solution results in a higher (8.46%) misclassification percentage. Additional investigations showed the lowest observed misclassification percentage of 5.39% to be simultaneously the lowest misclassification percentage for the C, AP, and JAPA groups, while the PND group required 20 components to reach its minimum level. It is therefore rather obvious that the component 17 carry a significant part of the between-group differences though it explains by itself an almost negligible proportion of the total variance (0.2%). Figure 4 also illustrates clearly the bias coming from the use of the resubstitution rule to estimate these misclassification percentages, which converge toward zero with increasing component numbers, while percentages calculated using cross-validation reach a plateau around 10.0% after 20 components.

KNN misclassifications results are between 1.5 and 2 times higher than LDA ones. The application of the broken stick model retained only the first four

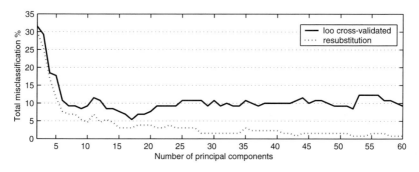

Figure 4. Evolution of the leave-one-out (loo) cross-validated total misclassification using linear discrimination with increasing numbers of principal components derived from Fourier harmonics (crania, see text).

components, resulting in at least two times more misclassified individuals, while Jolliffe's rule provided intermediate results sometimes as good as our optimized one. As a rule that will apply equally to the rest of the discrimination results, whether based on the cranium or the vault, misclassified Japanese crania key into the PND group, those from PND key within AP or JAPA (never within the C group), C within AP or PND, and AP within PND (cranium). In all cases the variation of the misclassification rates of LDA in relation to the number of components was greater than two times that observed with KNN: 11.1% compared to 3.9% for the cranium, 4.9% compared to 0.6% for the vault. These results tend effectively to confirm that KNN are less sensitive to the number of retained components than LDA discriminations.

Landmarks: The MANOVA using the landmark configuration of the complete cranium is also highly significant (Wilks' Lambda $= 0.0321$, $F = 17.501, df = 42/336, p < 10^{-4}$). The CVA yields overall similar results (Figure 5, 91.87%). The most important differences are found between the

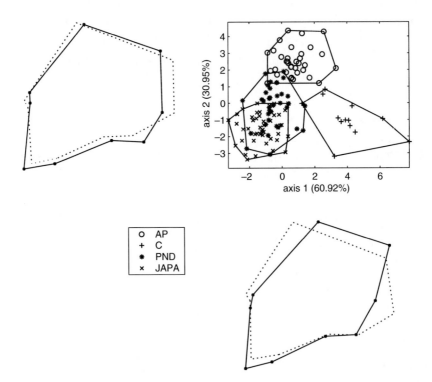

Figure 5.   Canonical variate analysis of the four groups using GPA results for landmarks (i.e., control points) of the crania. Abbreviations: see legend of Figure 3.

circumferentially deformed and the undeformed Peruvians (along the first canonical axis), while the second axis contrasts the undeformed and the AP group. This first plane (CV1 and CV2) shows more overlap than it does in the outline based analysis. The visualization along the two axes (Figure 5) show the pronounced frontal and occipital flattening in the C group, as well as the increased facial protrusion seen before. The AP group on the other hand, is characterized in this analysis by the occipital compression and a facial protrusion, but only a minor frontal flattening.

If on the average, the misclassification rates (Table 2, second column) for landmarks are almost twice as high as they are for outlines, the differences vary substantially between groups. Only the AP deformed group is equally well classified with both types of data. KNN misclassifications rates are again higher than LDA ones, but the difference on LDA results is less than with outlines.

Vaults

Outlines: This MANOVA was also highly significant (Wilks' Lambda = 0.1107, $F = 65.216$, df $= 9/399$, $p < 10^{-4}$). Results of CVA (Figure 6, 90.77% of variance) were generally consistent with those based on the cranium, but, as could be expected, separation of groups was lower. The first two canonical axes mostly isolate the C deformed group (Figure 6), AP remain close to the two undeformed groups. Undeformed Peruvians show much overlap with the other undeformed groups and are therefore not well identified.

The lowest cross-validated misclassification percentages (Table 2, third and fourth columns) are observed for the C deformed and the Japanese samples. As in the case of the cranium, misclassifications rates are higher with landmarks than with outlines, by a factor of three for the total percentages. KNN misclassifications are roughly equivalent to LDA ones, no matter what type of data was used, while application of the broken stick rule resulted in noticeably higher misclassification rates (Table 2, Sixth line).

Landmarks: When the groups comparison is based on vault landmarks only, results are overall comparable (Figure 7, 91.52% of variance). The MANOVA is still highly significant (Wilks' Lambda = 0.0837, $F = 20.428$, df $= 30/462$, $p < 10^{-4}$). The most distant group is, as in the case of outlines, the C deformed group on the canonical axis 1, followed by the contrast between the two deformed groups along the second axis. In terms of shape change in this landmark configuration, the visualization of the canonical axes indicates frontal

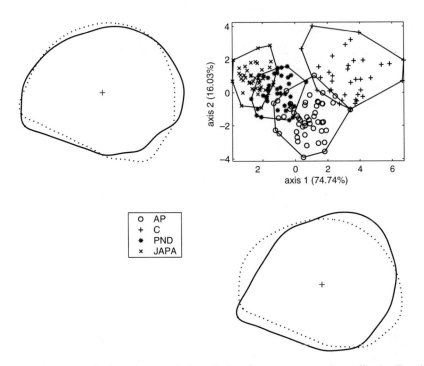

Figure 6. Canonical variate analysis of the four groups using elliptic Fourier parameters derived for the vaults. Abbreviations: see legend of Figure 3.

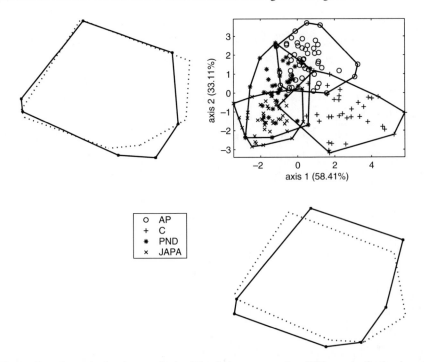

Figure 7. Canonical variate analysis of the four groups using GPA results for landmarks of the vaults. Abbreviations: see legend of Figure 3.

and occipital flattening as well as an upward and backward protrusion of the parietal portion of the vault in the two deformed groups. The second axis shows that the major differences lie in the posterior compression of the vault, as seen in the AP deformed group. The frontal portion of the vault also appears to be configured differently in the AP group, but whether this can be interpreted as frontal flattening remains unclear. Finally, there is a noticeable shift in the *foramen magnum* orientation that was also observed in the outline-based analysis.

DISCUSSION AND CONCLUSION

When it comes to quantitative analyses of shape variation, the advantages of geometric morphometric approaches have been demonstrated in the present volume and elsewhere (Rohlf, 2000; Rohlf and Marcus, 1993). Particularly the use of multiple points for Procrustes alignment (Dryden and Mardia, 1998; Rohlf, 1990) has been shown to improve morphometric studies compared to conventional registration methods, especially in anthropology (Frieß, 1999; Penin and Baylac, 1995; Penin and Baylac, 1999). However, the use of landmarks alone may not always be suitable for the biological shape under study. One of the goals of this chapter is to show how the advantages of GPA can be combined with an outline-based analysis (EFA), and to demonstrate this with an example. Studies of cranial shape variation, though often performed using landmark data, are sensitive to the location and number of landmarks used to describe the contour. The analyses of artificially deformed crania presented here stressed that deformed and undeformed groups are statistically more accurately characterized and distinguished with outlines than with landmarks only. Outlines not only provided a better coverage of the shape differences, they also allowed in most comparisons for lower cross-validated misclassification percentages.

These remarks are a reminder that if global comparisons are useful at a first step, in many instances a deeper analysis of shape differences is needed. Paradoxically, this may become a necessity when differences are large and involve many subregions of the objects. In such circumstances, one needs to know the particular contributions of each subregion. Our results clearly point out that artificial cranial deformation leads to modifications of the basic cranial architecture, represented by the landmarks used here. These modifications extend well beyond the frontal and occipital, that is, the portions to which the deforming apparatus is attached. The face and the basioccipital are shifted in different directions depending on the deformation type. While circumferential deformation leads to a somewhat flattened basioccipital region and a more protruding face (see Frieß and Baylac, 2003), anteroposterior deformation is associated with an increased flexion in this area at least in its posterior portion. Nevertheless, while the vault contour alone allows for low misclassification rates—even lower than when entire cranium is compared—in C and undeformed groups, the best discriminations for the AP group ask for the inclusion of the basioccipital and of the facial regions. Clearly, the shift in orientation of the *foramen magnum* to the clivus in AP deformed skulls contributes to their better classification, but it remains unclear why the same does not apply to the C deformation. It can be proposed that the changes in this region seen in the C deformation are less marked then they are in the AP group, for which they become a distinct characteristic. Conversely, when only vault characteristics are taken into account, the remarkable shape of the C deformed skulls improves their classification percentage relative to the AP deformed skulls, while the shape of the latter group is less well captured without the inclusion of the face and the occipital clivus. In addition, it can also be emphasized that, since landmarks proved to be as efficient as outlines in the case of AP deformations, it appears that it is the overall architecture of the cranium which is modified. The difference in this case is more global, while the C deformation, especially in the frontal and occipital portions is more of regional scale.

We believe that similar comparisons of results using landmarks and outlines enable us to gain additional insights in such analyses. Unlike landmarks, contours do not allow easy delineation of the relative importance of regions by simply discarding or adding subsets of points (Bookstein, 1991; see also Baylac and Penin, 1998 for an illustration). In this study this could be achieved only for a single large region, the vault. A further subsetting of the original outlines, limited for example to the face and/or the basioccipital region, would imply rather artificial and arbitrary closures of large open sections. In such a context, comparing results between outlines and control points, provides a form of subsetting that omits the portions of outlines located between landmarks. Although we used only the entire set of control points, further analyses would be possible by using subsets of it. This is a place where the landmark relaxation framework introduced by Bookstein (1997, 1998; see also Rohlf, 2003) could be particularly helpful, being more flexible in this sense than Fourier approaches.

Finally, the important issue of the dimensionality reduction of the shape space deserves some comments. Our selection rule maximizes an objective criterion applied to discrimination, and uses a cross-validation assessment that provide unbiased estimates (Kraznowski, 1998; Ripley, 1996). Its application retained relatively high numbers of components: from 10 to 16 with outlines (between 97% and 98% of the total variance) and from 6 to 10 (between 95% and 100% of variance) with landmarks. These numbers are much higher than the commonly used first three principal components dealt with most of the reported analyses of shape. They are also far more numerous, and by a factor from 2 to 10, than numbers resulting from the use of the classic broken stick rule. Its application, resulting in higher misclassification rates, by a factor from 1.5 to 5, clearly underestimates the shape information content. Jolliffe's rule in turn provides intermediate results more satisfying with the vaults than with the whole cranium. That high numbers of components may be necessary to achieve correct contour discriminations may simply follow from the use of PCs in the discrimination context. Jolliffe (1986, chapter 9.1.), among many authors, noticed that discriminations using principal component scores usually give large weights to small components. PCA of total covariance matrices indifferently mix the within- and between-group covariances, and there is no guarantee that the between-group separation will involve only high variance PCs. Small component numbers may be needed if the within and between-group covariances share similar directions, while large numbers will be required if both directions are independent (Jolliffe, 1986). The need of large PC numbers to recover fine patterns of the between-group variability may be also the consequence of the inflation of the within-group shape variances, particularly noticeable with outlines, which oversample the shape information. This inflation may be seen for instance, in the fact that the first PCA axes of Fourier coefficients rarely account for a significant proportion of the between-group differences (see e.g., the contrast between PCA and discriminant plots or results in Daegling and Jungers (2000), Ferson et al. (1985), McLellan and Endler (1998); similar results were observed with the present data set—being more acute with the cranium than with the somewhat simpler vault—as well as with the data set analyzed in Monti et al. (2001)). It seems that increasing the amount of details in contour coordinates inflates proportionally more the among-individual variability than the between-one. Since the substitution of their principal components to the original shape variables results in a complex mixing of the within- and the between-group variabilities, we believe that the proper way to reduce the dimensionality of the data space is less to select the best subspace representation, but to maximize the discrimination values between the groups using nonbiased estimates given by cross-validation procedures.

Initially, we had thought that KNN could be useful in the present context. Frequently considered as a powerful discrimination approach (Ripley, 1996), they provided only modest results with the present data set. If KNN results were found effectively less sensitive than LDA to the dimensionality of the shape space, this was at the expense of generally lower performances. In addition, their results are highly sensitive to the number of nearest-neighbors used in their calculations, which should be systematically investigated (Ripley, 1996).

In most biological fields the number of objects that may be used in quantitative analyses of shape is not extensible. In many cases, dimensionality reduction approaches appear therefore inevitable, particularly when shapes involve complex outlines or large sets of three-dimensional landmarks. Nevertheless, it seems to us that dimensionality reduction in morphometrics is a nontrivial task that has been too frequently solved by simple, subjective, yet rarely justified, decisions. Such a task should use reproducible and objectively defined approaches.

ACKNOWLEDGMENTS

We thank Dennis Slice for his active review and editorial process and one anonymous referree for his constructive comments. This article is a contribution of the "Service de Morphométrie" of the Systematic and Evolution Department of the Paris Museum, formerly "Groupe de travail Morphométrie et Analyses de Formes," and of the CNRS GDR 2474 "Morphométrie et évolution des formes."

REFERENCES

Antón, S. C., 1989, Intentional cranial vault deformation and induced changes of the cranial base and face, *Am. J. Phys. Anthropol.* 79:253–267.

Antón, S. C. and Weinstein, K. W., 1999, Artificial cranial deformation and fossil Australians revisited, *J. Hum. Evol.* 36:195–209.

Baylac, M. and Penin, X., 1998, Wing static allometry in *Drosophila simulans* males (Diptera, Drosophilidae) and its relationships with developmental compartments, *Acta Zool.* 44:97–112.

Baylac, M., Villemant, C., and Simbolotti, G., 2003, Combining geometric morphometrics to pattern recognition for species complexes investigations, *Biol. J. Linn. Soc.* 80:89–98.

Bookstein, F. L., 1991, *Morphometric Tools for Landmark Data*, Cambridge University Press, New York.

Bookstein, F. L., 1997, Landmark methods for forms without landmarks: Morphometrics of group differences in outline shape, *Med. Image. Anal.* 1:225–243.

Bookstein, F. L. and Mardia, K. V., 1998, Some new Procrustes methods for the analysis of outlines, Department of Statistics, Report 469, University of Leeds.

Brothwell, D. R., 1975, Possible evidence of a cultural practice affecting head growth in some late Pleistocene East Asian and Australian populations, *J. Archaeolog. Sci.* 2:75–77.

Cheverud, J. M., Kohn, L. A. P., Konigsberg, L. W., and Leigh, S. R., 1992, Effects of fronto-occipital artificial cranial vault modification on the cranial base and face, *Am. J. Phys. Anthropol.* 88:323–345.

Daegling, J. D. and Jungers, W. L., 2000, Elliptical Fourier analysis of symphyseal shape in great apes mandibles, *J. Hum. Evol.* 39:107–122.

Dembo, A. and Imbelloni, J., 1938, Deformaciones intencionales del cuerpo humano de caracter etnico. *J. Anesi.* Buenos Aires.

Dingwall, E. J., 1931, *Artificial Cranial Deformation. A Contribution to the Study of Ethnic Mutilations*, John Bale, Sons & Danielsson Ltd, London.

Dobigny, G., Baylac, M., and Denys, C., 2002, Geometric morphometrics, neural networks and diagnosis of sibling *Taterillus* species (Rodentia, Gerbillinae), *Biol. J. Linn. Soc.* 77:319–327.

Dryden, I. L. and Mardia, K. V., 1998, *Statistical Shape Analysis*, John Wiley & Sons, New York.

Falkenburger, F., 1938, Recherches anthropologiques sur la déformation artificielle du crâne, Paris.

Ferson, S., Rohlf, F. J., and Koehn, R. K., 1985, Measuring shape variation of two-dimensional outlines, *Syst. Zool.* 34:59–68.

Frieß, M., 1999, Some aspects of cranial size and shape, and their variation among later pleistocene hominids, *Anthropologie* (Brno) 37(3):231–238.

Frieß, M. and Baylac, M., 2003, Exploring artificial cranial deformation using elliptic Fourier analysis of Procrustes aligned outlines, *Am. J. Phys. Anthropol.* 122:11–22.

Ihaka, R. and Gentleman, R., 1996, R: A language for data analysis and graphics. *Computational and Graphical Statistics*, 5:299–314.

Jolliffe, I. T., 1986, *Principal Component Analysis*, Series in Statistics, Springer, New York.

Kohn, L. A. P., Leigh, S. R., Jacobs, S. C., and Cheverud, J. M., 1993, Effects of annular cranial vault modification on the cranial base and face, *Am. J. Phys. Anthropol.* 90:147–168.

Krzanowski, W. J., 1987, Cross-validation in principal component analysis, *Biometrics* 43:575–584.

Krzanowski, W. J., 1988, *Principles of Multivariate Analysis. A User's Perspective*, Clarendon Press, Oxford.

Kuhl, F. P. and Giardina, C. R., 1982, Elliptic Fourier features of a closed contour, *Computer Graphics and Image Processing*, 18:236–258.

Lestrel, P. E., 1997, *Fourier Descriptors and Their Applications*, Cambridge University Press, Cambridge.

McLellan, T. and Endler, J. A., 1998, The relative success of some methods for measuring and describing the shape of complex objects, *Syst. Biol.* 47:264–281.

Monti, L., Baylac, M., and Lalanne-Cassou, B., 2001, Elliptical Fourier analysis of the form of genitalia in two *Spodoptera* species and their hybrids (Lepidoptera: Noctuidae), *Biol. J. Linn. Soc.* 72:391–400.

Moss, J. L., 1958, The Pathogenesis of Artificial Cranial Deformation, *Am. J. Phys. Anthropol.* 16:269–286.

Penin, X. and Baylac, M., 1995, Analysis of the skull shape changes in Apes, using 3D procrustes superimposition, *Proceedings in Current Issues in Statistical Analysis of Shape*, pp. 208–210.

Penin, X. and Baylac, M., 1999, Comparaison tridimensionnelle des cranes de *Pan* et *Pongo* par superpositions procrusteennes, *C.R. Acad. Sci. Paris, Sciences de la Vie* 322:1099–1104.

Ripley, B. D., 1996, *Pattern Recognition and Neural Networks*, Cambridge University Press.

Rohlf, F. J., 1986, Relationships among eigenshape analysis, Fourier analysis and analysis of coordinates, *Math. Geol.* 18:845–854.

Rohlf, F. J., 1990, Fitting curves to outlines, in: *Proceedings of the Michigan Morphometric Workshop*, F. J. Rohlf, and F. L. Bookstein, eds., The University of Michigan Museum of Zoology, Ann Harbor, MI, pp. 167–178.

Rohlf, F. J., 1996, TpsDig. Ecology and Evolution, State University of New York at Stony Brook, http://life.bio.sunysb.edu/morph/.

Rohlf, F. J., 2000, Statistical power comparisons among alternative morphometric methods, *Am. J. Phys. Anthropol.* 111:463–478.

Rohlf, F. J., 2003, TPSRelw. Ecology and Evolution, State University of New York at Stony Brook, http://life.bio.sunysb.edu/morph/.

Rohlf, F. J. and Marcus, L. F., 1993, A revolution in morphometrics, *TREE* 8(4):129–132.

Rusakov, D., 1996, Dimension reduction and selection of landmarks, in: *Advances in Morphometrics: Proceedings of the 1993 NATO-ASI on Morphometrics*, L. F. Marcus, M. Corti, A. Loy, G. Naylor, and D. Slice, eds., Plenum Publishers, New York.

Sakamaki, Y. and Akimoto, S., 1997, Wing shape of Gelechiid moths and its functions: Analysis by elliptic Fourier transformation, *Annls Entomol. Soc. Am.* 90(4):447–452.

Tangchaitrong, K., Brearley-Messer, L., Thomas, C. D. L., and Townsend, G. C., 2000, Fourier analysis of facial profiles of young twins, *Am. J. Phys. Anthropol.* 113:369–379.

Trinkaus, E., 1983, *The Shanidar Neandertals*, Academic Press, New York.

CHAPTER SEVEN

Problems with Landmark-Based Morphometrics for Fractal Outlines: The Case of Frontal Sinus Ontogeny

Hermann Prossinger

INTRODUCTION

The functional role of the frontal sinuses in *H. sapiens* is currently inconclusively debated. Because in archaic *Homo* and in Australopithecines the frontal sinuses were much larger (Prossinger et al., 2000a), the debate has implications in human evolution studies. The two competing models about the functional role are (a) the spatial models of supraorbital torus formation (Hylander et al., 1991; Moss and Young, 1960; Ravosa, 1991; Shea, 1986) and (b) the masticatory stress hypothesis models (Bookstein et al., 1999; Demes, 1982, 1987; Prossinger et al., 2000b; Spencer and Demes, 1993).

Support for the masticatory stress hypothesis models is considered tenuous, so in this chapter we approach descriptions of the spatial models of supraorbital

---

Hermann Prossinger • Department for Anthropology, University of Vienna, Vienna, Austria.

Modern Morphometrics in Physical Anthropology, edited by Dennis E. Slice.
Kluwer Academic/Plenum Publishers, New York, 2005.

torus formation in a novel way. In two previous chapters, we began by investigating geometric properties of frontal sinus ontogeny (Prossinger, 2001; Prossinger and Bookstein, 2003). One difficulty is the enormous variability in volume, morphology, and extent of frontal sinuses (Anon et al., 1996; Fairbanks, 1990; Maresh, 1940; Szilvássy, 1982). Nonetheless, the ontogeny of the (statistically very noisy) cross-sections in *H. sapiens* could be modeled (Prossinger and Bookstein, 2003) with a sigmoid function written as

$$A(t) = K \frac{1}{1 + e^{-r(t-t_0)}} \tag{1}$$

where $A(t)$ is the frontal sinus cross-section as a function of time t, with K, r, and t_0 being parameters to be estimated from the data (K is the asymptotic value—the total cross-section of the adult frontal sinus; t_0 is the time at the point of inflection—the age of the individual when the rate of cross-sectional increase is greatest; r is a growth parameter—more properly: erosion parameter (Prossinger and Bookstein, 2003), which determines the rate of cross-section increase for each individual). Because we could show (Prossinger and Bookstein, 2003) that t_0 is in very good agreement with the sex-specific onset of puberty, we argued that it is the parameter r that varies considerably among individuals, thus resulting in the observed noisiness of the adult cross-section K.

The function $A(t)$ is a solution of the autocatalytic equation

$$A(t) = \frac{\mathrm{d}A}{\mathrm{d}t} = rA\left(1 - \frac{A}{K}\right) \tag{2}$$

This equation is the starting point of a suite of approaches we present here. It has been known that Equation (2) can lead to chaotic solutions (which are temporal fractals), if r is sufficiently large. Consequently, although r is small enough (in frontal sinus ontogeny) for $A(t)$ not to be a (temporal) fractal, perhaps the outline encompassing the area A is a (spatial) fractal. In order to pursue the viability of this thought, several methodological approaches must be developed. They are introduced here.

Fractals are objects with a "fractional dimension" (explained below) that have the same (geometric) structure on every scale of magnification. The frontal sinus outlines, if indeed fractal, are therefore not smooth curves, and we can then neither rely on a straightforward method of finding landmarks on them (a prerequisite for conventional geometric morphometrics), nor use the method of "sliding landmarks" (Bookstein et al., 1999) to statistically assess the shape between landmarks (points can not slide along a fractal). If we can validate the claim that frontal sinus outlines are fractal, then we must draw attention to the necessity of analyzing their shape with a methodology different from conventional geometric morphometrics. One method (which we apply here) of trying to identify points—"surrogate landmarks"—on a fractal outline uses SVD (singular value decomposition) of the data matrix of a population of outlines (explained below).

There is an intriguing physiological possibility why the solutions of the autocatalytic equation could be generated as fractals—a possibility hinted at in a previous publication (Prossinger and Bookstein, 2003): the expansion of the frontal sinus is a local erosion process involving osteoblasts and osteoclasts. We ask "How is it that a local process can lead to a large, biologically coherent structure?" In numerous studies, primarily in oncology, it has been shown that local processes can lead to long-range correlations that characterize statistical fractals (Nonnenmacher et al., 1993). We introduce one such local process, the percolation cluster model (PCM) and show how it can be adapted to simulate the ontogeny of frontal sinus outlines with fractal properties.

In summary, we show: first, that we can identify the outlines of a growth process with an ontogeny described by Equation (1) and Equation (2) as fractals; second, how to approximate these outlines with polygons and subsequently determine their fractal dimension; third, how singular value decomposition can identify points that take the role of landmarks for morphometric analysis; and, fourth, how fractals increasing in size may simulate frontal sinus ontogeny.

MATERIALS AND METHODS

Materials

One outline we use is from the frontal sinus of a (female) Neanderthal cranium, Krapina C (Gorjanović-Kramberger, 1899; Radovčić et al., 1988), which we obtain by flood-filling the CT-scan of the frontal bone (Prossinger, 2000a) and then projected the resulting virtual sinus onto an image plane so that a "Caldwell view" (Caldwell, 1918) results (Figure 1). The outline is digitized with Rohlf's program tpsDig (Rohlf, 2001).

Another set of outlines we use is from tracings made by Kritscher (1980) of the roentgenograms of the left and right lobes of 711 crania (housed at the Natural History Museum in Vienna), which had been catalogued with known sex and geographical (often including ethnic) provenience. All were of adults

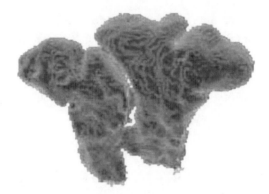

Figure 1. The "Caldwell" projection of a CT-scan of the frontal sinuses of Krapina C. The (small) squares visible in the image are the projections of the (cubic) voxels of the CT-scan.

Figure 2. The frontal sinus cross-section outlines traced by Kritscher (1980) from the roentgenograms of 24 Chinese skulls. The left lobes of the individuals are the right ones in the image. (The ★ symbol indicates the outline that is shown in the detailed analysis of Figures 3–6.)

at time of death (details in Kritscher, 1980). These outlines of 24 Chinese have also been digitized using tpsDig (Rohlf, 2001). See Figure 2.

The coordinates of the outline points (of Krapina C and the Chinese set) were imported into MATHEMATICA® programs and subjected to the mathematical and statistical analyses described in this article.

Methods

We find the center of mass of the (noncircular) disk enclosed by the outline and then find the polar coordinates of every point on the outline with the center of mass of the *disk* being the origin (Figure 3). The outline is then approximated by a polygon encoded in polar coordinates. We find those points between corners of the polygon that are integral multiples of some angle $\Delta\alpha$ ($\Delta\alpha = 5°$ in Figure 4 and $\Delta\alpha = 2°$ in Figure 5) by linear triangulation between the corners of the outline polygon (Figure 6). (The starting vector is the horizontal one pointing left, as defined by the Caldwell view.) We then construct a (column) vector of the radial components of the polar vectors from the center of mass to these interpolated points; there are 72 components if the angles are integral multiples of 5°, or n angles ($n = 360°/\Delta\alpha, n \in \mathbb{N}$) for some other $\Delta\alpha$.

Percolation Cluster Models Generate Fractals: The Percolation Cluster Model (PCM) algorithm (Gaylord and Wellin, 1994) is a generalization of the Eden Model algorithm (Eden, 1961): one first initializes a set of sites on a grid (called an initial perimeter) and then randomly chooses some number

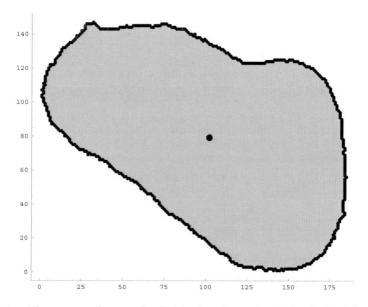

Figure 3. The center of mass of a digitized outline of an individual's right frontal sinus tracing. The center of mass is defined as the center of mass of all pixels inside (and including) the pixels of the outline. Coordinates are in pixel units; the center of mass of the (noncircular) disk is the large dot.

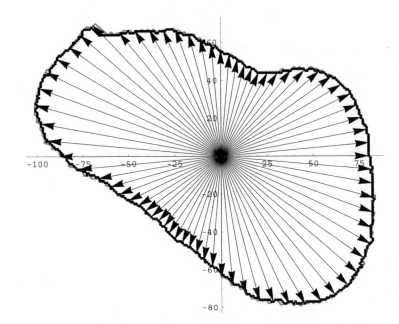

Figure 4. The polar vectors approximating the outline tracing of the right frontal sinus of the individual from Figure 3. The vectors are 5° apart, and are interpolated between nearest pixels of the digitized outline.

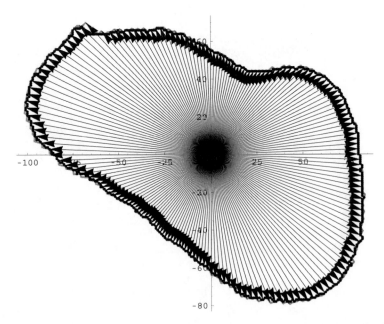

Figure 5. The polar vectors approximating the outline tracing of the right frontal sinus of the individual from Figure 3. The vectors are 2° apart, and are interpolated between nearest pixels of the digitized outline.

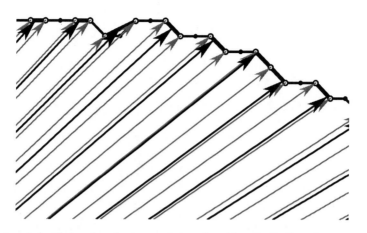

Figure 6. A detail showing the interpolation algorithm for finding the vectors shown in Figures 4 and 5. A pair of vectors (dark gray, small arrowheads) point to pixels of the digitized outline (open circles) nearest to integral multiples of $\Delta\alpha$, and every interpolation vector is linearly triangulated between such a pair (black vector, large arrowhead). The outline segments that straddle the integral multiples of $\Delta\alpha$ are drawn as thick line segments; the points of the outline that are not needed for the triangulation are drawn as black dots.

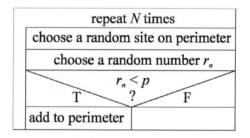

Figure 7. A Nassi-Schneidermann structogram of the Percolation Cluster Model algorithm. Using a probability $p < 1$ is a generalization of the Eden Model, in which $p = 1$. For useful simulations, N must be large; in this study, $N = 75,000$, $145,000$, and $175,000$ and $p = 0.55$ (see Figure 12).

r_n $(0 \leq r_n \leq 1)$. If this number is less than some predefined probability value p (chosen by the user before the execution of the program), then a randomly chosen site adjacent to—but outside—the perimeter is added to the perimeter. This iterative process is continued for a specified number of iterations N (Figure 7). After N iterations, the perimeter will be a fractal; its fractal dimension depends on p.

Detecting Fractality: Each outline is a sequence $r_\Phi(j)$ $(j = 1, \ldots, n)$ of radii beginning at the lag angle $\Phi(0 \leq \Phi \leq 2\pi)$. We are interested how the radial component r_Φ at lag Φ covaries with its neighbor j steps away. We look at the differences

$$\Delta r_\Phi(j) = r(j + \Phi) - r(\Phi) \qquad (3)$$

and calculate the variance of these differences at the position Φ, namely

$$\mathrm{Var}(\Delta r_\Phi) = \left\langle \Delta r^2 \right\rangle_\Phi = \frac{1}{n-1} \sum_{j=1}^{n} \left(\Delta r_\Phi(j) - \langle \Delta r \rangle \right)^2 \qquad (4)$$

where $\langle \Delta r \rangle$ denotes the (arithmetic) mean over all j (observing that this average is independent of Φ). The standard deviation $\left\langle \Delta r^2 \right\rangle_\Phi^{1/2}$ is a function of Φ. If it varies as $\left\langle \Delta r^2 \right\rangle_\Phi^{1/2} \sim \Phi^F$, then the graph $\ln\left(\left\langle \Delta r^2 \right\rangle_\Phi^{1/2} \right)$ vs $\ln(\Phi)$ should be a straight

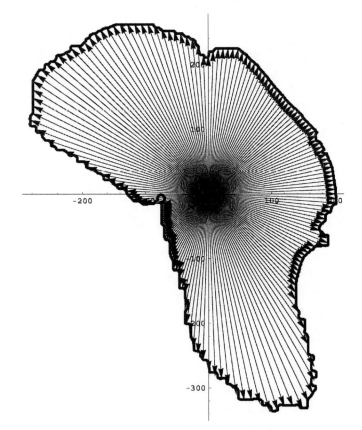

Figure 8. The polar vector interpolation outline of the right frontal sinus lobe of Krapina C. The interpolation method is the same as shown in Figures 3–6 for one Chinese specimen. Coordinates are in pixel units.

line with slope F. If F is integer, then the curve is smooth (a one-dimensional geometric object); if not, then it is a measure of the fractal dimension of the outline (Baumann et al., 1997). Geometric outlines that derive from biological morphologies are statistical fractals; they do not have a fixed F for all Φ, but only over some range. For outlines to be considered statistical fractals, then the linear region should be over some reasonable interval $[\Phi_A, \Phi_B]$, as will be demonstrated in the case of Krapina C. See Figure 8.

RESULTS

The Fractal Dimension of the Krapina C Outline

To exemplify the methodology, we first determine the fractal dimension of the frontal sinus outline of Krapina C. Figure 1 shows a projection of the CT-scan of her frontal sinuses and Figure 8 the vectors, $\Delta\alpha = 2°$ apart, that approximate the digitized outline of her right lobe. Figure 9 is the graph of $\ln\left(\langle\Delta r^2\rangle_\Phi^{1/2}\right)$ vs $\ln(\Phi)$ derived from the polygon approximation to this outline. A linear region (lag angles Φ_2 to Φ_9) has a well-defined slope $F = 0.66147 (P < 8 \times 10^{-12}; r^2 = 0.9997)$.

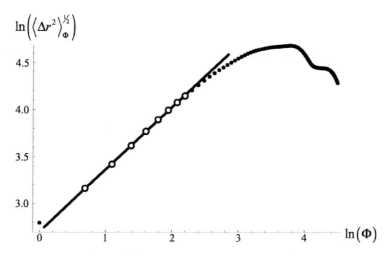

Figure 9. The $\ln\left(\langle\Delta r^2\rangle_\Phi^{1/2}\right)$ vs $\ln(\Phi)$ graph for the left frontal sinus outline of Krapina C when interpolated by vectors that are 2° apart (as in Figure 8). The points used for the linear interpolation are represented with open circles. There is a clear linear region between the second and the ninth points (inclusive; corresponding to a scaling range 4°–18°). This observed linearity indicates a fractal outline for the left lobe. The slope of the linear region is 0.661467, so $\langle\Delta r^2\rangle_\Phi^{1/2} \sim \Phi^{0.661467} (r^2 = 0.9997)$.

In this linear region $\langle \Delta r^2 \rangle_\Phi^{1/2} \sim \Phi^{0.661467}$, so the outline must be a fractal (Bassingthwaite et al., 1994; Czirok, 2001). If the outline were constructed as a self-similar fluctuation over a straight line (i.e., the deviations r are perpendicular to the abscissa, rather than from a point origin), then the Hurst exponent H would be $H = E + 1 - F = 1.343455$ ($E = 1$ is the Euclidean dimension; Bassingthwaite, 1994). However, in our case, r is the radial component of a polar vector, and the fractal dimension of a closed outline does not have the straightforward relationship $H = E + 1 - F$, so we cannot (at present) compute the Hurst exponent.

THE RIGHT FRONTAL SINUS OUTLINES
FROM A CHINESE SAMPLE

Figure 5 shows the 180 polar vector polygon for $\Delta\alpha = 2°$ for one Chinese individual's right lobe outline. We repeat this vector polygon interpolation process for all 23 other right frontal sinus outlines and construct a (singular, rectangular) matrix

$$A = \begin{pmatrix} r_{11} & r_{1j} & r_{1\,24} \\ & \cdot & \\ & \cdot & \\ \cdots & \cdots & r_{ij} & \cdots & \cdots \\ & \cdot & \\ & \cdot & \\ r_{180\,1} & r_{180j} & r_{180\,24} \end{pmatrix} \tag{5}$$

where each column vector $(r_{1j} \ldots \ldots r_{ij} \ldots \ldots r_{180j})^T$ is the (2°) vector polygon of the jth individual's outline, scaled to its Centroid Size. We calculate the Singular Value Decomposition (SVD) of A (Leon, 1998) namely,

$$A = \begin{pmatrix} \cdots & \vec{u}_k & \cdots \end{pmatrix} \cdot \begin{pmatrix} \sigma_1 & & \\ & \cdot & \\ & & \cdot \\ & & & \sigma_{24} \\ 0 & \cdot & \cdot & 0 \\ & \cdot & \\ 0 & \cdot & \cdot & 0 \end{pmatrix} \cdot \begin{pmatrix} \cdots & \vec{v}_l & \cdots \end{pmatrix}^T . \tag{6}$$

There are 24 singular values σ_j because 24 individuals were used for constructing the decomposition. The matrix $A_1 = \sigma_1 \vec{u}_1 \vec{v}_1^{\mathrm{T}}$ can be obtained by setting all singular values to zero except the first one. This matrix A_1 is the first approximation of A with respect to the Frobenius norm (Leon, 1998). The column vectors of this matrix A_1 are the 24 outlines that are the closest fit for the 24 individuals. In fact, they are practically identical, as can be seen in Figure 10. SVD thus produces an average outline of the 24 (fractal) frontal sinus outlines.

If one calculates $A_2 = \sigma_1 \vec{u}_1 \vec{v}_1^{\mathrm{T}} + \sigma_2 \vec{u}_2 \vec{v}_2^{\mathrm{T}}$ by setting $\sigma_j = 0$ (for $j = 3 \ldots 24$), then one obtains a better approximation of A (in the Frobenius norm sense). In this case, the 24 column vectors $\vec{r}_j^{\mathrm{T}}$ of A_2 represent a very interesting approximation of these 24 frontal sinus outlines, as can be seen in Figure 11. We point out (and take advantage of) a remarkable feature: Each first approximation outline must intersect the second approximation outline in 4 points; however, these 4 intersection points of all 24 outlines are very close together; they are (almost)

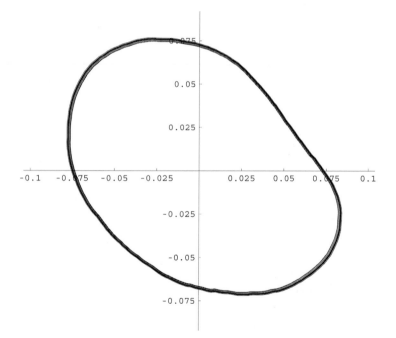

Figure 10. The outlines of the 24 Chinese left frontal sinuses using only the first singular value of the SVD method. The outlines generated by this approximation are almost indistinguishable and are the means of the outlines in the Frobenius sense.

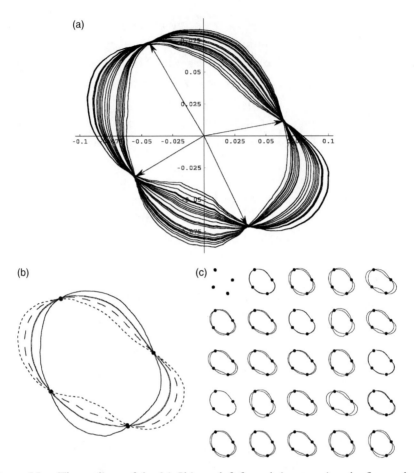

Figure 11. The outlines of the 24 Chinese left frontal sinuses using the first and the second singular values of the SVD method (see text). (a) The outlines using the first singular value cross the outline generated by this (second) approximation at 4 points, and these 4 points have a common direction (indicated by the four arrows) for all 24 outlines. We use these four points on every outline for geometric morphometric analyses of landmarks; we know of no other method to find landmarks for fractal outlines. The directions are at the angles $12°$, $123°$, $212°$, and $298°$, measured counterclockwise from the positive abscissa direction. We note that the 4 points of every outline are very close to the "homologous points" of all other 23 outlines; (b) The outlines of three individuals obtained by the SVD approximation methodology. In the first approximation, all three contours almost indistinguishable. The second outlines are represented by a solid line, a short dashed line and a long dashed line. Observe how the intersection points for all three outlines (black dots) are almost at the same position; (c) A graphic array of all 24 first and second approximations. The intersection points are black dots. In the upper lefthand corner, the $4 \times 24 = 96$ "surrogate landmarks" are drawn; we observe that they lie very close together in four regions, all having the angular directions mentioned in (a).

common to all outlines (as exemplified in Figures 11b–c). We think this is a noteworthy discovery: by choosing the angles of each of these four directions (12°, 123°, 212°, and 298°; see Figures 11a and 11c), we can extract the four radial components of four position vectors on the frontal sinus outline of each individual. We define the four points obtained this way as *surrogate land-marks*. If we multiply the respective column vectors $\vec{r}_j^{\mathrm{T}}$ of A_2 by their respective Centroid Size, we obtain a scatter of $4N$ landmarks (in this sample of Chinese, $N = 24$), which can be analyzed by geometric morphometrics.

DISCUSSION

In Figure 12, we show the first step used to simulate a frontal sinus ontogeny: the three fractal outlines are for $N = 75,000$, $N = 145,000$, and $N = 175,000$ iterations of the PCM algorithm (with $p = 55\%$). In a second step, we use the thin-plate spline interpolation (Bookstein, 1997) to warp these three fractals to the frontal sinus outline polygon obtained from either the tracing of its roent-genogram or the "Caldwell" projection of a CT-scan. In Figure 13, we apply this procedure to the outline of Krapina C and claim we have thereby simulated Krapina C's ontogeny. There are, however, two unresolved issues: one, how to smooth a PCM-modeled fractal so as to compare the model's fractal dimen-sion with the biological specimen's; and two, the development of a statistical test of when fractal dimensions can be considered statistically indistinguishable. (i.e., What initial perimeter configurations and what probabilities p in the PCM algorithm are statistically significantly different?)

A fractal dimension is a measure of self-similar scaling (Barabási and Stanley, 1995; Czirok, 2001). There are various methods of finding the fractal dimension (box counting, correlation analysis as presented here, etc.). The transformation rules from a fractal dimension obtained by one method to one obtained by another are known only for fractals grown on either a straight or a planar substrate (Barabási and Stanley, 1995). Because our biologically defined perimeters have a closed topology, we have currently no method for finding the equivalent transformation rules. Likewise, we cannot determine the relation of these fractal dimensions with those of a random walk on a closed path (Bookstein, personal communication). These topics necessitate further investigations, because the results promise insights into possible physiological processes that control the ontogenetic parameters K, t_0, and r in the sigmoid function (Equation 1).

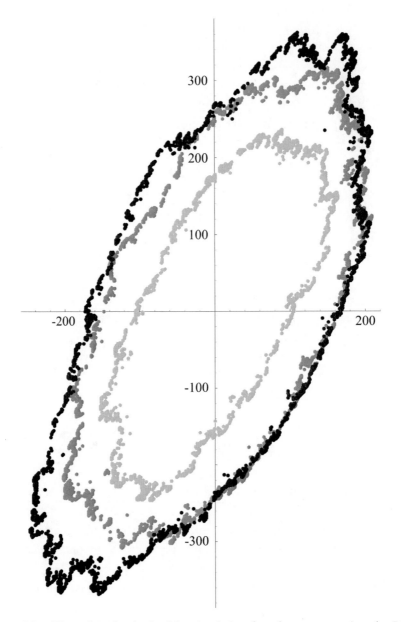

Figure 12. Three fractals obtained by simulating fractal ontogeny using the PCM algorithm. The number of iterations (N in Figure 7) is 75,000 (light gray dots), 145,000 (medium gray dots), and 175,000 (black dots) respectively; the probability is $p = 0.55$. The number of points on each (fractal) perimeter is 1,706, 2,268, and 2,549, respectively.

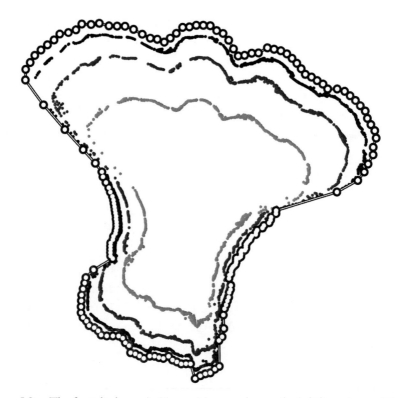

Figure 13. The fractals shown in Figure 12 warped onto the left frontal sinus lobe 2°
polygon of Krapina C (open circles connected by a double-line) using a thin-plate spline
(Bookstein, 1997). The 2° polygon perimeters of the three fractals from Figure 12 were
splined onto the left lobe 2° polygon outline with a scaling factor proportional to the
number of perimeter points (the largest fractal being arbitrarily chosen to be 95% of
the final outline shown in Figure 1). The successive increases in area within the fractal
contours generated by the PCM algorithm is then determined by the relative increase
in the number of points on the perimeter (Figure 12). The resulting fractal contours
(same gray level coding as in Figure 12) generate a remarkably suggestive model for the
ontogeny $A(t)$ of this frontal sinus lobe.

The fact that the SVD method identifies surrogate landmarks is intriguing.
Combining the four surrogate landmarks of the left and the four of the right
lobes offers the possibility of analyzing left/right directional and fluctuating
asymmetries as well as allometric scaling of the outline coordinates relative to
the supraorbital browridge. Heretofore, geometric morphometric analysis of
outlines was not possible for rough, fractal-like outlines. The SVD method
presented here opens up such an opportunity.

CONCLUSIONS AND IMPLICATIONS

One method of analyzing outline shapes is to use identifiable points, commonly called landmarks. If, however, the outlines contain neither characteristic geometric features (such as extremes of curvature) nor homologous points (such as where a suture crosses a ridge curve) then the SVD method we have introduced here allows the identification of points that may be considered surrogate homologues with respect to morphometric analysis. The SVD method is particularly useful if the outlines are extremely variable contours, as the fractals presented here are. We note that the points found by the SVD method cannot be identified by visual inspection of the outline; they are carriers of geometric outline information which, we postulate, includes biological ontogeny information.

The landmarks obtained by the SVD method of polar coordinates for outlines of one set of lobes (the left, say) may not be subjected to a generalized Procrustes fit: because of the use of integral multiples of the angle $\Delta\alpha$ as a determination of the column vector components, we may not rotate the outlines; only superposition to a common center of mass is permissible. In the case of geometric morphometrics analysis of both lobes together, standard Procrustes fitting procedure is possible.

Because the fractal dimension of a biological outline is a statistical measure of its texture, we point out that the determination of the fractal dimension and these surrogate landmarks (using SVD on fractals) implies texture feature analysis. (The biological smoothing process of such structures is yet to be determined.) It is highly suggestive to use the SVD method for outlines that are characterized/dominated by biological texture features—smoothness is not a prerequisite for the SVD method, whereas it is for sliding semilandmark algorithms (Bookstein et al., 1999). Algorithms that let semilandmarks slide need fixed points to constrain their sliding range. If the fractal outlines are approximated by "$\Delta\alpha$-polygons," then one can implement the sliding semilandmark algorithms by using the four surrogate landmarks (identified with the SVD method) as fixed points that constrain the sliding of semilandmarks that are intermediate to these four.

This study introduces several methodologies needed to investigate the morphology and ontogeny of noisy, fractal outlines without landmarks: (a) how to determine the fractal dimension of biological outlines that have long-range correlations despite the local nature of the processes; (b) how to determine surrogate landmarks of a sample of outlines by using singular value decomposition of polygon approximations to the outlines; (c) how to use the PCM algorithm to simulate the local processes (in the case of frontal sinuses: erosion processes); and (d) how to use thin-plate spline warping of fractal perimeters (obtained by the PCM algorithm) to simulate frontal sinus outline ontogeny.

ACKNOWLEDGMENTS

I thank H. Kritscher (Natural History Museum, Vienna) for permission to use his outline tracings and Jakov Radovčić (Croatian Natural History Museum, Zagreb) for permission to use the CT-scan of the Krapina C Neanderthal. Furthermore, I thank F. L. Bookstein (University of Michigan, Ann Arbor) for suggestive, supportive, and encouraging comments, as well as P. O'Higgins (University of York, York) for insightful debates. Thanks also to P. Gunz and P. Mitteröcker (University of Vienna, Vienna) who kindly gave me permission to use their thin-plate spline interpolation modules programmed in MATHEMATICA®.

REFERENCES

Anon, J. B., Rontal, M., and Zinreich, S. J., 1996, *Anatomy of Paranasal Sinuses*, Thieme Verlag, New York.

Barabási, A. L. and Stanley, H. E., 1995, *Fractal Concepts in Surface Growth*, Cambridge University Press, Cambridge, UK.

Bassingthwaite, J. B., Liebovitch, L. S., and West, B. J., 1994, *Fractal Physiology*, Oxford University Press, New York.

Baumann, G., Dollinger, J., Losa, G. A., and Nonnenmacher, T. F., 1997, Fractal analysis of landscapes in medicine, in: *Fractals in Biology and Medicine*, G. A. Losa, D. Merlini, T. F. Nonnenmacher, and E. R. Weibel, eds., Vol. 2, Birkhäuser Verlag, Basel, pp. 97–113.

Bookstein, F. L., 1997, *Morphometric Tools for Landmark Data: Geometry and Biology*, Cambridge University Press, Cambridge, UK.

Bookstein, F. L., Schaefer, K., Prossinger, H., Seidler, H., Fieder, M., Stringer, C., Weber, G., Arsuaga, J. L., Slice, D. E., Rohlf, F. J., Recheis, W., Mariam, A. J., and Marcus, L. F., 1999, Comparing frontal cranial profiles in archaic and modern Homo by morphometric analysis, *Anat. Rec.* (*New Anat.*) 257:217–224.

Caldwell, E. W., 1918, Skiagraphy of the accessory sinuses of the nose, *Am. J. Radiol.* 5:569–574.

Czirok, A., 2001, Patterns and correlations, in: *Fluctuations and Scaling in Biology*, T. Vicsek, ed., Oxford University Press, Oxford, UK, pp. 48–116.

Demes, B., 1982, The resistance of primate skulls against mechanical stress, *J. Hum. Evol.* 11:687–691.

Demes, B., 1987, Another look at an old face: Biomechanics of the Neanderthal facial skeleton reconsidered, *J. Hum. Evol.* 16:297–303.

Eden, M., 1961, A two-dimensional growth process, in: *Biology and Problems of Health*, Proc. 4th Berkeley Symposium on Mathematics, Statistics and Probability, Vol. 4, F. Neyman, ed., University of California Press, Berkeley, pp. 223–239.

Fairbanks, D. N. F., 1990, Embryology and anatomy, in: *Pediatric Otolaryngology*, C. D. Bluestone, S. E. Stool, and M. D. Scheetz, eds., Saunders, Philadelphia, pp. 605–631.

Gaylord, R. J. and Wellin, P. R., 1994, *Computer Simulations with MATHEMATICA®: Explorations in Complex Physical and Biological Systems*, Springer Verlag, New York, pp. 44–55.

Gorjanović-Kramberger, D., 1899, Der paläolithische Mensch und seine Zeitgenossen aus dem Diluvium von Krapina in Kroatien, *Mitt. Anthrop. Ges. Wien* 29:65–68.

Hylander, W. L., Picq, P. G., and Johnson, K. R., 1991, Function of the supraorbital region of primates, *Arch. Oral Biol.* 36:273–281.

Kritscher, H., 1980, Stirnhöhlenvariationen bei den Rassen des Menschen und die Beziehungen der Stirnhöhlen zu einigen anthropometrischen Merkmalen, doctoral thesis, University of Vienna (unpublished thesis).

Leon, S. J., 1998, *Linear Algebra with Applications*, Prentice Hall, Upper Saddle River, NJ.

Maresh, M. M., 1940, Paranasal sinuses from birth to late adolescence as observed in routine postanterior roentgenograms, *Am. J. Dis. Child.* 60:55–78.

Moss, M. L. and Young, R. W., 1960, A functional approach to craniology, *Am. J. Phys. Anthrop.* 18:281–292.

Nonnenmacher, T. F., Losa, G. A., and Weiberl, E. R., eds., 1993, *Fractals in Biology and Medicine*, Vol. 1, Birkhäuser, Basle.

Prossinger, H., 2001, Sexually dimorphic ontogenetic trajectories of frontal sinus cross sections, *Coll. Antropol.* 25:1–11.

Prossinger, H., Wicke, L., Seidler, H., Weber, G. W., Recheis, W., and Müller, G., 2000a, The CT scans of fossilized crania with encrustations removed allow morphological and metric comparisons of paranasal sinuses, *Am. J. Phys. Anthropol.* 30(Suppl):254.

Prossinger, H., Bookstein, F., Schäfer, K., and Seidler, H., 2000b, Reemerging stress: supraorbital torus morphology in the midsagittal plane? *Anat. Rec. (New Anat.)* 261:170–172.

Prossinger, H. and Bookstein, F. L., 2003, Statistical estimators of frontal sinus cross section ontogeny from very noisy data, *Journ. Morph.* 257:1–8.

Radovčić, J., Smith, F. H., Trinkaus, E., and Wolpoff, M. H., 1988, *The Krapina Hominids: An Illustrated Catalog of Skeletal Collections*, Mladost, Zagreb.

Ravosa, M. J., 1991, Interspecific perspective on mechanical and nonmechanical models of primate circumorbital morphology, *Am. J. Phys. Anthrop.* 86:369–396.

Rohlf, J., 2001, tpsDig: A program for digitizing landmarks and outlines for geometric morphometric analysis, Department of Ecology and Evolution, State University of New York, Stony Brook, NY.

Shea, B. T., 1986, On skull form and the supraorbital torus in primates, *Curr. Anthropol.* 26:257–260.

Spencer, M. A. and Demes, B., 1993, Biomechanical analysis of masticatory system configuration in Neandertals and Inuits, *Am. J. Phys. Anthrop.* 91:1–20.

Szilvássy, J., 1982, Zur Variation, Entwicklung und Vererbung der Stirnhöhlen. *Ann. Naturhist. Mus. Wien* 84A:97–125.

CHAPTER EIGHT

An Invariant Approach to the Study of Fluctuating Asymmetry: Developmental Instability in a Mouse Model for Down Syndrome

Joan T. Richtsmeier, Theodore M. Cole III, and Subhash R. Lele

INTRODUCTION

Aneuploidy refers to the condition where the number of chromosomes within an organism is not an exact multiple of the haploid number. Examples of aneuploidy include monosomy (a single chromosome instead of a pair exists for a given chromosome) and trisomy (three copies of a chromosome are present for a given chromosome). Trisomy 21 (Ts21) or Down syndrome (DS) is the

Joan T. Richtsmeier • Department of Anthropology, The Pennsylvania State University, University Park, PA 16802 and Center for Craniofacial Development and Disorders, The Johns Hopkins University, Baltimore, MD 21205. **Theodore M. Cole III** • Department of Basic Medical Science, School of Medicine, University of Missouri—Kansas City, Kansas City, MO 64108.
Subhash R. Lele • Department of Mathematical and Statistical Sciences, University of Alberta, Edmonton, Alberta, Canada T6G 2G1.

Modern Morphometrics in Physical Anthropology, edited by Dennis E. Slice.
Kluwer Academic/Plenum Publishers, New York, 2005.

most frequent live-born aneuploidy in humans, occurring in approximately one in 700 live births. The cause of Ts21 is most commonly nondisjunction during meiosis (Antonarakis, 1991; Antonarakis et al., 1992, 1993), but little is known about the mechanisms responsible for developmental anomalies associated with the DS phenotype.

Two ideas have been articulated about the cause of anomalies associated with Ts21. The first is that specific genes on Chr 21, when occurring in triplicate, cause the production of particular phenotypes (Delabar, 1993; Korenberg, 1991; Korenberg et al., 1990, 1994). The second is that Ts21 phenotypes result from a generalized genetic imbalance that causes amplified *developmental instability* (DI) produced by altered responses to genetic and environmental factors to which all individuals are exposed. This idea was proposed by Hall (1965) and supported by Shapiro and others (Greber-Platzer et al., 1999; Shapiro, 1975, 1983, 2001), who suggested that the observation of increased variability in linear measurements of many features in DS, as compared to unaffected individuals, supported this idea. The developmental mechanisms that underlie DI remain largely unexplained, however (Hallgrímsson and Hall, 2002).

Bilateral symmetry is a phylogenetically widespread characteristic of many complex organisms (Palmer, 1996). In those organisms that tend toward bilateral symmetry, there is a midline plane that divides the body into right and left halves (Figure 1). Midline symmetry is secured by ontogenetic and phylogenetic mechanisms, so that the breaking of symmetry is a relatively rare event and, therefore, of interest to biologists. Fluctuating asymmetry (FA), the variance of deviations from perfect symmetry, has been proposed and is widely used as a measure of DI (Palmer and Strobeck, 1992; Polak, 2003). Since a single genome controls the development of both the left and right sides, and the environment is typically the same for both sides, the expectation is that the two sides of an organism are replicates, or mirror images of each other. Deviations from symmetry are thought to represent the effects of random perturbations during development.

It is commonly held that the development of the organism is driven by a plan that includes perfect symmetry for traits that occur bilaterally. Even in a stable environment, however, small random perturbations of biological processes produce phenotypic deviations from the ideal. These perturbations, commonly called *developmental noise*, result in part from the accumulation of the products of stochastic gene expression mechanisms (see Kirschner and Gerhart, 1998;

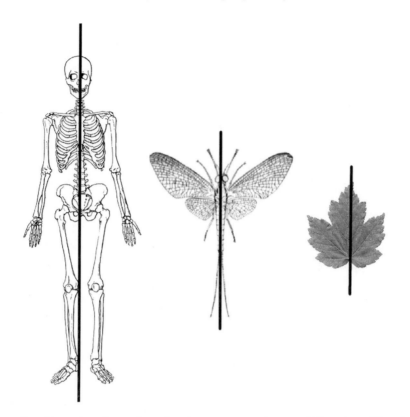

Figure 1. Examples of organisms showing symmetry along a midline plane. The wings of the mayfly are an example of "matching symmetry," where symmetry is observed in separate bilateral structures. The leaf is an example of "object symmetry," where symmetry is seen within a single structure centered on a midline plane. The human skeleton is composed of anatomical units showing both matching symmetry (e.g., upper and lower limbs) and object symmetry (e.g., skull, vertebrae).

McAdams and Arkin, 1997). *Developmental stability* is the suppression of phenotypic variation within individuals and refers to the capacity for developmental trajectories to resist accidents and perturbations during growth.

In the comparative studies of right and left sides of an organism, the underlying developmental assumption is that organisms possess some sort of homeostatic mechanisms that control the development of traits that occur bilaterally (Van Valen, 1962). These mechanisms, though poorly understood, determine the organism's developmental stability. According to Klingenberg (2002), developmental noise can cause differences between body sides. These responses are mediated by the organism's DI, defined as the organism's tendency to produce a morphological change in response to developmental perturbations. Developmental instability and developmental stability are, therefore, two sides of the same coin; the former referring to the organism's phenotypic response to perturbations, the latter to the organism's capacity to buffer these insults through homeostatic mechanisms that inhibit the expression of a phenotypic response (Klingenberg, 2003). Many questions about these homeostatic mechanisms remain unanswered, and little is known about the developmental basis for asymmetry.

Though departure from symmetry is a property of the individual, patterns of asymmetry in a particular trait are studied at the level of the population or sample (Palmer, 1994). When departure from symmetry is quantified as the difference between similar measures on the left and right sides (L–R) in a population, three basic types of asymmetry are defined on the basis of the frequency distribution of the (L–R) measure. Small, subtle deviations from perfect symmetry, which do not show a tendency to a specific side (nondirectional), characterize FA. The pattern of L–R symmetry in a sample of individuals exhibiting FA, shows a unimodal distribution with a mean of zero and with variation symmetrically distributed around the mean (Figure 2a). Evidence for a positive correlation between FA and DI comes from the results of studies that show FA to increase as environmental and/or genetic "stress" increases (Møller and

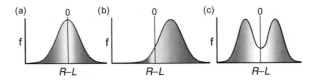

Figure 2. Asymmetry may be characterized by the distribution of asymmetry values within a population (after Palmer and Strobeck, 1986; Van Valen, 1962). In this figure, "asymmetry" is quantified as the signed arithmetic difference between right and left measurements (L–R) of a single dimension. (a) In *fluctuating asymmetry*, deviations from symmetry are small and randomly distributed as to side, so the distribution of L–R is unimodal and centered at zero. This definition assumes that the underlying "ideal" form is perfectly symmetric (i.e., L–R = 0). (b) *Directional asymmetry* describes a measurement that is generally larger on the same side in all members of a population, and the mean value is something other than zero (L–$R \neq 0$). (c) *Antisymmetry* refers to a measurement that is usually asymmetric, but larger on the right in some individuals, and larger on the left in others. In this case, the distribution of (L–R) is bimodal or platykurtic (adapted from Palmer, 1994).

Swaddle, 1997; Palmer and Strobeck, 1992; Zakharov, 1992), but others argue that the relationship between FA and DI is weak (Fuller and Houle, 2002). *Directional asymmetry* (DA) describes a pattern where the difference between sides is biased as to side (i.e., one side tends to be consistently larger across individuals in a population). An example of DA occurs in the bill of the wry-billed plover (*Anarhynchus frontalis*), which is always bent to the right at the tip by up to 12° (Neville, 1976). DA need not favor a single side for all characters within an organism, but can favor the left side for some traits and the right side for different traits. Distributions for characters showing DA in a population are unimodal with a mean that is different from (either greater or less than) zero (Figure 2b). *Antisymmetry* describes a pattern of bilateral variation in a sample where the difference between sides is consistent, but nondirectional. A common example is the fiddler crab (*Uca pugnax*) where the crusher claw is always larger than the cutter claw, but it is just as likely that the right claw be the crusher claw as it is that the crusher claw be on the left. Because the left or the right side may be predominant in cases of antisymmetry, the distribution that describes antisymmetry in a sample is bimodal and centered on zero (Figure 2c).

Since we are interested in DI as a basis for the production of the DS phenotypes, this study is concerned with FA. Most simply, FA can be thought of as a metric that compares corresponding measures from the right and left sides of organisms within a sample. Most analyses aim to determine whether or not differences in the magnitude of FA exist in two samples. Although our ultimate goal is to understand the development of phenotypes in humans with DS, humans provide a less than ideal study subject since genetic background cannot be controlled and collection of data from certain developmental time points is not possible. To our advantage, several informative mouse models for DS have been developed (Davisson et al., 1993; Sago et al., 1998) and are useful in the study of DI in aneuploidy as demonstrated by FA.

THE QUANTITATIVE STUDY OF FA

The use of FA as an indicator of increased levels of DI has been broadly reviewed (e.g., DeLeon, 2004; Møller and Swaddle, 1997; Polak, 2003). Traditional methods for studying FA were described fully by Palmer (1994). Superimposition for the purpose of studying asymmetry was briefly introduced by Bookstein (1991: 267–270) and applied in a more fully developed context by Auffray et al. (1996). The Procrustes approach was later revised by

Smith et al. (1997), thus formally linking the study of FA with geometric morphometrics (see Bookstein, 1991; Marcus et al., 1996; Richtsmeier et al., 1992, 2002a for reviews of geometric morphometrics). The Procrustes approach to FA was extended by Klingenberg and McIntyre (1998) to a two-factor ANOVA design for the purpose of estimating and testing the different components of asymmetry. Mardia et al. (2000) elaborated on the formal statistics of symmetry of shapes using Procrustes superimposition.

Procrustean approaches fall within the class of geometric morphometrics called superimposition methods (Richtsmeier et al., 2002a). Superimposition methods involve the translation, rotation, and scaling of landmark data from two or more objects into the same coordinate space according to a specified rule. With two objects, one object is designated as the "reference," and the other is designated as the "target." The displacements necessary to take the landmarks in the reference to their new locations in the superimposed target are used to characterize the differences between the two landmark sets. With more than two objects, variation in form is described relative to the iteratively computed sample mean.

We have chosen to develop an alternate approach to the study of asymmetry for the following reasons. When using Procrustes, the researcher chooses a particular criterion for superimposing the two sides. For example, the least-squares criterion (where, after reflection, the forms are superimposed so that the sum of the squared distances between corresponding landmarks on the two forms are minimized) leads to the Generalized Procrustes superimposition. This is currently the most commonly used strategy for superimposition (Klingenberg et al., 2002). Alternatively, the generalized resistant fit algorithm (Chapman, 1990; Rohlf and Slice, 1990; Siegel and Benson, 1982) uses repeated medians to calculate the best fit between two mean forms and attributes differences to a small number of landmarks, instead of spreading it over the whole object, as is done using the least-squares approach. The use of different fitting criteria for matching gives different superimpositions (Richtsmeier et al., 2002a; Rohlf and Slice, 1990). This means that localized differences between two objects or the local measures of variation among objects in a sample will vary depending upon the superimposition criterion used. Results are, therefore, affected by the scientist's arbitrary choice of a superimposition criterion. The choice of superimposition criteria is rarely consciously made by the researcher, but instead is integrated into the software program. The crucial point is that the superimposition scheme used in analysis can change results of an analysis by shifting the location of maximum differences from one biological location to another, or spread the effects of a shifted biological locus to unaffected, neighboring biological loci. Moreover, the data cannot inform us of which superimposition is the correct one (Richtsmeier et al., 2002b).

What follows statistically from these observations is that, due to the nuisance parameters of rotation and translation, neither the mean nor the variance–covariance matrix can be estimated consistently from data using Procrustes. Lele and Richtsmeier (1990) first recognized this problem. It was further explained by Lele (1991, 1993) and proven mathematically by Lele and McCulloch (2002). Walker (2001) published similar findings. If the variance–covariance structure cannot be estimated correctly using Procrustes, the development of models that decompose Procrustes variance structures in order to separate components of symmetric variation among individuals from that within individuals seems ill-advised.

AN ALTERNATE APPROACH TO THE ASSESSMENT OF FA

Our approach to the study of FA is based on Euclidean distance matrix analysis (EDMA; Lele and Richtsmeier, 2001). Suppose we have an object that is described by a collection of landmarks in three dimensions. In contrast with most other landmark-based morphometric methods, EDMA does not require placement of the observations under study into an arbitrary coordinate system in order to describe or compare them. Instead the coordinate data are rewritten as a matrix of interlandmark distances. These distances remain the same, no matter how the objects are positioned or oriented. This property is called *coordinate-system invariance* (Lele and McCulloch, 2002; Lele and Richtsmeier, 2001).

Before we can study FA in a sample, we must first be able to describe DA, because measurement of the former is dependent on the latter. Our algorithm for the analysis of DA is described in terms of a single, left–right pair of linear distance between landmarks. However, the steps of the algorithm are applied to every left–right distance pair. For each individual in a sample, a form matrix is computed, consisting of all unique interlandmark distances. The linear distances that occur bilaterally are paired, one being from the left side of the organism (L) and other from the right (R). For each individual i, we define the (signed) asymmetry of a distance pair as $(L-R)_i$. If $(L-R)_i = 0$, then individual i

is perfectly symmetric for that pair of distances. Asymmetric individuals will either have $(L-R)_i > 0$ or $(L-R)_i < 0$, depending on which side is larger. The *sample distribution* of $(L-R)$ contains information about both DA and FA. The mean of the sample, $\overline{(L-R)}$, measures DA. The amount of dispersion (variation) in the sample (the measurement of which is described below) is a measurement of FA.

Our bootstrap-based algorithm for measuring DA was developed by Cole (2001) and is an extension of work by O'Grady and Antonyshyn (1999). Programs are available from the Richtsmeier laboratory website: http://oshima.anthro.psu.edu. The approach is reviewed briefly here. Again, for the sake of clarity, we are describing the algorithm in terms of a single pair of left–right distances. However, in practice, the bootstrapping procedure is not applied independently to each distance pair. Instead, entire individuals (i.e., with linear distances calculated from the complete set of landmarks used in the study) are resampled randomly and with replacement, so that information about the covariances among measurements is retained in the data.

Preliminary step: We describe the DA in a sample for each linear distance pair by calculating the mean of $(L-R)$, calling it $\overline{(L-R)}$. If this mean were exactly zero, then the sample would be symmetric *on the average*, even though each of the individuals in the sample might be asymmetric to some degree. In such a case, there would be no DA for the sample. However, if $\overline{(L-R)}$ were, in fact, different from zero (and it would be likely to be at least slightly different for any real sample), we would then want to know how far it must be from zero before we would consider the DA in the sample to be significant. To determine this, we use the remainder of the algorithm to construct a confidence interval for $\overline{(L-R)}$ using the bootstrap.

Step 1: Denote the size of sample X as n_X. Construct a bootstrap *pseudosample*, called X*, by selecting n_X individuals from X randomly and with replacement. This is a typical resampling strategy for nonparametric bootstrapping (Davison and Hinkley, 1997; Efron and Tibshirani, 1993). Use this pseudosample to compute a *bootstrap estimate* of the mean asymmetry, calling it $\overline{(L-R)}^*$.

Step 2: Repeat Step 1 M times where M is some large number (e.g., 1,000 or more) generating a new random pseudosample each time. The result is a distribution of M estimates of $\overline{(L-R)}^*$, the pseudosample means.

Step 3: Sort the vector of M bootstrap estimates of $\overline{(L-R)}^*$ in ascending order: $\overline{(L-R)}_{[1]}\ldots\overline{(L-R)}_{[M]}$. Truncate the sorted vector to obtain a bootstrap estimate of the marginal confidence interval for $\overline{(L-R)}$. For a $100(1-\alpha)\%$ confidence interval, the lower bound will be $\overline{(L-R)}^*_{[(M)(\alpha/2)]}$ and the upper bound will be $\overline{(L-R)}^*_{[(M)(1-\alpha/2)]}$. For example, when $M = 1{,}000$ and 90% confidence intervals are desired (where $\alpha = 0.10$), the estimates of the lower and upper bounds will be $\overline{(L-R)}^*_{[50]}$ and $\overline{(L-R)}^*_{[950]}$, respectively. This method for obtaining a confidence interval by truncating a sorted vector of bootstrap estimates is called the percentile method (Davison and Hinkley, 1997; Efron and Tibshirani, 1993).

Step 4: Evaluate the DA of the sample by determining whether the confidence interval includes zero, which is the expected value of $\overline{(L-R)}$ when there is no DA. If the interval *excludes* zero, then the null hypothesis is rejected, and we conclude that there is a significant degree of DA—a "handedness"—in the sample as a whole for the distance being considered.

If significant DA is found for one or more linear distance pairs, we must decide whether the *biological* interpretation of DA should be a part of the analysis. For some studies, an understanding of DA patterns may be of primary importance. For example, our original application of EDMA to the study of asymmetry (Cole, 2001) was a study of children affected with unilateral coronal craniosynostosis (a problem of antisymmetry, although it was treated as DA problem after reflections of some observations in the sample; Figure 3). In this case, we were interested not only in identifying which specific distances were asymmetric (as the result of premature suture fusion), but also in identifying the "handedness" of each asymmetric distance (relative to the side of suture fusion). However, if there is no rationale for a biological investigation of DA, we might consider DA to be a nuisance that confuses our measurement of FA. Whether the DA in a measurement is significant or not, it must be accounted for before FA can be accurately measured (Palmer and Strobeck, 1986). Otherwise, we run the risk of confusing FA and the "total" asymmetry in a sample.

There are many different ways to quantify FA in paired distances (e.g., Palmer and Strobeck, 1986). One simple way is to express the asymmetries of individuals as absolute deviations from the sample mean. To simplify further discussion, we introduce additional notation. Let us use A_i to represent the absolute value of the difference of individual i's left–right asymmetry from the

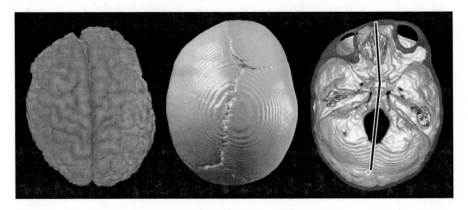

Figure 3. An example of DA where the midline is directly and visibly affected. This figure shows a superior view of (from left to right) the neural surface of the brain, the superior surface of the neurocranium, and the endocranial base with a line showing the midsagittal plane of an individual with premature closure of the left coronal suture. Although asymmetry of midline structures is obvious in this case, the potential for direct effects of asymmetry on midline points should not be ignored in analyses of fluctuating asymmetry.

sample mean:

$$A_i = \left| (L - R)_i - \overline{(L - R)} \right|$$

Because $\overline{(L - R)}$ represents a measure of DA for the sample, the distribution of A is a measure of the amount of FA for the sample (i.e., the subtraction of $\overline{(L - R)}$ means that the directional-asymmetry component of the total asymmetry has been removed).

We now present our algorithm for comparing levels of FA in two samples. Our null hypothesis is that there is no difference between samples in the amount of FA. This algorithm is similar to the DA algorithm in that it uses the bootstrap.

Preliminary step: Suppose we are comparing two samples called X and Y. Calculate the $L - R$ means for both samples, calling them $\overline{(L - R)}_X$ and $\overline{(L - R)}_Y$, respectively. For sample X, calculate the A statistic for each individual in the sample. Recall that $A_i = \left| (L - R)_i - \overline{(L - R)} \right|$ and that this is calculated for each distance pair. Calculate the sample mean and call it $\overline{A}_X$. Similarly, calculate the sample mean of A for sample Y, calling it $\overline{A}_Y$. These means are measures of FA within their respective samples, with larger values of A indicating greater degrees of asymmetry. Our null hypothesis is that the amount of FA is the same in the two samples, or H_0: $\overline{A}_X - \overline{A}_Y = 0$. We can give the difference in means a new name, D. We use the bootstrap to construct a confidence interval that will determine whether D is significantly different from zero. This is an application of Hall and Martin's (1988) bootstrap-based two-sample test.

Step 1: Denote the size of sample X as n_X. Similarly, denote the size of sample Y as n_Y. As with the DA algorithm, we will use a nonparametric bootstrap approach to resampling. Construct a bootstrap pseudosample called X* by sampling n_X individuals from X randomly and with replacement. Similarly, construct a pseudosample called Y* by sampling n_Y individuals randomly and with replacement from Y. Compute the means of A from the bootstrap samples and call them $\overline{A}_X{}^*$ and $\overline{A}_Y{}^*$. Then call the differences in bootstrap means, $D^* = \overline{A}_X{}^* - \overline{A}_Y{}^*$.

Step 2: Repeat Step 1 M times where M is some large number (e.g., 1,000 or more), generating new random pseudosamples each time. The result is a vector of M bootstrap estimates (D^*) for the difference between sample means (D).

Step 3: Sort the M bootstrap estimates of D in ascending order: $D^*_{[1]} \ldots D^*_{[M]}$. Truncate the vector to obtain $100(1 - \alpha)\%$ confidence intervals, as described above for the DA algorithm.

Step 4: If the bootstrap confidence interval for D excludes zero, we reject the null hypothesis and conclude that there is a significant difference in the amount of FA in the two samples. If the null hypothesis is rejected and $D = \overline{A}_X - \overline{A}_Y > 0$, we conclude that there is a greater amount of FA in sample X. Conversely, if the null hypothesis is rejected and $D < 0$, we conclude that there is a greater amount of FA in sample Y.

This algorithm may be applied in cases of both "matching" and "object" symmetry (Mardia et al., 2000). For matching symmetry, where there are no landmarks that belong to the midline plane by definition (e.g., insect wings), we examine asymmetry in all possible distances that are present bilaterally. Because we are using interlandmark distances, our measurements of asymmetry are coordinate-system invariant and are not affected by arbitrary locations or orientations of the left- and right-side structures. For "object" symmetry, there may be landmarks that lie in the midline plane by definition (e.g., midsagittal landmarks on the skull; see Figure 3). With our approach, we can include these landmarks if appropriate. The inclusion of midline landmarks allows us to examine asymmetry in bilateral distances that have midline landmarks at one end. We do not consider distances *between* midline landmarks in our analyses because they are not paired. Note that our method makes no assumptions about the midline points being coplanar; information about any distortion in the midline plane will be contained in comparisons of all the bilateral distances. As with considerations of matching symmetry, our use of interlandmark distances ensures coordinate-system invariance so that the orientations and positions of the individual observations do not affect the results.

Finally, we should mention that there is another potential factor that can confuse our consideration of FA, particularly in comparisons between samples: variation in scale (Palmer and Strobeck, 1986). Suppose we are studying a sample of humans affected with a particular genetic disorder, and we want to compare the degree of FA in these humans with a genetically engineered mouse model of the same disorder (e.g., DS). Because measurements of human skulls are absolutely much larger than the corresponding measurements on the skulls of mice, we would expect the $\overline{A}$ statistics to be larger in humans, even if the actual degree of FA is the same. This is because of the well-known positive association between the means and variances of linear distances (Lande, 1977; Palmer and Strobeck, 1986). If we want to compare levels of FA in samples of organisms that differ substantially in size, we need to incorporate a scale-adjustment, so that we will explicitly examine *relative* FA. Further discussion of this problem, along with some proposed solutions, is found in Palmer and Strobeck (1986).

COMPARATIVE ANALYSIS OF FA IN ANEUPLOIDY

Using Animal Models to Study DS

As noted previously, two distinct schools of thought have emerged to explain why the inheritance of three copies of Chr 21 genes results in disruption of normal patterns of development. The *amplified DI* hypothesis holds that the correct balance of gene expression in pathways regulating development is disrupted by dosage imbalance of the hundreds of genes on Chr 21 (Shapiro, 1975, 1983, 1999, 2001). Support for this hypothesis includes: (a) the observation that features seen in DS are nonspecific, occurring in other trisomic conditions (Hall, 1965; Shapiro, 1983) and in the population at large (albeit at much lower frequency); and (b) measures of significantly increased individual phenotypic variation among individuals with Ts21 compared to euploid individuals (Kisling, 1966; Levinson et al., 1955; Roche, 1964, 1965; Shapiro, 1970). The amplified DI hypothesis states that DS phenotypes, and the increased variation noted in DS populations, result from a disruption of an evolutionarily achieved balance of genetic programs regulating development and recognizes that pathways disrupted by Ts21 involve many more genes than those on Chr 21 (Reeves et al., 2001).

The other ideas proposed to explain why Ts21 disrupts normal patterns of development are summarized by the *gene dosage effects* hypothesis that argues for a more specific relationship between particular genes and specific individual DS traits (Delabar, 1993; Korenberg et al., 1994). The gene dosage effects hypothesis holds that dosage imbalance of a specific gene or small group of genes from Chr 21 is responsible for specific individual DS traits.

It is clear that the debate surrounding these hypotheses (Pritchard and Kola, 1999; Reeves et al., 2001; Shapiro, 1999) cannot be resolved by continued study of adult DS individuals. A joint focus on the mechanism of gene action (e.g., Saran et al., 2003) and the phenotypic consequences for development is needed to understand the etiology of this complex disorder. A comprehensive explanation of the etiology of DS features should consider *developmental* consequences of aneuploidy and not only the direct overexpression of the triplicated genes or the phenotypic consequences of this overexpression as manifest in the adult (Reeves et al., 2001). Since the biological processes underlying these two hypotheses and the data needed to sufficiently test them cannot be evaluated using human data, experimental organisms are required. Mouse strains with segmental trisomy 16 have been studied as genetic models of DS (Baxter et al., 2000; Davisson et al., 1993; Neville, 1976; Reeves et al., 1995; Richtsmeier et al., 2000; Sago et al., 1998, 2000). Ts1Cje is a segmental trisomy 16 model that arose as a fortuitous translocation of mouse Chromosome 16 (Chr 16) in a transgenic mouse line (Sago et al., 1998). These mice are at dosage imbalance for a segment of mouse Chr 16 corresponding to a human Chr 21 region that spans 9.8 Mb and contains 79 of the 225 genes in the Chr 21 gene catalog (Hattori et al., 2000). The genetic insult in Ts1Cje mice and in another segmental trisomy 16 model, Ts65Dn, has been shown to correspond closely to that of segmental Ts21 in human beings (Reeves et al., 1995; Sago et al., 2000). Although species differences need to be kept in mind when complex characters are compared in mouse and human, a detailed, three-dimensional analysis of the skull of segmentally trisomic mice and their normal littermates demonstrated direct parallels between the human DS craniofacial phenotype and that of the Ts1Cje and Ts65Dn mouse skull (Richtsmeier et al., 2000, 2002b). When compared statistically to the skulls of normal littermates, the segmentally trisomic Ts65Dn and Ts1Cje craniofacial skeletons showed an overall reduction in size, a disproportionately reduced midface, maxilla and mandible, and reduced interorbital breadth. Since the effects of gene dosage imbalance on conserved genetic pathways are expected to be similar in mice and human beings, mice with segmental trisomy provide the experimental basis to investigate corresponding developmental processes disrupted by the analogous trisomy in mouse and human. Analysis of prenatal mice is ongoing (Richtsmeier et al., 2002c, 2003).

Analysis of FA in Aneuploid and Euploid Ts1Cje Mice

If differences in developmental stability between euploid and aneuploid mice are the basis for, or contribute to, the craniofacial anomalies of development previously quantified, then differences in the measures of FA should also be evident in a comparison of the euploid and aneuploid samples. Moreover, if aneuploidy results from amplified DI, we predict that measures of FA should be increased in aneuploid as compared to euploid mice. The analysis presented here uses three dimensional coordinates of landmark data collected using the Reflex microscope from adult, segmentally trisomic Ts1Cje mice ($N = 15$) and unaffected littermates ($N = 12$) (Figure 4). Landmarks collected multiple times from each specimen were 27 in number. Measurement error studies were done following details given previously (Richtsmeier et al., 1995). When it was determined that measurement error was minimal and comparable to previous studies using the Reflex microscope (Richtsmeier et al., 2000), an average was computed from two data collection trials.

As noted by Palmer and Strobeck (1986) and Palmer (1994), the calculation of FA is particularly sensitive to measurement error. The method we present here does not, as yet, include an integrated test of FA over measurement error. This may be important because measurement error can contribute directly to measures of FA and can be responsible, at least in part, for differences in FA between measures and between samples. Because the data sets used here were initially collected to study difference in shape, we estimated the precision of each landmark separately. Precision refers to the average absolute difference between repeated measures of the same individual (Kohn and Cheverud, 1992). Three-dimensional coordinates of landmarks were collected several times from 10 mouse skulls with the skull remaining in the same position for each trial. The average variance along the x, y, and z axes for all landmarks

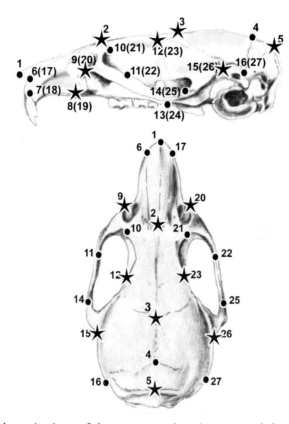

Figure 4. Schematic views of the mouse cranium (upper panel: lateral view; lower panel: superior view) showing landmarks collected using the Reflex microscope. Landmark number and label are given. For bilateral landmarks, the number of the right-sided landmark is shown in parentheses on the lateral view. Landmarks marked by a star were used as an endpoint for linear distances used in the current analysis of FA. *Cranial landmarks*: 1, nasale; 2, nasion; 3, bregma; 4, intersection of parietal and interparietal bones; 5, intersection of interparietal and occipital bones at the midline; 6(17), anterior-most point at intersection of premaxillae and nasal bones; 7(18), center of alveolar ridge over maxillary incisor; 8(19), most inferior point on premaxilla–maxilla suture; 9(20), anterior notch on frontal process lateral to infraorbital fissure; 10(21), intersection of frontal process of maxilla with frontal and lacrimal bones; 11(22), intersection of zygomatic process of maxilla with zygoma (jugal), superior surface; 12(23), frontal–squamosal intersection at temporal crest; 13(24) intersection of maxilla and sphenoid on inferior alveolar ridge; 14(25), intersection of zygoma (jugal) with zygomatic process of temporal, superior aspect; 15(26) joining of squamosal body to zygomatic process of squamosal; 16(27) intersection of parietal, temporal, and occipital bones.

ranged from 0.002–0.061 mm. Further refinements of the method presented here will include integrated measures of FA and measurement error.

From the group of all possible linear distances among the landmarks, we used 18 paired distances to determine if the degree of FA is increased in aneuploid mice. Distances were chosen on the basis of their contribution to significant cranial dysmorphology in Ts1Cje mice (Richtsmeier et al., 2002a). Mean directional asymmetries were computed for left- and right-paired distances in the aneuploid and euploid Ts1Cje samples (X_i and Y_i), and the between-sample difference between measures of absolute asymmetry, A_i, were calculated for corresponding linear distances. As stated previously, the null hypothesis is that for each measure, the two samples show similar magnitudes of absolute asymmetry. Therefore, the expected value of the between-sample difference for measures of absolute asymmetry for corresponding linear distances is zero. In our application, the measure of absolute asymmetry for each linear distance in the euploid sample was subtracted from the corresponding measure in the aneuploid sample, so that values > 0 indicate greater asymmetry in the aneuploid sample for a given distance, while values < 0 indicate greater asymmetry in the euploid sample. The measures of A_i for each sample are given in Figure 5.

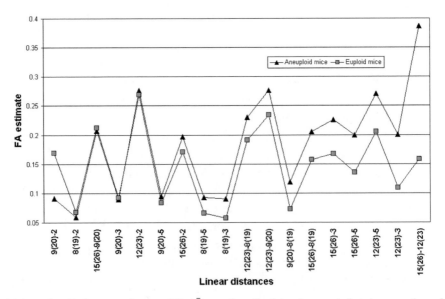

Figure 5. Estimates of mean FA ($\bar{A}_i$ as described in the text) for the samples of aneuploid and euploid Ts1Cje adult mouse crania for linear distances between landmarks given in Figure 4. Landmarks on the right side of the skull are given in parentheses.

Figure 6 provides a summary of the difference in means, D, for all linear distances considered in the Ts1Cje mouse model for DS. Fourteen of the 18 paired linear distances indicate a larger degree of asymmetry in the aneuploid sample ($D_i > 0$ on the right of Figure 6), two show approximately equal measures of asymmetry in the two samples, and two linear distances show a higher degree of FA in the euploid Ts1Cje sample ($D_i < 0$ on the left side of Figure 6). Remember that the distribution of D is a measure of difference in FA for the sample, because the DA component of total asymmetry has been removed in a previous step of the algorithm. Of those linear distances that indicate a larger degree of asymmetry in the aneuploid sample, confidence interval testing shows that four of these differences are significant (marked by arrows on Figure 6). None of

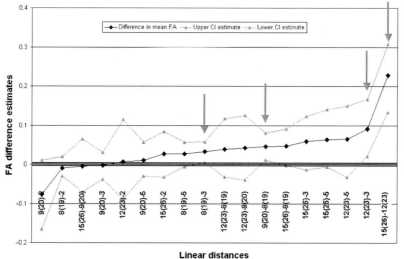

Figure 6.   Graph of comparison of measures of FA in aneuploid and euploid Ts1Cje mice. The y-axis is the measure of the difference in absolute asymmetries between aneuploid and euploid samples for all linear distances considered. The x-axis represents the 18 paired linear distances. Linear distances are defined by the landmarks used as endpoints; landmarks on the right side of the skull are given in parentheses. The estimates of the difference in FA between the two samples are shown as black diamonds. Estimates of the lower and upper bounds of the confidence interval ($\alpha = 0.10$; 1,000 resamples) for each linear distance appear as gray triangles. Those measures of difference in fluctuating asymmetry that show a significant difference in asymmetry in the euploid and aneuploid sample (i.e., 0 is not included in the confidence interval) are marked with a gray arrow.

the measures that are more asymmetric in the euploid sample are shown to be significant by confidence interval testing.

When these results are used to identify the anatomical locations that show significant differences in FA between aneuploid and euploid samples, certain landmarks are shown to be involved more frequently than others (Figure 7). Landmarks that contribute disproportionately to a greater degree of asymmetry in the trisomic mice are located on the intersection of the premaxilla and maxillary bones (landmarks 8 and 19) and the neurocranium at bregma (landmark 3). Overall increased FA in the Ts1Cje aneuploid skull is not limited to a specific

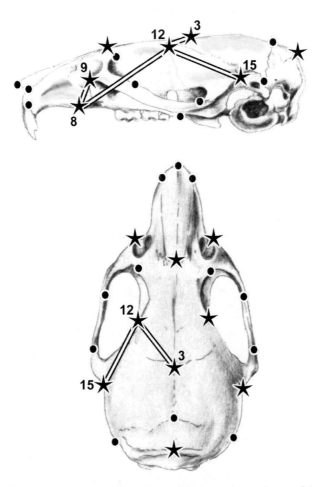

Figure 7. Graphic depiction of those linear distances shown by confidence intervals (see Figure 6) to be relatively more asymmetric in the aneuploid sample are depicted on one side of the skull only.

bone or functional unit of the skull, but is distributed across the skull. Further tests of the significance of the contribution of specific landmarks to FA can be conducted following methods for the detection of influential landmarks outlined by Lele and Richtsmeier (1992).

A previous comparison of the skulls of Ts1Cje euploid and aneuploid mice (Richtsmeier et al., 2002b) found that the Ts1Cje aneuploid mice have a relatively shorter skull along the rostro-caudal axis, with the primary reduction being located on the bones of the face (i.e., premaxilla, maxilla, anterior frontal) and a marked reduction in interorbital distance. The Ts1Cje aneuploid neurocranium was also reduced along the mediolateral axis, but to a lesser degree. Localization of the linear distances that show significantly increased FA in the aneuploid sample (Figure 7) indicates increased FA local to the premaxilla and maxilla of the aneuploid mouse—an area that was shown previously to be significantly dysmorphic in Ts1Cje. However, linear distances on the neurocranium that show statistically significant increased FA in the aneuploid sample in this study were not previously identified as significantly different when compared to their euploid littermates (Richtsmeier et al., 2002b).

As discussed previously, two viable, though not mutually exclusive, hypotheses have been proposed to explain the phenotypes associated with trisomy. We suggested earlier that if increased DI were the cause of dysmorphic features of the skull in Ts1Cje mice, then measures of FA should be increased in aneuploid mice relative to their euploid littermates. We have found some support for this hypothesis in that a majority (78%) of linear distances show higher values of FA in aneuploid mice, though only four of these measures (28%) are significant. However, these linear distances do not correspond with measures of shape that were previously shown to be significantly different from normal (Richtsmeier et al., 2002b).

To fully test the amplified DI hypothesis, an increased understanding of the processes that underlie DI is required coupled with additional analyses using larger sample sizes and additional age groups. Unpublished analyses of morphological integration of Ts1Cje and Ts65Dn mice, find that crania and postcrania of the adult aneuploid mice show *increased* morphological integration as compared to their normal littermates (Hill et al., 2003; Richtsmeier et al., 2002c, 2003). Although only a conjecture at this point, we can envision a developmental scenario where gene action in aneuploidy affects cellular processes in such a way that localized phenotypic dysmorphology of the skeleton results. This dysmorphology may be subtle, but significant enough that developmental adjustments need to be made to ensure adequate structural stability and proper function. The adjustments could include regions that are not directly affected by dysmorphogenesis. The result is a predictable phenotype composed of localized areas of heightened dysmorphology that can be characterized according to a quantifiable distribution. The phenotypic targets of dysmorphogenesis and the adjustments that need to be made in adjoining tissues combine to produce a typical phenotype (like the characteristic DS facial appearance) that manifests itself at the individual level as a "characteristic" phenotype, but at the population level as one of increased phenotypic variability. The increased variability comes from both the actual distribution of effects on localized structures and the requisite and customized adjustments made by adjoining tissues in response to the primary dysmorphology. If the processes responsible for impacting skull growth in trisomy operate in ways similar to what is described above, this could explain the combined findings of increased phenotypic variability, increased morphological integration, and localized increases in FA in samples of aneuploid mice. Theory and methods from evolutionary biology that account for the coordination of developmental modules (e.g., Klingenberg, 2003; Wagner, 1995) will be useful in the evaluation of these ideas.

SUMMARY AND CONCLUSIONS

We have presented a novel method for statistical comparison of FA. The advantages of our method include:

1. the straightforward inclusion of three-dimensional data;
2. the lack of superimposition, so that the user does not need to arbitrarily select a fitting criterion;
3. identification of significant differences in FA by bootstrap confidence intervals;
4. presentation of local measures of FA, enabling identification of the affected anatomical structures and the proposal of testable developmental hypotheses.

The results of our analysis of FA in the Ts1Cje mouse provide preliminary support for the amplified DI hypothesis and provide the basis for a model of the interplay of dysmorphology and FA in aneuploidy that can be further explored in studies of development.

ACKNOWLEDGMENTS

Programs for the EDMA approach to FA are available for free download from http://oshima.anthro.psu.edu (Richtsmeier lab website) and from http://3dlab.umkc.edu (Cole lab website). We thank our collaborators Roger H. Reeves and Charles J. Epstein for providing animals for this research. Ann Zumwalt collected some of the landmark data. Kristina Aldridge prepared figures 1, 2, 3, 6, and 7. We thank Dennis Slice for inviting us to contribute to this volume and for his efforts in bringing these papers together. This presentation was greatly improved and clarified due to helpful comments provided by Dennis Slice, Kristina Aldridge, and an anonymous reviewer. This work was supported in part by NSF-SBER004903; PHS awards F33DEO5706, P60 DE13078, HD 24605, HD 38384, and HD 34198.

REFERENCES

Antonarakis, S. E., 1991, Parental origin of the extra chromosome in trisomy 21 as indicated by analysis of DNA polymorphisms. Down Syndrome Collaborative Group, *N. Engl. J. Med.* 324:872–876.

Antonarakis, S. E., Petersen, M. B., McInnis, M. G., Adelsberger, P. A., Schinzel, A. A., Binkert, F. et al., 1992, The meiotic stage of nondisjunction in trisomy 21: Determination by using DNA polymorphisms, *Am. J. Hum. Genet.* 50:544–550.

Antonarakis, S. E., Avramopoulos, D., Blouin, J. L., Talbot, C. C., Jr., and Schinzel, A. A., 1993, Mitotic errors in somatic cells cause trisomy 21 in about 4.5% of cases and are not associated with advanced maternal age, *Nat. Genet.* 3:146–150.

Auffray, J.-C., Alibert, P., Renaud, S., Orth, A., and Bonhomme, F., 1996, Fluctuating asymmetry in *Mus musculus* subspecific hybridization: Traditional and Procrustes comparative approach, in: *Advances in Morphometrics*, L. Marcus, M. Corti, A. Loy, G. J. P. Naylor, and D. E. Slice, eds., Plenum Press, New York, pp. 275–283.

Baxter, L. L., Moran, T. H., Richtsmeier, J. T., Troncoso, J., and Reeves, R. H., 2000, Discovery and genetic localization of Down syndrome cerebellar phenotypes using the Ts65Dn mouse, *Hum. Mol. Genet.* 9:195–202.

Bookstein, F., 1991, *Morphometric Tools for Landmark Data: Geometry and Biology*, Cambridge University Press, Cambridge.

Chapman, R., 1990, Conventional Procrustes approaches, in: *Proceedings of the Michigan Morphometrics Workshop*, F. J. Rohlf and F. L. Bookstein, eds., University of Michigan Museum of Zoology, Ann Arbor, pp. 251–267.

Cole, T. M. III, 2001, Chapter 7. Further applications of EDMA, in: *An Invariant Approach to the Statistical Analysis of Shapes*, S. Lele and J. T. Richtsmeier, eds., Chapman and Hall/CRC Press, London, pp. 263–284.

Davison, A. C. and Hinkley, D. V., 1997, *Bootstrap Methods and their Applications*, Cambridge University Press, Cambridge.

Davisson, M., Schmidt, C., Reeves, N., Irving, E., Akeson, E., Harris, B. et al., 1993, Segmental trisomy as a mouse model for Down Syndrome, *Prog. Clin. Biol. Res.* 384:117–133.

Delabar, J.-M., Theophile, D., Rahmani, Z., Chettouh, Z., Blouin, J.-L., Prieut, M., Noel, B., and Sinet, P.-M., 1993, Molecular mapping of twenty-four features of Down syndrome on Chromosome 21, *Eur. J. Hum. Genet.* 1:114–124.

DeLeon, V. B., 2004, *Fluctuating Asymmetry in the Human Craniofacial Skeleton: Effects of Sexual Dimorphism, Stress, and Developmental Anomalies*, PhD Thesis, Center for Functional Anatomy & Evolution, Johns Hopkins University.

Efron, B. and Tibshirani, R., 1993, *An Introduction to the Bootstrap.* Chapman & Hall, New York.

Fuller, R. and Houle, D., 2002, Detecting genetic variation in devleopmental instability by artificial selection on fluctuating asymmetry, *J. Evol. Biol.* 15: 954–960.

Greber-Platzer, S., Schatzmann-Turhani, D., Wollenek, G., and Lubec, G., 1999, Evidence against the current hypothesis of "gene dosage effects" of trisomy 21: ets-2, encoded on chromosome 21 is not overexpressed in hearts of patients with Down syndrome, *Biochem. Biophys. Res. Commun.* 254:395–399.

Hall, B., 1965, Delayed ontogenesis in human trisomy syndromes, *Hereditas* 52:334–344.

Hall, P. and Martin, M., 1988, On the bootstrap and two-sample problems, *Austral. J. Stat.* 30A:179–192.

Hallgrímmson, B. and Hall, B., 2002, Modularity within and among limbs: Implications for evolutionary divergence in fore- and hind limb morphology in primates, *Am. J. Phys. Anthropol. Suppl.* 34:81 (abstract).

Hattori, M., Fujiyama, A., Taylor, T. D., Watanabe, H., Yada, T., Park, H. S. et al., 2000, The DNA sequence of human chromosome 21, *Nature* 405:311–319.

Hill, C., Reeves, R. H., Epstein, C. J., Valeri, C. J., Lindsay, E., Baxter, L. L. Cole, T. M., Richtsmeier, JT, 2003, Developmental instability and skeletal phenotypes in Down syndrome, *Am. J. Phys. Anthropol. Suppl.* 36:114 (abstract).

Kirschner, M. and Gerhart, J., 1998, Evolvability, *Proc. Natl. Acad. Sci. USA* 95:8420–8427.

Kisling, E., 1966, *Cranial Morphology in Down's Syndrome: A Comparative Roentgen-cephalometric Study in Adult Males*, Munksgaard, Copenhagen.

Klingenberg, C., 2003, A developmental perspective on developmental instability: Theory, models and mechanisms, in: *Developmental Instability: Causes and Consequences*, M. Polak, ed., Oxford University Press, New York, pp. 427–442.

Klingenberg, C. and McIntyre, G., 1998, Geometric morphometrics of developmental instability: Analyzing patterns of fluctuating asymmetry with Procrustes methods, *Evolution* 52:1363–1375.

Klingenberg, C. P., Barluenga, M., and Meyer, A., 2002, Shape analysis of symmetric structures: Quantifying variation among individuals and asymmetry, *Evolution* 56:1909–1920.

Kohn, L. and Cheverud, J., 1992, Calibration, validation, and evaluation of scanning systems: Anthropmetric imaging system repeatability, in: *Electronic Imaging of the Human Body Workshop*, CSERIAC, Dayton, OH.

Korenberg, R., 1991, Down syndrome phenotypic mapping. Progress in clinical Biology, *Research* 373:43–53.

Korenberg, J. R., Kawashima, H., Pulst, S. M., Ikeuchi, T., Ogasawara, N., Yamamoto, K. et al., 1990, Molecular definition of a region of chromosome 21 that causes features of the Down syndrome phenotype, *Am. J. Hum. Genet.* 47:236–246.

Korenberg, J., Chen, X., Schipper, R., Sun, Z., Gonsky, R., Gerwehr, S. et al., 1994, Down syndrome phenotypes: The consequences of chromosomal imbalance, *Proc. Natl. Acad. Sci. USA* 91:4997–5001.

Lande, R., 1977, On comparing coefficients of variation, *Syst. Zool.* 26:214–217.

Lele, S., 1991, Some comments on coordinate free and scale invariant methods in morphometrics, *Am. J. Phys. Anthropol.* 85:407–418.

Lele, S., 1993, Euclidean distance matrix analysis (EDMA) of landmark data: Estimation of mean form and mean form difference, *Math. Geol.* 25:573–602.

Lele, S. and McCulloch, C., 2002, Invariance and morphometrics, *J. Am. Stat. Assoc.* 971:796–806.

Lele, S. and Richtsmeier, J., 1990, Statistical models in morphometrics: Are they realistic? *Syst. Zool.* 39:60–69.

Lele, S. and Richtsmeier, J. T., 1992, On comparing biological shapes: Detection of influential landmarks, *Am. J. Phys. Anthropol.* 87:49–65.

Lele, S. and Richtsmeier, J., 2001, *An Invariant Approach to the Statistical Analysis of Shapes*, Chapman and Hall/CRC Press, London.

Levinson, A., Friedman, A., and Stamps, F., 1955, Variability of mongolism, *Pediatrics* 16:43.

Marcus, L., Corti, M., Loy, A., Naylor, G. J. P., and Slice, D. E., 1996, *Advances in Morphometrics*, Plenum, New York.

Mardia, K., Bookstein, F., and Moreton, I., 2000, Statistical assessment of bilateral symmetry of shapes, *Biometrika* 87:285–300.

McAdams, H. H. and Arkin, A., 1997, Stochastic mechanisms in gene expression, *Proc. Natl. Acad. Sci. USA* 94:814–819.

Møller, A. and Swaddle, J., 1997, *Asymmetry, Developmental Stability, and Evolution*, Oxford University Press, Oxford.

Neville, A., 1976, *Animal Asymmetry*, Edward Arnold, London.

O'Grady, K. F. and Antonyshyn, O. M., 1999, Facial asymmetry: Three-dimensional analysis using laser scanning, *Plast. Reconstr. Surg.* 104:928–940.

Palmer, A., 1994, Fluctuating asymmetry: A primer, in: *Developmental Instability: Its Origins and Implications*, T. Markow, ed., The Netherlands, Kluwer, Dordrecht, pp. 335–364.

Palmer, A. R., 1996, From symmetry to asymmetry: Phylogenetic patterns of asymmetry variation in animals and their evolutionary significance, *Proc. Natl. Acad. Sci. USA* 93:14279–14286.

Palmer, A. and Strobeck, C., 1986, Fluctuating asymmetry: Measurement, analysis, patterns, *Ann. Rev. Ecol. Syst.* 17:391 421.

Palmer, A. and Strobeck, C., 1992, Fluctuating asymmetry as a measure of developmental stability: Implications of non-normal distributions and power of statistical tests, *Acta Zool. Fenn* 191:57–72.

Polak, M., 2003, *Developmental Instability: Causes and Consequences*, Oxford University Press, Oxford.

Pritchard, M. A. and Kola, I., 1999, The "gene dosage effect" hypothesis versus the "amplified developmental instability" hypothesis in Down syndrome, *J. Neural Transm. Suppl.* 57:293–303.

Reeves, R., Irving, N., Moran, T., Wohn, A., Kitt, C., Sisodia, S. et al., 1995, A mouse model for Down syndrome exhibits learning and behavior deficits, *Nat. Genet.* 11:177–184.

Reeves, R. H., Baxter, L. L., and Richtsmeier, J. T., 2001, Too much of a good thing: Mechanisms of gene action in Down syndrome, *Trends Genet.* 17:83–88.

Richtsmeier, J., Cheverud, J., and Lele, S., 1992, Advances in anthropological morphometrics, *Ann. Rev. Anthropol.* 21:231–253.

Richtsmeier, J. T., Paik, C. H., Elfert, P. C., Cole, T. M., and Dahlman, H. R., 1995, Precision, repeatability, and validation of the localization of cranial landmarks using computed tomography scans, *Cleft Palate Craniofac. J.* 32:217–227.

Richtsmeier, J. T., Baxter, L. L., and Reeves, R. H., 2000, Parallels of craniofacial maldevelopment in Down syndrome and Ts65Dn mice, *Dev. Dyn.* 217:137–145.

Richtsmeier, J., DeLeon, V., and Lele, S., 2002a, The promise of geometric morphometrics, *Yearb. Phys. Anthropol.* 45:63–91.

Richtsmeier, J. T., Zumwalt, A., Carlson, E. J., Epstein, C. J., and Reeves, R. H., 2002b, Craniofacial phenotypes in segmentally trisomic mouse models for Down syndrome, *Am. J. Med. Genet.* 107:317–324.

Richtsmeier, J., Leszl, J., Hill, C., Aldridge, K., Aquino, V. et al., 2002c, Development of skull dysmorphology in Ts65Dn segmentally trisomic mice, *Am. J. Hum. Genet. Suppl.* 72:280 (abstract).

Richtsmeier, J. T., T. M. Cole III, Leszl, J. M., Hill, C. A., Budd, J. L., and Reeves, R. H., 2003, Developmental instability of the skull in aneuploidy, *European Society for Evolutionary Biology, 9th Congress*, Leeds, August (abstract).

Roche, A., 1964, Skeletal maturation rates in mongolism, *Am. J. Roentgenol.* 91:979–987.

Roche, A., 1965, The stature of mongols, *J. Ment. Defic. Res.* 9:131–145.

Rohlf, F. J. and Slice, D. E., 1990, Extensions of the procrustes method for the optimal superimposition of landmarks, *Syst. Zool.* 39:40–59.

Sago, H., Carlson, E., Smith, D., Kilbridge, J., Rubin, E., Mobley, W. et al., 1998, Ts1Cje, a partial trisomy 16 mouse model for Down syndrome, exhibits learning and behavioral abnormalities, *Proc. Natl. Acad. Sci. USA* 95:6256–6261.

Sago, H., Carlson, E., Smith, D. J., Rubin, E. M., Crnic, L. S., Huang, T. T. et al., 2000, Genetic dissection of region associated with behavioral abnormalities in mouse models for Down syndrome, *Pediatr. Res.* 48:606–613.

Saran, N. G., Pletcher, M. T., Natale, J. E., Cheng, Y., and Reeves, R. H., 2003, Global disruption of the cerebellar transcriptome in a Down syndrome mouse model, *Hum. Mol. Genet.* 12(16):2013–2019.

Shapiro, B., 1970, Prenatal dental anomalies in mongolism: Comments on the basis and implications of variability, *Ann. NY Acad. Sci.* 171:562–577.

Shapiro, B., 1975, Amplified developmental instability in Down syndrome, *Ann. Hum. Genet.* 38:429–437.

Shapiro, B., 1983, Down syndrome—a disruption of homeostasis, *Am. J. Med. Genet.* 14:241–269.

Shapiro, B. L., 1999, The Down syndrome critical region, *J. Neural. Transm. Suppl.* 57:41–60.

Shapiro, B. L., 2001, Developmental instability of the cerebellum and its relevance to Down syndrome, *J. Neural Transm. Suppl.* 61 11–34.

Siegel, A. and Benson, R., 1982, A robust comparison of biological shapes, *Biometrics* 38:341–350.

Smith, D., Crespi, B., and Bookstein, F., 1997, Fluctuating asymmetry in the honey bee, *Apis mellifera*: Effects of ploidy and hybridzation, *J. Evol. Biol.* 10:551–574.

Van Valen, L., 1962, A study of fluctuating asymmetry, *Evolution* 16:125–142.

Wagner, G., 1995, Adaptation and the modular design of organisms, in: *Advances in Artificial Life*, A. Moran, J. Merelo, and P. Chacon, eds., Springer-Verlag, Berlin, pp. 317–328.

Walker, J., 2001, Ability of geometric morphometric methods to estimate a known covariance matrix, *Syst. Biol.* 49:686–696.

Zakharov, V., 1992, Population phenogenetics: Analysis of developmental stability in natural populations, *Acta Zool. Fenn* 191:7–30.

PART TWO

Applications

CHAPTER NINE

Comparison of Coordinate and Craniometric Data for Biological Distance Studies

Ashley H. McKeown and Richard L. Jantz

INTRODUCTION

In physical anthropology, biological distance studies have employed quantitative traits observed on the human body or skeleton to measure the degree of population divergence among groups separated by time and/or geography (Buikstra et al., 1990). These studies estimate phenotypic distances among human populations from observed morphological variation in order to infer evolutionary history. Craniometrics, or measurements designed to quantify craniofacial morphology, have been a popular and effective tool for biological distance studies (e.g., Heathcote, 1986; Howells, 1973; Jantz, 1973; Relethford, 1994), and research has demonstrated a considerable genetic component to cranial form (Devor, 1987; Sparks and Jantz, 2002).

Ashley H. McKeown and Richard L. Jantz • Department of Anthropology, 250 South Stadium Hall, University of Tennessee, Knoxville, TN 37996.

Modern Morphometrics in Physical Anthropology, edited by Dennis E. Slice.
Kluwer Academic/Plenum Publishers, New York, 2005.

An alternate approach to quantifying size and shape is the use of Cartesian coordinates of cranial landmarks. Since these coordinates are collected in two or three dimensions, they simultaneously incorporate more information than one-dimensional linear measurements, and, therefore, provide more powerful statistical analysis (Rohlf and Slice, 1990). Collection of coordinate data is relatively simple with digitizing equipment that downloads data directly into a computerized format, eliminating the need for multiple calipers and manual recording. Additionally, repatriation of archaeological skeletal material necessitates thorough documentation, and coordinate data offer a way to archive the cranium in a format that permits the reconstruction of a substantial portion of the original form. For a more in-depth discussion of the benefits of coordinate data collection, see Ousley and McKeown (2001).

As the statistical tools necessary for the analysis of coordinate data, known as geometric morphometry, have become increasingly accessible, coordinate data are seeing wider application including investigations of craniofacial variation and its relationship to biological distances among human populations (e.g., McKeown, 2000; Ross et al., 1999). Using both craniometric and coordinate data observed on the same population samples, this chapter compares the two types of data by evaluating results from the same statistical analyses designed to discern patterns of biological distance among the groups under investigation.

This comparison is designed to evaluate the two different types of data employed to answer the same questions about populations that lived in the past. Craniometric data sets are not simply a mindless list of inter-landmark distances (which could be calculated from the three-dimensional coordinates), but represent an attempt to quantify cranial size and shape through a series of specific linear measurements. Some of these measurements are not replicated in the coordinate data set because their endpoints are type III landmarks (endpoints of maximal or minimal dimensions). Also, the collection protocol for the craniometric data set was different from the coordinate data set, and the data sets incorporate any errors or biases associated with the particular method and instruments used. Hence, this study seeks to demonstrate the utility of coordinate data for anthropological questions, such as biological distance among past populations, and to compare it to the current standard for quantitative studies of craniofacial morphology, craniometrics.

MATERIALS AND METHODS

Samples

Craniometric and coordinate data observed on specimens from archaeological components attributed to the protohistoric and postcontact Arikara of South Dakota were analyzed using the same statistical tests. Male and female adult specimens from 11 components of 8 sites in the Middle Missouri region were employed in this comparative study (Table 1). Sample sizes for the components vary between the data sets for the most part due to the repatriation of specimens since the craniometric data sets were recorded. In most cases, the sample size difference is not very large and likely has little effect on the study presented here. The largest discrepancy is for the samples from Mobridge Feature 2 (109 craniometric specimens and 85 coordinate specimens); nevertheless, the sample available for the coordinate data set is large enough to provide a reasonable estimate of the sample's phenotypic variability. This may not be true for the coordinate data set from Swan Creek, where 37 specimens comprise the craniometric sample while only 12 individuals were available for inclusion in the coordinate sample. Temporally, the sites cover approximately 250 years of Arikara history and are distributed along the southward flowing Missouri River

Table 1. Coalescent tradition components with sample sizes for craniometric and coordinate data sets

Site name/component	Site number	Cultural affiliation	Dates	Sample sizes	
				Craniometric	Coordinate
Nordvold 2 & 3 (ND2&3)	39CO32–3	EC[a], La Roche	1550–1675	30	27
Anton Rygh (RY)	39CA4	EC, La Roche	1600–1650	25	23
Mobridge F1 (MBF1)	39WW1A	EC, La Roche	1600–1650	31	31
Sully A (SLA)	39SL4A	EC, La Roche	1650–1675	28	17
Sully D (SLD)	39SL4D	EC, La Roche	1650–1675	26	18
Mobridge F2 (MBF2)	39WW1B	PCC[b], Le Beau	1675–1700	109	85
Sully E (SLE)	39SL4E	PCC, Le Beau	1675–1700	18	21
Swan Creek (SC)	39WW7	PCC, Le Beau	1675–1725	37	12
Larson (LA)	39WW2	PCC, Le Beau	1679–1733	128	128
Cheyenne River (CR)	39ST1	PCC, Bad River	1740–1795	16	18
Leavenworth (LW)	39CO9	Historic Arikara	1802–1832	52	44

Notes:
[a] Extended coalescent.
[b] Postcontact coalescent.
Sources: Blakeslee, 1994; Jantz, 1997; Key, 1983.

Figure 1. Map of the Missouri River trench in South Dakota showing the locations of the archaeological sites associated with the samples included in this study. Nordvold 2 & 3 is located just south of the historic villages of Leavenworth.

(Figure 1). The sites fall into three cultural and temporally defined periods within the Coalescent Tradition: Extended Coalescent, Postcontact Coalescent, and Disorganized Coalescent or Historic Arikara. These cultural and temporal periods are archaeological constructs based on changing material culture and chronology and are further subdivided into phases and foci based on similar criteria. All sites from the Extended Coalescent have been assigned to the late prehistoric and protohistoric La Roche focus, and are the earliest in the sample. The Postcontact Coalescent sites are subdivided into two contemporaneous phases, Le Beau and Bad River, distinguished by differences in material culture and geography. Le Beau sites are generally located on the left bank of the Missouri River, and Bad River sites are located on the right bank of the Missouri River (Hoffman and Brown, 1967). Cheyenne River is the only site included that is attributed to the Bad River phase. Mobridge and Sully are considered multicomponent sites with evidence for both La Roche and Le Beau occupations. The Disorganized Coalescent, or Historic period, is represented by a single site, Leavenworth. The villages at Leavenworth were visited and documented by numerous European and European-American explorers and fur traders, including Lewis and Clark in 1804.

Craniometric and Landmark Data

The craniometric data set contains 40 linear measurements, most of which are defined by Howells (1973), and are shown in Table 2. Beyond the standard interlandmark distances and chords, subtenses and radii are also included. The craniometric data were collected by four observers: W. W. Howells, P. Lin, P. Key, and R. L. Jantz, with the majority being collected by Key for his 1983 craniometric study of Plains Indians.

The coordinate data set contains 30 landmarks recorded in three dimensions with a MicroScribe-3DX. Both midline and bilateral landmarks were observed in an effort to capture the entire craniofacial morphology. These data were recorded by the first author on collections housed at the National Museum of Natural History, Smithsonian Institution, and the University of Tennessee, Knoxville. The cranial landmarks collected as three-dimensional coordinates are listed in Table 3 and are visually represented in Figure 2.

The landmarks chosen for the coordinate data set are commonly used as endpoints for standard craniometric measurements. The craniometric variables employed in this study were selected in an attempt to parallel the geometric information contained in the three-dimensional coordinate data. Hence, the two data sets should approximate one another with regard to recording cranial dimensions and contain similar information about phenotypic variability.

Table 2. Craniometric variables employed in this study

Glabella–occipital length	Biorbital breadth	Nasion radius
Nasion–occipital length	Dacryon subtense	Subspinale radius
Basion–nasion length	Malar length, inferior	Prosthion radius
Basion–bregma height	Malar length, maximum	Dacryon radius
Maximum cranial breadth	Cheek height	Zygoorbitale radius
Minimum frontal breadth	Frontal chord	Frontomolare radius
Bizygomatic breadth	Frontal subtense	Ectoconchion radius
Biauricular breadth	Parietal chord	Zygomaxillare radius
Biasterionic breadth	Parietal subtense	Molar alveolus radius
Nasio-dacryal subtense	Occipital chord	Bregma radius
Bimaxillary breadth	Occipital subtense	Lambda radius
Bimaxillary subtense	Foramen magnum length	Opisthion radius
Bifrontal breadth	Foramen magnum breadth	Basion radius
Naso-frontal subtense		

Table 3.   Cranial landmarks employed in this study

1. Subspinale	11. Zygoorbitale, L	21. Pterion, L
2. Nasion	12. Zygoorbitale, R	22. Pterion, R
3. Dacryon, L	13. Zygomaxillare anterior, L	23. Asterion, L
4. Dacryon, R	14. Zygomaxillare anterior, R	24. Asterion, R
5. Frontomalare anterior, L	15. Alare, L	25. Lambda
6. Frontomalare anterior, R	16. Alare, R	26. Midparietal
7. Posterior frontomalare, L	17. Midfrontal	27. Zygomatic root, L
8. Posterior frontomalare, R	18. Frontotemporale, L	28. Zygomatic root, R
9. Ectoconchion, L	19. Frontotemporale, R	29. Basion
10. Ectoconchion, R	20. Bregma	30. Posterior occipital

Note: L, Left; R, Right.

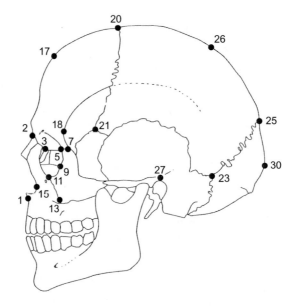

Figure 2.   Lateral view of skull with landmark locations identified. The numbers indicate the corresponding landmark name listed in Table 3. Bilateral landmarks are represented by the left antimere, and all midline landmarks are represented except basion.

Statistical Analysis

Similar analytical techniques were employed to derive biological distance matrices and to explore the implications of the pattern of distances between components. However, due to the fact that the landmark coordinates are recorded with respect to arbitrary axes, the tools of geometric morphometry must be used to map the data into a common coordinate system so that they can be used in traditional statistical analysis (Bookstein, 1996; Slice, 1996). Specifically, the three-dimensional coordinate-based configurations are registered via Procrustes analysis. This procedure translates and rotates the configurations in such a way that the sum of the squared distances across all landmarks is minimized and scales each configuration to a common size.

Size was removed from the craniometric data according to the method described by Darroch and Mosimann (1985). Shape variables were calculated by dividing each variable by the geometric mean (the nth root of the product of n variables). Using SAS system software for Windows Version 6.12 (copyright © 1989–1996 SAS Institute, Inc.), principal component analysis was conducted on the shape variables, and the first 28 principal components representing 95% of the total variation were used for canonical discriminant analysis that produced the squared Mahalanobis distance matrix (D^2).

The three-dimensional configurations were subjected to Procrustes analysis via the General Least Squares procedure in GRF-ND (Slice, 1994). Again using SAS, the residuals from this procedure were used for principal component analysis, and initially the first 51 principal components representing 95% of the overall variation were utilized as variables in the canonical discriminant analysis. The squared Mahalanobis distance matrix (D^2) generated from all 51 principal components contained exceptionally high distances for the component Sully E. This was an unexpected result as previous craniometric and coordinate-based investigations of the Sully components had failed to indicate that this component was so divergent, especially from the other two components at the same site. Additionally, the craniometric distance matrix generated for this study did not produce similarly large distances for Sully E. This suggests that some unknown factor was contributing to the relatively large distances for this particular sample. After discerning that principal components 27 and 30 were contributing disproportionately to the differences between Sully E and the other samples, the corresponding eigenvectors were scaled by factors of ±0.1 and added to the mean configuration. Plots of these hypothetical configurations representing the extremes along principal components 27 and 30 revealed that in morphological terms they represent vault asymmetry, which apparently occurs at a much higher frequency in the Sully E sample than the other samples. The asymmetry is not clearly patterned and affects highly variable vault landmarks (e.g., asterion). The craniometric data did not have comparably large distance values for Sully E because the one-dimensional nature of craniometric data does not capture such information, unless the data collection protocol is designed to detect asymmetry. The random nature of the asymmetry represented by principal components 27 and 30 is likely due to unknown factors, and these components were removed from the canonical data analysis for the coordinate data. All statistics, including the squared Mahalanobis distance matrix, were recalculated.

Mantel tests (matrix correlation), as developed by Mantel (1967) and generalized by Manly (1986) and Smouse et al. (1986), were used to investigate the factors patterning the biological variation and to evaluate the congruence of dissimilarity information contained in the biological distance matrices. Using a program written by the second author, pairwise and three-way matrix correlations were computed to look for correspondence between the biological distance matrices with temporal and geographic parameters. In order to compare the biological distances to the geographic and temporal distances, a Mahalanobis distance matrix (D) was computed for each data set by calculating the square root of each distance in the squared Mahalonobis distance matrix. Each biological Mahalanobis distance matrix (D) was compared individually to the temporal and geographic distance matrices. Then the biological distance matrices were compared to the temporal matrix with the geographic held constant and the geographic matrix with the temporal held constant. Additionally, the Mahalanobis distance matrices from the craniometric and coordinate data sets were directly compared. In all cases, 999 permutations were used to assess statistical significance by determining the number of comparisons involving a randomly rearranged matrix that produced a correlation value as large or larger than the observed correlation.

In order to visualize the correspondence of the two biological distance matrices, the principal coordinate ordinations were derived from each squared Mahalanobis biological distance matrix (D^2) and matched via Procrustes analysis. Ten principal coordinate dimensions were calculated for each biological distance matrix using NTSYS-PC resulting in two data sets containing 11 ten-dimensional configurations. These two sets of principal coordinate ordinations were matched using Procrustes analysis without scaling (Rohlf and Slice, 1990). In this case, the Least Squares procedure in GRF-ND (Slice, 1994) was used and the configurations were not scaled.

RESULTS

Matrix Correlation Analysis

If the craniometric and coordinate data sets are providing similar information about phenotypic divergence among the population samples, then the overall pattern of distances in the biological distance matrices for the two data sets should be similar. The squared Mahalanobis distance matrices are shown in Table 4, and Table 5 presents the results of the Mantel tests. Both the craniometric and coordinate biological distance matrices are positively correlated with the geographic distance matrix, even when the temporal distance matrix is taken into consideration. Although both correlations are significant at $p < 0.01$, the correlation for the coordinate data is stronger with only one random permutation generating a higher correlation than the observed correlation. According to the model developed by Konigsberg (1990), a positive correlation between the biological and geographic distance matrices indicates that the morphological variation conforms to a model of isolation by geographic distance, with increasing similarity between samples as the geographic distances between them decreases. Both the craniometric and coordinate biological distance matrices have very low and statistically insignificant correlations with the temporal distance matrix. These results indicate congruence between the data sets, and suggest that the craniofacial variation observed is geographically patterned and not the product of a temporal trend. The stronger correlations and the paucity of random permutations that exceed the correlation generated by the actual comparison suggest that the coordinate data more effectively captured the shape information relevant to the pattern of distances observed. Despite the similarity of these patterns, the correlation between the Mahalanobis distance matrices for the two data sets is relatively low ($r = 0.5568$, $p < 0.01$), suggesting that the coordinate data contain phenotypic variation not present in the craniometric data.

Principal Coordinates

The first two principal coordinate ordinations for the craniometric and coordinate data sets are shown in Figures 3 and 4, respectively. Both figures depict three loosely formed clusters incorporating the same site components with the second principal coordinate axis possibly representing temporal variation. The components from Sully cluster loosely as do Rygh and Feature 1 from Mobridge. These clusters are expected based on their geographic and temporal commonalities. Both graphs suggest that the majority of burials from Sully E may not be associated with a Le Beau occupation, but may instead be La Roche. This tentative observation is supported by a review of the burial artifacts indicating that Sully E does not contain a higher frequency of European grave

Table 4. Matrix of squared Mahalanobis distances (D^2) for craniometric and coordinate data[a]

	CR	LA	LW	MBF1	MBF2	ND2&3	RY	SC	SLA	SLD	SLE
CR	0	6.3395	7.0275	5.7523	3.0226	6.3033	5.0662	7.7398	5.7779	6.5964	8.4997
LA	7.8985	0	3.3353	3.1255	2.3448	4.0628	2.3210	4.6438	4.5515	6.3406	5.2233
LW	7.9316	5.4075	0	6.6234	3.1178	3.1459	5.1121	5.7300	4.2893	5.2711	5.8559
MBF1	11.7366	5.9435	8.7586	0	4.4004	8.1718	2.7540	6.9394	5.4498	8.6395	5.9346
MBF2	6.0971	3.6050	5.5160	6.6760	0	2.0126	3.0091	3.2078	3.1780	3.1068	4.8568
ND2&3	4.8150	6.6121	5.8607	10.1115	4.7423	0	6.2278	6.0942	4.7105	4.3870	6.1352
RY	13.7600	7.4868	11.2515	5.1692	6.7246	11.8614	0	4.5743	4.9595	6.8087	4.9310
SC	7.2800	6.8032	9.2376	9.6844	6.1359	5.4242	11.0275	0	5.7735	5.9521	7.5419
SLA	8.1269	8.7334	10.3574	9.7532	5.2768	6.5571	9.5114	9.3541	0	2.5171	3.0705
SLD	7.3702	11.351	13.4635	12.2820	7.8540	7.9517	11.087	11.2974	5.584	0	4.2023
SLE	13.0692	16.263	17.2992	17.9573	11.0328	11.3752	18.6386	18.0611	8.11016	12.8846	0

Note:
[a] Craniometric distances are above the diagonal and coordinate distances are below the diagonal.

Table 5.   Results of the matrix correlation analysis

Matrix comparison	Controlling for	Correlation	Random > Observed	p-value
Craniometric with geographic		0.3278	7	0.008
Craniometric with temporal		0.0661	442	0.443
Craniometric with geographic	Temporal	0.3320	11	0.012
Craniometric with temporal	Geographic	0.1072	353	0.354
Coordinate with geographic		0.4604	1	0.002
Coordinate with temporal		0.0014	434	0.435
Coordinate with geographic	Temporal	0.4442	0	0.001
Coordinate with temporal	Geographic	0.0480	329	0.330
Craniometric with coordinate		0.5568	0	0.001

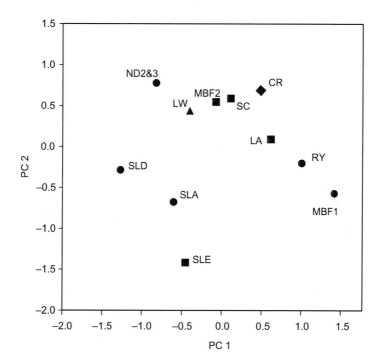

Figure 3.   Principal coordinate ordinations for the craniometric data set; filled circle—La Roche, filled square—Le Beau, filled diamond—Bad River, filled triangle—Historic Arikara.

goods than either Sully A and D (Billeck, W., 2003, personal communication). The remaining cultural components, Nordvold 2 & 3, Larson, Swan Creek, Mobridge Feature 2, Cheyenne River, and Leavenworth, form the third loose cluster [although the more central location of Rygh in the craniometric plot makes the distinction between the early North (Rygh and Mobridge F1) and

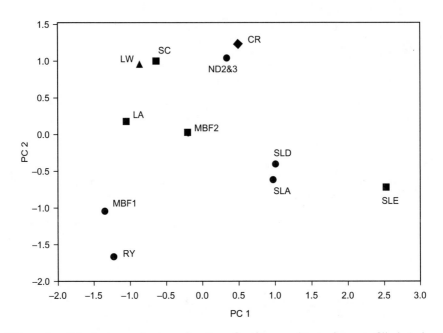

Figure 4. Principal coordinate ordinations for the coordinate data set; filled circle—
La Roche, filled square—Le Beau, filled diamond—Bad River, filled triangle—Historic
Arikara.

the later sites (including Le Beau, Bad River, and Historic) clusters less distinct].
With the exception of Cheyenne River, these sites share geographic proximity
at the northern end of the South Dakota stretch of the Missouri River. The
inclusion of Cheyenne River in this cluster suggests that the Missouri River
probably did little to inhibit gene flow between the Le Beau and Bad River
populations of the Postcontact period. Although Nordvold 2 & 3 has traditionally been attributed to the late prehistoric period, it is clear that a majority
of the burials are similar to the Postcontact period samples implying that the
cultural and temporal assessment of this site may be erroneous.

The major differences between these figures involve which axis separates
Sully E from the other Sully components and the more distinct clustering in the
ordinations based on the coordinate data. This is particularly evident in Figure 3
where Larson is located intermediate between the early North cluster of Rygh
and Mobridge and the later sites. It is likely that a greater degree of variation
captured by the coordinate data is responsible for the better separation among
the clusters, which generally conform to expectations based on geographic and
temporal parameters.

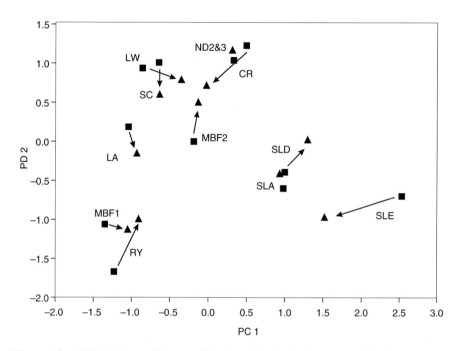

Figure 5. Principal coordinate ordinations for both data sets after least-squares Procrustes analysis; filled triangle—craniometric data, filled square—coordinate data. Arrows indicate the paired samples.

Figure 5 depicts the first two principal coordinate ordinations for the samples from the craniometric and coordinate biological distance matrices after least-squares matching. Although the sample locations for each data type do not correspond exactly, there is enough similarity to argue for congruence between the two data sets. The cultural components remain in the same general vicinity and reflect the same relative positions. Again, greater group separation is associated with the coordinate data. It is of note that the two samples with the greatest disparity in sample size (Mobridge F2 and Swan Creek) both show a reasonable degree of congruence across the two data sets suggesting that the sample size differences do not have an appreciable effect on this analysis.

DISCUSSION AND CONCLUSIONS

The similarity in the biological distance matrices for the craniometric and coordinate data sets is indicated by the congruent results from the Mantel tests and the direct correlation between the two matrices. It is clear that the phenotypic variation captured by both data sets conforms to a model of isolation by geographic distance. Reassignment of components Sully E and Nordvold 2 & 3 to appropriate cultural and temporal phases may permit the identification of the temporal trend that is suggested by the second principal coordinate axis of each plot. The coordinate data produced more powerful statistics with higher correlations and lower probabilities of spurious results, suggesting that the coordinate data provides more information about morphological variation among the population samples.

The principal coordinate ordinations also present similar results and the least-squares match between the two sets of principal coordinates indicates relatively close agreement. Again, the coordinate data provide a clearer picture of the variation with more distinct clusters. This is likely due to the greater variation present in the coordinate data, which is evident in the wider distribution of the principal coordinate ordinations. An unexpected finding that was only detected in the coordinate data was the greater expression of bilateral asymmetry among the crania from Sully E. This had not been identified in previous craniometric studies as those data were not designed to record this form of variation. This example illustrates the greater flexibility of coordinate data for analytical purposes, in addition to its easy mode of collection and ability to archive the two- or three-dimensional form.

For the purposes of biological distance studies, where identification and exploration of morphological variation allow inferences about the relationships among populations, coordinate data clearly offer considerable advantages. In this case, three-dimensional coordinate data captured more variation than the craniometric data, producing distinct patterns associated with geography among closely related subdivisions of a larger population. Given that the possible visual representations of morphological variation are not employed in this study due to space considerations, it is evident that coordinate data provides a wealth of new tools to the physical anthropologist.

ACKNOWLEDGMENTS

The first author would like to acknowledge the William M. Bass III Endowment for providing funds for the purchase of the Microscribe 3DX used to collect the coordinate data. The authors would also like to thank the volume editor and two anonymous reviewers for their constructive comments on earlier drafts of this chapter.

REFERENCES

Blakeslee, D. J., 1994, The archaeological context of human skeletons in the Northern and Central Plains, in: *Skeletal Biology in the Great Plains*, D. W. Owsley and R. L. Jantz, eds., Smithsonian Institution Press, Washington, DC, pp. 9–32.

Bookstein, F., 1996, Combining the tools of geometric morphometrics, in: *Advances in Morphometrics*, L. F. Marcus, M. Corti, A. Loy, G. J. P. Naylor, and D. E. Slice, eds., Plenum Press, New York, pp. 131–152.

Buikstra, J. E., Frankenburg, S. R., and Konigsberg, L. W., 1990, Skeletal biological distance studies in American physical anthropology: Recent Trends, *Am. J. Phys. Anthropol.* 82:1–7.

Darroch, J. N. and Mosimann, J. E., 1985, Canonical and principal components of shape, *Biometrika* 72:241–252.

Devor, E. J., 1987, Transmission of human craniofacial dimensions, *J. Craniofac. Genet. Dev. Biol.* 7:95–106.

Heathcote, G. M., 1986, *Exploratory Human Craniometry of Recent Eskaleut Regional Groups from Western Arctic and Subarctic of North America*, BAR International Series 301, BAR, Oxford.

Hoffman, J. and Brown, L., 1967, The bad river phase, *Plains Anthropol.* 12:323–343.

Howells, W. W., 1973, *Cranial Variation in Man: A Study by Multivariate Analysis of Patterns of Difference among Recent Human Populations*, Peabody Museum, Harvard University, Cambridge, MA.

Jantz, R. L., 1973, Microevolutionary change in Arikara crania: A multivariate analysis, *Am. J. Phys. Anthropol.* 38:15–26.

Jantz, R. L., 1997, Cranial, postcranial, and discrete trait variation, in: *Bioarcheology of the North Central United States*, D. W. Owsley and J. Rose, eds., Arkansas Archeological Survey Research Series, pp. 240–247.

Key, P., 1983, *Craniometric Relationships among Plains Indians*, Department of Anthropology, University of Tennessee, Knoxville.

Konigsberg, L. W., 1990, Analysis of prehistoric biological variation under a model of isolation by geographic and temporal distance, *Hum. Biol.* 62:49–70.

Manly, B., 1986, Randomization and regression methods for testing for associations with geographical, environmental, and biological distances between populations, *Res. Popul. Ecol.* 28:201–218.

Mantel, N., 1967, The detection of disease clustering and a generalized regression approach, *Cancer Res.* 27:209–220.

McKeown, A. H., 2000, *Investigating Variation Among Arikara Crania Using Geometric Morphometry*, Unpublished Ph.D. dissertation, Department of Anthropology, University of Tennessee, Knoxville.

Ousley, S. D. and McKeown, A. H., 2001, Three dimensional digitizing of human skulls as an archival procedure, in: *Human Remains: Conservation, Retrieval and Analysis*, E. Williams, ed., BAR International Series 934, Archaeopress, Oxford, UK, pp. 173–184.

Relethford, J. H., 1994, Craniometric variation among modern human populations, *Am. J. Phys. Anthropol.* 95:53–62.

Rohlf, F. J. and Slice, D. E., 1990, Extensions of the Procrustes method for optimal superimposition of landmarks, *Syst. Zool.* 39:40–59.

Ross, A. H., McKeown, A. H., and Konigsberg, L. W., 1999, Allocation of crania to groups via the "New Morphometry," *J. Forensic. Sci.* 44:584–587.

Slice, D. E., 1994, *GRF-ND: Generalized rotational fitting of N-dimensional data.* Department of Ecology and Evolution, State University of New York at Stony Brook.

Slice, D. E., 1996, Three-dimensional generalized resistant fitting and the comparison of least-squares and resistant-fit residuals, in: *Advances in Morphometrics*, L. F. Marcus, M. Corti, A. Loy, G. J. P. Naylor, and D. E. Slice, eds., Plenum Press, New York, pp. 179–199.

Smouse, P., Long, J., and Sokal, R., 1986, Multiple regression and correlation extensions of the Mantel test of matrix correspondence, *Syst. Zool.* 35:627–632.

Sparks, C. S. and Jantz, R. L., 2002, A reassessment of human cranial plasticity: Boas revisited. *Proc. Natl. Acad. Sci. USA* 99(3):14636–14639.

CHAPTER TEN

Assessing Craniofacial Secular Change in American Blacks and Whites Using Geometric Morphometry

Daniel J. Wescott and Richard L. Jantz

INTRODUCTION

Over the past 150 years, American crania have undergone striking changes (Angel, 1976, 1982; Jantz, 2001; Jantz and Meadows Jantz, 2000; Jantz and Moore-Jansen, 1988; Moore-Jansen, 1989; Ousley and Jantz, 1997; Smith et al., 1986). The most notable changes are increases in vault height, base length, total length, as well as a narrowing of the vault and face (Jantz and Meadows Jantz, 2000; Moore-Jansen, 1989). The single dimension showing the greatest change is vault height or the basion to bregma dimension (Jantz, 2001). Furthermore, changes in cranial shape are more pronounced than increases in size, and modifications in the vault are greater than those in the face (Jantz and Meadows Jantz, 2000).

Even though cranial secular change is well-documented, little is known about the nature of the anatomical transformations that have occurred, the proximate

Daniel J. Wescott • University of Missouri—Columbia, Department of Anthropology, 107 Swallow Hall, Columbia, MO 65211,   **Richard L. Jantz** • University of Tennessee, Department of Anthropology, 252 South Stadium Hall, Knoxville, TN 37996.

Modern Morphometrics in Physical Anthropology, edited by Dennis E. Slice.
Kluwer Academic/Plenum Publishers, New York, 2005.

231

causes, or the ultimate causes. Analyses of cranial secular trends using traditional craniometric data have been unable to resolve the question of whether changes in vault height are due to alterations in the inferior or superior vault. Angel (1982) showed that cranial base height (porion to basion) significantly increased from the 19th to the 20th century and that this increase was proportionately greater than increases in vault height (basion to bregma). Moore-Jansen (1989), on the other hand, suggested that changes in vault height are primarily due to increases in the superior vault because the transmeatal axis to bregma dimension increased proportionately more than the transmeatal axis to basion dimension. A deeper understanding of these anatomical changes will allow us to better pinpoint when these changes occur during growth and development (proximate causes) and how environmental and genetic factors (ultimate causes) influence cranial morphology.

Documenting the exact anatomical modifications responsible for secular change in Americans is difficult using traditional craniometrics, but the use of landmark data in the form of Cartesian coordinates allows for a much easier depiction of complex anatomical variation and may provide a better understanding of the proximate and ultimate causes responsible for the observed secular change. Previously, we undertook a preliminary analysis of cranial change using three-dimensional Cartesian coordinate data to detect the anatomical changes responsible for secular trends in American Blacks and Whites (Wescott and Jantz, 2001). Our results demonstrated that the change in vault height was primarily due to a downward extension of basion, but the study was hampered by small sample size, particularly of recent crania. In this study, we examine the issue of secular change in American Blacks and Whites over the last 150 years using a large sample of two-dimensional (2-d) Cartesian coordinates. By examining changes in the relative position of cranial landmarks, and not just changes in dimensions between landmarks, we hope to gain a better understanding of secular change in the United States over the last 150 years. We describe here the anatomical changes responsible for the secular trends in cranial morphology, discuss the nature of these changes, and discern some of their proximate and ultimate causes.

MATERIALS AND METHODS

The examination of secular change using 2-d coordinates was stimulated by the realization that we could reconstruct them from Howells' (1973) traditional measurements using the import truss feature of Morpheus et al. (http://life.bio.sunysb.edu/morph/morpheus/) (Slice, 1998). The import truss feature uses an iterative algorithm to fit measurements sharing a sufficient number of landmarks (Carpenter et al., 1996). Howells' (1973) measurements contain considerable redundancy, allowing for a reasonable number of landmarks to be reconstructed. We used measurements from 13 landmarks (Figure 1) and were able to reconstruct the coordinates with little error.

Samples from 644 Black and White crania of both sexes were available for landmark reconstruction (Table 1). The crania had previously been measured following Howells' (1973) definitions. The 19th century material comes from the Terry and Todd collections, while the 20th century material was drawn primarily from the Forensic Data Bank (Jantz and Moore-Jansen, 1988; Ousley and Jantz, 1997). Sex and race for all crania were known from premortem records and not estimated from morphology. In nearly all cases, birth year was known or could be calculated from age and date of death. In a few instances, age was estimated from skeletal remains and used with date of death to calculate year of birth. This would result in only a few years of potential error for only a handful of crania. Table 2 shows the mean year of birth and range for each race/sex group. Data were available for each race/sex group from the 1830s to the 1970s.

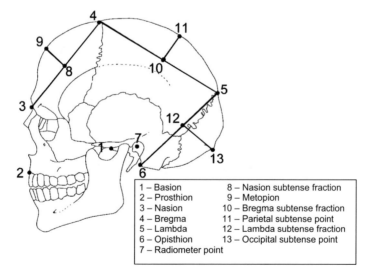

Figure 1. Cranial landmarks reconstructed from traditional craniometric measurements. Black lines illustrate subtenses and subtense fractions.

Table 1.   Sample size and origin by century of birth

Group	1850–1899[a]		1900–1975[a]		Totals
	T/T[b]	FDB[c]	T/T	FDB	
White male	97	0	15	105	217
White female	80	0	31	54	165
Black male	80	0	22	30	132
Black female	82	0	30	18	130
Totals	339	0	98	207	644

Notes:
[a] Birth years.
[b] T/T = Terry and Todd anatomical collections.
[c] FDB = Forensic Data Bank.

Table 2.   Sample size, mean birth year (MBY) and range (BYR) by race/sex

Group	N	MBY	BYR
White male	217	1907.7	1843–1977
White female	165	1909.6	1856–1975
Black male	132	1894.2	1848–1970
Black female	130	1899.1	1835–1970

Analysis was carried out using the thin-plate spline regression program (TPSREGR). The program executes the Procrustes superimposition, resolves the coordinate space into 10 partial warps and two uniform components (for these data), and performs a multivariate regression of the partial warps onto a dependent variable. In this case, the independent variable is year of birth.

RESULTS

Significant changes are seen in all race/sex groups by the multivariate regression of partial warps onto year of birth (Table 3). The nature of the changes reflected in the skull can be seen by means of vectors showing direction of change at each landmark in relation to year of birth (Figure 2). The most noticeable change is a relative movement of basion inferiorly and posteriorly. Lambda moves anterosuperiorly and metopion moves slightly up and posterior. There is virtually no change at prosthion, nasion, or bregma. These results clearly show that the secular change in American crania is concentrated on the base and posterior aspect of the skull.

Table 3. Multivariate regression of partial warps on year of birth

Group	Wilks' lambda	p-value	Permutations
White male	0.570	<0.001	1/1000
White female	0.604	<0.001	1/1000
Black male	0.654	<0.001	1/1000
Black female	0.571	<0.001	1/1000

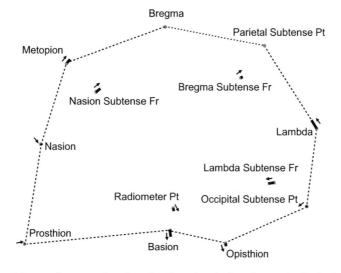

Figure 2. Means of vectors showing the direction (→) and magnitude (━) of change through time at each landmark.

The consistency and regularity of this change can be readily observed by dividing the sample into five 25-year birth cohorts (approximately one generation) starting in 1850 and ending in 1975. The mean coordinates of the inferior landmarks for each birth year cohort are shown in Figure 3 for Whites. The relative downward and backward movement of basion and opisthion and upward movement of lambda are readily apparent. There is an almost identical picture for Blacks (Figure 4). The same downward movement of the cranial base and upward movement of lambda are apparent. Prosthion exhibits more variation in Blacks than in Whites, but the change does not seem to be concordant with year of birth.

We also obtained the principal component scores of the Procrustes residuals and used them to compute distances and canonical variate scores by birth

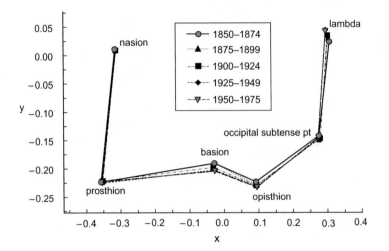

Figure 3.   Mean coordinates of inferior landmarks for the five 25-year cohorts (Whites).

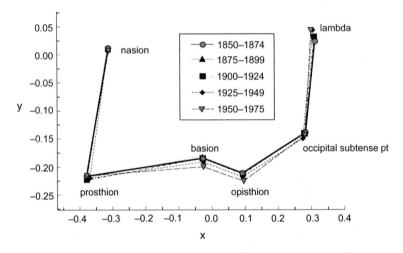

Figure 4.   Mean coordinates of inferior landmarks for the five 25-year cohorts (Blacks).

year cohort and race. Figure 5 illustrates the first two canonical vectors, which together account for 93.2% of variation. The first axis, accounting for nearly 83% of variation, separates Blacks from Whites. The secular changes within each population are primarily reflected on the second axis, which accounts for just over 10% of the variation. This axis shows that Blacks and Whites are proceeding along an approximately parallel course of secular change, which is oblique to the axis that separates them.

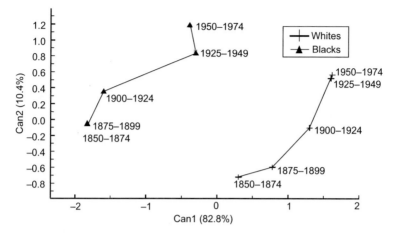

Figure 5. First two canonical vectors illustrating the secular trend in Blacks and Whites.

DISCUSSION

Previous research (Jantz, 2001; Jantz and Meadows Jantz, 2000; Jantz and Moore-Jansen, 1988; Moore-Jansen, 1989) has shown that the crania of both Black and White Americans have become higher, longer, and narrower through time. We demonstrate that most of these changes are associated with the cranial base. Changes in the superior vault and face are minimal. Our results have several implications for the understanding of craniofacial secular changes and the meaning of morphometric variation in the skull. We will summarize the implications as follows: (a) nature of changes, (b) plasticity vs genetic variation, (c) proximate causes, and (d) ultimate causes.

Nature of Changes

Vault height (basion to bregma) shows greater secular change than any other single dimension. Hypotheses concerning the nature of this change that emerged from traditional craniometric data centered on questions of whether this change involved the superior or inferior vault. That is, whether basion moves down or bregma moves up. Moore-Jansen (1989) finds that the transmeatal to bregma dimension increased proportionately more than the transmeatal to basion dimension. Therefore, he suggests that most of the secular change in the basion to bregma dimension is associated with an increase in vault height superior to the transmeatal axis. In other words, the position of bregma is moving superiorly over time more than basion is moving inferiorly. However, a significant increase in cranial base height (porion to basion) since the 19th century has been documented in Americans (Angel, 1982). In fact, Angel (1982) argues that proportionately, changes in cranial base height are more important to the increase in the basion to bregma dimension than are changes above the transmeatal axis.

At first, these two studies appear to be contradictory. However, by examining the movement of landmarks and not just the dimensions between landmarks, it is clear why both Angel (1982) and Moore-Jansen (1989) reached these conclusions. The present results convincingly show that change in American crania is limited to the inferior vault. Basion, opisthion and the transmeatal axis all "move" in an inferior and slightly posterior direction relative to the other vault landmarks, while bregma moves scarcely at all (Figure 2). The coordinate data demonstrate that the entire base moves inferiorly, which explains Moore-Jansen's (1989) observation of a relatively greater increase between basion and bregma. Furthermore, the coordinate data show that basion moved inferiorly more than the transmeatal axis over time, which explains Angel's (1982) observation of significant changes in cranial base height. Therefore, the landmark data clearly explain both researchers' conclusions and strongly support Angel's (1982) hypothesis that the cranial base has undergone more secular change than the superior vault.

Plasticity vs Genetic Variation

Over 90 years after Boas' (1912, 1940) classic study demonstrating changes in the descendants of immigrants, the extent to which cranial morphology reflects genetic variation remains controversial (Armelagos and Goodman, 1998; Sparks, 2001; Sparks and Jantz, 2002). The present results will not settle the issue, but may provide some additional insights. One could argue that significant changes in cranial morphology over the span of just 150 years are most likely due to plasticity. But, if this is true, they are confined to within-group change. The parallel course of secular change in Blacks and Whites, combined with their failure to converge upon a common morphology (see Figure 5), supports the hypothesis that genetic variation between the two groups is reflected in cranial morphology, despite exposure to a common environment. American Blacks and Whites undergo similar secular trends, but genetic differences between them are maintained.

Proximate Causes

Most of the secular change in American crania appears to occur early (during infancy and childhood) in the growth period (Jantz, 2001; Jantz and Meadows Jantz, 2000). The most dramatic changes are a result of modification in the growth of the posterior cranial base, which reaches adult size early in life. Moreover, the correlations of cranial vault morphology with year of birth are stronger than those for long bones or the face (Jantz and Meadows Jantz, 2000), further suggesting that most of the craniofacial changes occur early in life. Since the cranial vault follows a more rapid neural growth pattern than the long bones or face, there are fewer opportunities for catch-up growth (Skuse, 1998).

Basicranial dimensions and brain size primarily influence the shape of the adult cranial vault (Lieberman et al., 2000). Both the cranial base and the neurocranium follow a neural growth trajectory and, therefore, act as an integrated unit. The cranial base develops endochondrally and is the first region of the skull to reach adult size (Enlow, 1990). As a result, cranial base dimensions influence the overall shape of the neurocranium (Howells, 1973; Lieberman et al., 2000). Lieberman et al. (2000) found that maximum cranial breadth moderately correlates with the ratio of endocranial volume and cranial base breadth. Furthermore, cranial base breadth and endocranial volume appear to slightly influence the length and height of the neurocranium. Lieberman et al. (2000) also found that, in general, as cranial breadth narrows, vault size increases vertically and posteriorly. In other words, individuals with large brains and narrow cranial bases have slightly higher and longer cranial vaults. This corresponds to the secular trends we see in American crania. Over the last 150 years, American crania have become narrower and brain size has increased by nearly 150 cc (Jantz and Meadows Jantz, 2000). However, Lieberman et al. (2000) ascertained that less than 40% of the variation in neurocranium length and height is explained by cranial base breadth and endocranial volume, and we have noticed that among Great Plains Native Americans, cranial base breadth influences the height of the vault but not necessarily the length. Howells (1973) also discovered that occipital length (occipital chord) affects cranial vault height. Furthermore, the breadth of the cranial base may constrain facial breadth (Enlow, 1990), but since the face follows an intermediate somatic/neural growth pattern, it is not greatly influenced by cranial base dimensions. Lieberman et al. (2000) discovered that individuals with a narrow cranial base tend to have a narrower face than individuals with a wide cranial base, but the correlation is relatively low. These results suggest that while many of the changes we see in American craniofacial morphology are directly associated with changes in the cranial base, other factors must also be operating.

Secular trends in stature are associated with growth rate allometry (Meadows Jantz and Jantz, 1999), and this is probably also true for the cranium since vault shape changes more than vault size (Kouchi, 2000). Even small increases in the rate of cranial growth would lead to significant shape changes (Kouchi, 2000). Similar to the United States, an increase in vault height has been observed among the Japanese since 1900. However, unlike the secular trends observed in the United States, increases in vault height are associated with lateral expansion (larger vault breadth) and not an increase in vault length. In both populations, vault height increases, but in the United States vault breadth narrows and in Japan it broadens. Kouchi (2000) contends that vault breadth is the key characteristic in both groups. With enlargement of the brain, the preferential increase in growth rate appears to be in a lateral direction among the Japanese and in a posteroinferior direction among American Blacks and Whites. Kouchi (2000) argues this may be due to differences in facial morphology.

Ultimate Causes

Along with genetic changes, improved health and nutrition (Angel, 1982; Angel et al., 1987; Cameron et al., 1990; Jantz and Meadows Jantz, 2000; Kouchi, 2000) and biomechanical responses to a more processed diet (Carlson and Van Gerven, 1977; Larsen, 1997) have been put forward as ultimate causes of craniofacial alterations over time. In reality, craniofacial morphology is probably a reflection of all these factors and numerous others, but changes in infant health and nutrition appear to be the most credible explanation for the craniofacial secular changes in American populations (van Wieringen, 1986).

Angel (1982) argues that cranial base height is sensitive to growth stress and that secular change in this dimension is due to improved health and nutrition. Undernutrition, according to Angel (1982), affects the cranial base's ability to support the weight of the brain. As a result, downward growth of the cranial base is inhibited and it becomes flattened. Larsen (1997), however, argues that the "relationship between cranial base height and nutritional quality may be more apparent than real" because cranial base cartilage is resistant to compressive loading, as are other primary cartilages (e.g., limb bones). There does, however, appear to be a strong relationship between a flat cranial base and other indicators of childhood growth stress (Angel, 1982; Cameron et al., 1990; Jantz and Meadows Jantz, 2000). Angel (1982) found a strong correlation between cranial base height and pelvic inlet depth and stature in Americans. Likewise, Jantz and Meadows Jantz (2000) found cranial height to follow the same general secular pattern as long bone length, but cranial height was much more pronounced. That is, there is a stronger correlation with year of birth. We agree with Angel (1982) that the relationship between cranial base height, nutrition, and health are real.

While there appears to be a strong relationship between cranial base height, nutrition, and health, changes in vault shape are probably not because cranial base growth is inhibited by the weight of the brain when health and nutrition are poor. Instead, as health and nutrition have improved, the brain has become larger and the vault has increased in height (Miller and Corsellis, 1977). To accommodate these modifications, there have likely been allometric changes in the direction of growth (Kouchi, 2000). Among American Blacks and Whites, increases in brain size and cranial vault height due to improved health and nutrition appear to be associated with a decreased growth rate in vault breadth and an increased growth rate in vault length.

Several studies (Corruccini and Whitley, 1981; Goose, 1962, 1972; Lundström and Lysell, 1953; Lysell, 1958) suggest the dental arch has become narrower with an increased diet of softer, more highly processed foods, and Larsen (1997) suggests that this functional model explains the recent narrowing of the face in Americans. While dietary changes may actually affect facial breadth, improvements in health and nutrition among Americans over the past 150 years far exceed changes in diet over the same period. Furthermore, it is unlikely that functional demands would affect cranial base dimensions. Since most of the secular changes in Americans are associated with the cranial base, there would have to be substantial differences in the functional demands of infants and young children for the biomechanical model to be plausible.

In a broader context, it is possible to view the secular changes in the cranial vault as morphological change in response to the widespread demographic transition occurring in industrial societies. Boldsen (2000) argues that most of the world populations of modern *Homo sapiens* have undergone three demographic transitions, each associated with morphological change. The first demographic transition is associated with the Neolithic and is accompanied by skeletal gracilization. The second transition occurs during the Middle Ages and is accompanied by brachycephalization. The third transition is characterized by marked decrease in mortality followed by decrease in fertility. Stature increase associated with industrialization is just part of a suite of extensive skeletal changes, including changes in cranial morphology as shown in the present study.

Boldsen (2000) contends that selection is responsible for brachycephalization resulting from the second demographic transition, and it seems equally likely that there is a genetic component in the more recent trends caused by changes in mortality occurring over the past couple of centuries. In the past 150 years alone, infant mortality has gone from as high as 160 per 1,000 births to less than 10 per 1,000. Mortality changes such as these have the potential to effect considerable genetic change, especially in growth potential.

CONCLUSIONS

American crania have changed significantly in the past 150 years. In the sagittal plane, most of the change is associated with a downward movement of the cranial base, especially at basion. Secular change in American crania is proximately related to a decrease in cranial base breadth and an increase in cranial capacity, and ultimately a reflection of improved infant growth due to better health and nutrition. The environment of 20th century Americans has no parallel in history. Activity levels are at an all-time low, and diet has improved to the point where overnutrition has surpassed undernutrition as our most serious malnutrition problem (Flegal et al., 1998). In addition, epidemic infectious diseases are now mostly controlled, and mortality is at an all-time low (Armstrong et al., 1999). It is not surprising that there is a biological response to this unparalleled environmental change.

REFERENCES

Angel, J. L., 1976, Colonial to modern skeletal change in the U.S.A., *Am. J. Phys. Anthropol.* 45:723–736.
Angel, J. L., 1982, A new measure of growth efficiency: Skull base height, *Am. J. Phys. Anthropol.* 58:297–305.

Angel, J. L., Kelley, J. O., Parrington, M., and Pinter, S., 1987, Life stresses of the free black community as represented by the First African Baptist Church, Philadelphia, 1823–1841, *Am. J. Phys. Anthropol.* 74:213–229.

Armelagos, G. J. and Goodman, A. H., 1998, Race, racism, and anthropology, in: *Building A New Biocultural Synthesis*, A. H. Goodman and T. L. Leatherman, eds., University of Michigan Press, Ann Arbor, pp. 359–378.

Armstrong, G. L., Conn, L. A., and Pinner, R. W., 1999, Trends in infectious disease mortality in the United States during the 20th century, *JAMA* 281:61–66.

Boas, F., 1912, Changes in bodily form of descendants of immigrants, *Am. Anthropol.* 14:530–563.

Boas, F., 1940, Changes in bodily form of descendants of immigrants, in: *Race, Language, and Culture*, F. Boas, ed., Free Press, New York, pp. 60–75.

Boldsen, J. L., 2000, *Human Demographic Evolution—Is it Possible to Forget about Darwin?* Research Report No. 14, Danish Center for Demographic Research, Odense University.

Cameron, N., Tobias, P. V., Fraser, W. J., and Nagdee, M., 1990, Search for secular trends in calvarial diameters, cranial base height, indices, and capacity in South African Negro crania, *Am. J. Hum. Biol.* 2:53–61.

Carlson, D. S. and Van Gerven, D. P., 1977, Masticatory function and post-Pleistocene evolution in Nubia, *Am. J. Phys. Anthropol.* 46:495–506.

Carpenter, K. E., Sommer III, H. J., and Marcus, L. F., 1996, Converting truss interlandmark distances into Cartesian coordinates, in: *Advances in Morphometrics*, L. F. Marcus, M. Corti, A. Loy, G. J. P. Naylor, and D. E. Slice, eds., pp. 103–111.

Corruccini, R. S. and Whitley, L. D., 1981, Occlusal variation in a rural Kentucky community, *Am. J. Orthod.* 79:250–262.

Enlow, D. H., 1990, *Facial Growth*, 3rd ed., W. B. Saunders, Philadelphia.

Flegal, M. D., Carroll, R. J., Kuczmarski, R. J., and Johnson, C. L., 1998, Overweight and obesity in the United States: Prevalence and trends, 1960–1994, *Int. J. Obes. Relat. Metab. Disord.* 22:39–47.

Goose, D. H., 1962, Reduction of palate size in modern populations, *Arch. Oral Biol.* 7:343–350.

Goose, D. H., 1972, Maxillary dental arch width in Chinese living in Liverpool, *Arch. Oral Biol.* 17:231–233.

Howells, W. W., 1973, Cranial variation in man. Papers of the Peabody Museum of Archaeology and Ethnology, Volume 67, Cambridge.

Jantz, R. L., 2001, Cranial change in Americans: 1850–1975, *J. Forensic Sci.* 46:784–787.

Jantz, R. L. and Meadows Jantz, L., 2000, Secular change in craniofacial morphology, *Am. J. Hum. Biol.* 12:327–338.

Jantz, R. L. and Moore-Jansen, P. H., 1988, Report of Investigations No. 47, University of Tennessee, Knoxville.

Kouchi, M., 2000, Brachycephalization in Japan has ceased, *Am. J. Phys. Anthropol.* 112:339–347.

Larsen, C. S., 1997, *Bioarchaeology: Interpreting Behavior from the Human Skeleton,* Cambridge University Press, New York, p. 19.

Lieberman, D. E., Pearson, O. M., and Mowbray, K. M., 2000, Basicranial influence on overall cranial shape, *J. Hum. Evol.* 38:291–316.

Lundström, A. and Lysell, L., 1953, An anthropological examination of a group of Mediaeval Danish skulls, *Acta Odontol. Scandinavica* 11:111–128.

Lysell, L., 1958, A biometrical study of occlusion and dental arches in a series of Medieval skulls from northern Sweden, *Acta Odontol. Scandinavica* 16:177–203.

Meadows Jantz, L. and Jantz, R. L., 1999, Secular change in long bone length and proportion in the United States, 1800–1970, *Am. J. Phys. Anthropol.* 110:57–67.

Miller, A. K. H. and Corsellis, J. A. N., 1977, Evidence for a secular increase in human brain weight during the past century, *Ann. Hum. Biol.* 4:253–257.

Moore-Jansen, P. H., 1989, A multivariate craniometric analysis of secular change and variation among recent North American populations, Ph.D Dissertation, University of Tennessee, Knoxville.

Ousley, S. D. and Jantz, R. L., 1997, The Forensic Data Bank: Documenting skeletal trends in the United States, in: *Forensic Osteology,* 2nd ed., Charles C. Thomas, K. J. Reichs, ed., Springfield, IL, pp. 441–458.

Skuse, D. H., 1998, Growth and phychosocial stress, in: *The Cambridge Encyclopedia of Human Growth and Development,* S. J. Ulijaszek, F. E. Johnston, and M. A. Preece, eds., Cambridge University Press, New York, pp. 341–342.

Slice, D. E., 1998, *Morpheus et al.: Software for Morphometric Research,* Revision 01-30-98, Department of Ecology and Evolution, State University of New York, Stony Brook, NY.

Smith, B. H., Garn, S. M., and Hunter, W. S., 1986, Secular trend in face size, *Angle Orthod.* 56:196–204.

Sparks, C. S., 2001, Reassessment of cranial plasticity in man: A modern critique of changes in bodily form of descendants of immigrants, M.A. Thesis, University of Tennessee, Knoxville.

Sparks, C. S. and Jantz, R. L., 2002, A reassessment of human cranial plasticity: Boas revisited. *PNAS* 99:14636–14639.

van Wieringen, J. C., 1986, Secular growth changes, in: *Human Growth*, 2nd ed., vol. 3, F. Falkner and J. M. Tanner, eds., Plenum Press, New York, pp. 307–331.

Wescott, D. J. and Jantz, R. L., 2001, Examining secular change in craniofacial morphology using three-dimensional coordinate data, *Am. Acad. Forensic Sci. Proc.* 7:262–263.

CHAPTER ELEVEN

Secular Trends in Craniofacial Asymmetry Studied by Geometric Morphometry and Generalized Procrustes Methods

Erin H. Kimmerle and Richard L. Jantz

INTRODUCTION

Previous work on the secular trends of growth, size, and shape of American craniofacial form over the past 200 years documents significant morphological change and increased variability. For example, Jantz and Meadows Jantz (2000) demonstrated that the American cranial vault has become longer and narrower over time. While the face was less affected by secular change, modifications were noted in its height and width. Interestingly, secular change in shape

Erin H. Kimmerle and Richard L. Jantz • Department of Anthropology, University of Tennessee, 250 South Stadium Hall, Knoxville, TN 37996-0720.

Modern Morphometrics in Physical Anthropology, edited by Dennis E. Slice.
Kluwer Academic/Plenum Publishers, New York, 2005.

was more pronounced than that determined for size. Increased phenotypic variance, like that noted by Jantz and Meadows Jantz (2000), is compounded by many influential genetic factors (i.e., homozygosity and directional selection) and their interaction with nutritional stress or parasitic load. Such genetic and environmental factors disrupt the development process and alter the phenotype. One well-studied manifestation of this is the asymmetrical development of bilateral traits. Developmental stability and canalization, on the other hand, are buffering mechanisms that provide an individual with the ability to overcome this developmental noise (Sciulli et al., 1979; Siegel et al., 1977; Waddington, 1957). As buffering mechanisms act to overcome developmental noise and stabilize development, they modify the gene-to-phenotype relationship (Rutherford, 2000). This leads to the question of whether periods of rapid morphological change affect such buffering mechanisms and are accompanied by increased asymmetry.

Traditional morphometric techniques have been used to measure asymmetry in a wide range of traits from soft to hard tissue, such as dermatoglyphics (Jantz and Brehme, 1993), dentition (Hershkovitz et al., 1993), and various skeletal elements (Farkas and Cheung, 1981; Peck et al., 1991). More recently, geometric morphometric techniques have been applied to the study of size and shape asymmetry of skeletal and soft tissue structures (Klingenberg et al., 2001; Mardia et al., 2000).

Geometric morphometrics, as defined by Slice et al. (1996), is the study of size and shape based on the multivariate analysis of Cartesian coordinate data and has been applied to the study of developmental stability by a number of researchers (Klingenberg and McIntyre, 1998; Klingenberg et al., 2001; Leamy et al., 2000; Richtsmeier, 1987). Typically, methods of geometric morphometry include the estimation of mean shapes and descriptions of sample variation in shape through the use of geometric principles, such as Procrustes distances (Slice et al., 1996). Since Procrustes methods inherently allow for shape and size variables to be analyzed separately, they are quite useful for the study of developmental stability through the measurement of asymmetry. The multivariate statistical procedures needed for analysis of developmental stability are easily applied to coordinate data.

While a comprehensive comparison between traditional methods of calculating asymmetry and geometric morphometrics has been made by Auffray and colleagues (1996), a few comments on the differences between these approaches, as they relate to asymmetry, are worthwhile. Beneficially, Procrustes methods allow for analysis of size and shape variables using three-dimensional (3D) visualization, in depth analysis of directionality, and in our experience, lower interobserver or repeated measurement error. Further, in some past studies of craniofacial landmarks (Peck et al., 1991), construction of a midline was needed for structures, such as the face, from which one could measure left and rights points. The creation of such a midline often leads to bias or inaccuracies because asymmetry usually involves the measure of very small deviations between left and right sides. Procrustes methods as used in this analysis allow for the elimination of a defined midline.

MATERIALS AND METHODS

The purpose of this chapter is to investigate the secular change of developmental stability in the craniofacial region of Americans through measurement of fluctuating and directional asymmetry using the generalized Procrustes analysis (Gower, 1985) of paired craniofacial landmark coordinates in the manner of Klingenberg and McIntyre (1998); refer also to Mardia and coworkers (2000) for more recent methods. Previous work on the secular trends of growth, size, and shape of American cranial form leads to the question of whether periods of rapid morphological change are accompanied by increased fluctuating asymmetry in American craniofacial morphology. To test for fluctuating and directional asymmetry and their association with rapid morphometric change, 3D coordinate data from 526 American males and females of African and European descent who were born between the years 1820–1980 A.D. were analyzed (Table 1). All craniofacial data used in this investigation were collected from skeletons of identified individuals, ensuring that accurate birth date or minimally, the years of birth are known.

The skeletal remains in this analysis originated from several national collections: The Robert J. Terry Anatomical Skeletal Collection located at the National Museum of Natural History, Smithsonian Institution in Washington, D.C.; the William M. Bass Donated Collection housed at the University of Tennessee, Knoxville; the Forensic Data Bank that includes forensic cases from throughout the United States, but is maintained by the Forensic Anthropology Center also at the University of Tennessee, Knoxville; and the Civil War Collection housed at the National Museum of Health and Medicine of the Armed Forces Institute of Pathology in Washington, D.C.

The majority of data were collected from the R. J. Terry Anatomical Collection representing skeletons that were retrieved from the medical cadavers of people who lived and died in St. Louis, Missouri, between the years

Table 1.   Sex and ancestral distribution by decade of birth

YOB	White males	White females	Black males	Black females	Total sample
1820–1829	5	—	—	1	6
1830–1839	13	1	1	1	16
1840–1849	14	5	8	7	34
1850–1859	10	10	9	10	39
1860–1869	12	14	12	15	53
1870–1879	10	12	10	13	45
1880–1889	11	11	10	12	44
1890–1899	10	10	10	11	41
1900–1909	10	12	18	15	55
1910–1919	12	11	12	10	45
1920–1929	19	17	7	7	50
1930–1930	21	6	7	2	36
1940–1949	12	4	4	2	22
1950–1959	12	3	2	—	17
1960–1969	10	0	6	3	19
1970–1979	—	—	3	—	3
1980–1989	—	1	—	—	1
Total	181	117	119	109	526

of 1820–1940 A.D. The birth places for these individuals varies nationwide. The W. M. Bass Donated sample consists of 127 skeletons retrieved from the donated cadavers of people who predominantly lived and died in East Tennessee, between the years of 1900–1980 A.D. Data collected from the Forensic Data Bank came from 27 forensic cases investigated at various universities and morgues throughout the United States. These individuals were born between the years of 1840–1980 A.D. Finally, the Civil War Collection consists of the crania of 28 Civil War veterans who died in battle and were born between the years of 1820–1840 A.D. in various parts of the United States.

Cartesian coordinate data were collected using a Microscribe-3DX 3D digitizer for the seven bilateral facial landmarks (dacryon, frontomalare anterior, frontomalare temporale, zygomaxillare, zygoorbitale, zygion, and asterion). The landmarks (Figure 1) were defined using W. W. Howells (1973) standard definitions.

While investigations into the buffering mechanisms of individuals may lead to greater understanding of observed variation, its precise measurement is more elusive. There are two different types of asymmetry commonly recognized that will be discussed here. Directional asymmetry is the propensity for a particular side of a trait to develop more than the other. In directional asymmetry, the

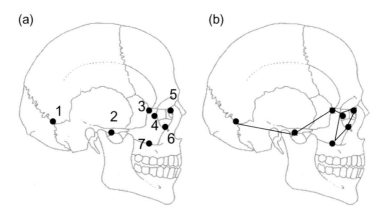

Figure 1. (a) Craniofacial landmarks: 1, asterion; 2, zygion; 3, frontomalare temporale; 4, frontomalare anterior; 5, dacryon; 6, zygoorbitale; 7, zygomaxillare. (b) The second panel illustrates the geometric abstraction of landmarks that is used throughout this study.

mean right minus left (R − L) trait values have a normal distribution with the mean value deviating from zero (Palmer and Strobeck, 1992), and it is typically argued to be genetically based. This differs from fluctuating asymmetry, which is the measure of random deviations from a normal distribution of right minus left values with a mean value of zero (Palmer and Strobeck, 1992) and is thought to result from disruptions to the buffering mechanisms during growth (i.e., environmental and genetic noise). The (R − L) landmark distributions were also checked for a third pattern, antisymmetry, the tendency of a random side to significantly deviate in size or shape, through scatter plots. None was noted.

There are many ways to measure fluctuating asymmetry. Most commonly, the absolute value of right-minus-left measurements, |R − L|, is used to compare individuals, and the variance of |R − L| is the measure of asymmetry at the level of the population. In contrast, directional asymmetry is measured as the signed differences between left and right sides. Various methods, such as a two-way (multivariate) analysis of variance models, test whether significant levels of directional and fluctuating asymmetry are present (Klingenberg and McIntyre, 1998).

Procrustes methods of superimposition offer one approach to the analysis of shape, the geometric properties of an object invariant to orientation, location, and scale (Slice et al., 1996). While there have been several studies investigating the asymmetry of shape, the overwhelming body of research in the area of developmental stability focuses on asymmetry in terms of size. In this study, size and shape are analyzed as separate variables.

Figure 2 illustrates the construction of shape asymmetry variables used in this analysis for a single individual. Landmark coordinates for the left and right structures of the individual are superimposed using Procrustes superimposition (see Chapter 14). The data are individually scaled to unit centroid size, reflected, and optimally (in the least-squares sense) translated and rotated to achieve a best fit. Once so superimposed, individual coordinate differences between the right and left structures provide a detailed multivariate description of shape asymmetry. To produce a univariate summary of this asymmetry, we compute the square root of the sum of squared coordinate differences (Procrustes distance) between the two configurations.

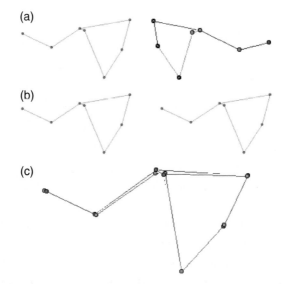

Figure 2.   This figure illustrates the construction of shape asymmetry variables used in this analysis for a single individual. Landmark coordinates for the left and right structures of the individual are superimposed using Procrustes superimposition. In this example, using only two specimens, ordinary Procrustes analysis (OPA) is demonstrated. The entire sample of right and left sides from all individuals are fitted through GPA so that differences between left and right sides may be computed. (a) Paired homologous craniofacial landmarks, (b) both sides scaled to centroid size and one side reflected, (c) paired configurations superimposed and rotated for maximum fit. The square root of the sum of the differences between the left and right landmark coordinates is the measure of shape asymmetry.

The illustration is for two specimens and uses ordinary Procrustes analysis (OPA). To jointly process the samples, right and left data for multiple individuals were combined and subjected to generalized Procrustes analysis (GPA). This will produce slightly different numerical results since GPA optimally superimposes individual data to an interactively computed sample mean, while OPA produces optimal pairwise superimpositions (see Slice, 2001 and Chapter 14). The multivariate shape asymmetry data from a GPA were used in the MANOVAs testing Side, Individual, and their interaction following Klingenberg and McIntyre (1998), while the individual pairwise distances (after GPA) were used for the regression analyses.

Centroid size was used for analyses of size asymmetry. The sizes of right and left structures were used for the ANOVA analyses, while the absolute value of their difference was used for the regressions.

Two-way ANOVA models with repeated measures were used to test the main effects of Side, Individual, and the interaction between Side and Individual on size (ANOVA) and shape (MANOVA) for both sex and ancestral subgroups (as defined by Klingenberg and McIntyre, 1998; Palmer and Strobeck, 1986). The Side variable was a fixed factor, representing the signed difference between the right and left configurations, and was a measure of the directional asymmetry component. The Individual was a random factor and was a measure of interindividual variation. The interaction term between Side and Individual was the measure of fluctuating asymmetry. Note that the degrees of freedom for the shape analysis were calculated for the 3D data as the number of landmarks times the number of dimensions, minus seven for the number of translations, rotations, and scaling (Bookstein, 1991). Finally, repeated measures were not available for the entire sample. Therefore, the individuals for whom repeated measures were available ($n = 19$) were analyzed and included in the generalized Procrustes superimposition. The variance between the two repeated measures was assumed to be consistent for the entire sample and was applied to the overall model so that a direct test of the presence of fluctuating asymmetry was obtained.

Polynomial regression was used to assess secular change in the craniofacial region of American White and Black male and female subgroups. To assess the secular trend of fluctuating asymmetry among the total sample and the four subgroups, the size and shape asymmetry variables (centroid size differences and shape distances) were regressed separately on the year of birth by polynomial regression, including linear and quadratic terms. Cubic terms were also tested, but the results were consistent with the quadratic terms. They are not presented here. Birth-years were divided into decade cohorts ranging from 1820–1990. The midpoint of each decade was used as the birth-year cohort term in the polynomial regression analysis. The mean fluctuating asymmetry scores from individuals born within each decade were used for each decade cohort. Since the sample sizes of each decade vary between one and fifty-seven, Weighted Least Squares (WLS) analysis was used to weigh the analysis for the sample size of each decade. Procrustes and statistical analyses and the creation of plots were performed using the programs Morpheus et al. (Slice, 1998) and SPSS, Inc. (SYSTAT, 1998).

RESULTS

The Presence and Types of Size Asymmetry

Table 2 presents a two-way ANOVA model with repeated measures used to test the main effects of Side and Individual and the interaction between the two on the total size variation. The degrees of freedom (df), sum of squares (SS), mean

Table 2. Total size variation. Two-way ANOVA tests for directional and fluctuating asymmetry

	df	SS	MS	F
White males				
Side	1	7.22	7.22	1.40
Individual	180	7879.02	43.77	8.47[a]
Individual × Side	180	930.07	5.167	11.52[a]
Measurement	38	17.05	0.4487	
White females				
Side	1	37.15	37.15	6.83[a]
Individual	116	4411.76	38.03	6.99[a]
Individual × Side	116	631.03	5.44	12.13[a]
Measurement	38	17.05	0.4487	
Black males				
Side	1	21.00	21.00	4.09[b]
Individual	118	5574.08	47.24	9.19[a]
Individual × Side	118	606.41	5.139	11.45[a]
Measurement	38	17.05	0.4487	
Black females				
Side	1	19.43	19.43	4.46[b]
Individual	108	4780.79	44.27	10.16[a]
Individual × Side	108	470.77	4.359	9.72[a]
Measurement	38	17.05	0.4487	

Notes:
[a] $p < 0.001$.
[b] $p < 0.05$.

Table 3. Total shape variation. Two-way MANOVA tests for directional and fluctuating asymmetry

	df	SS	MS	F
White males				
Side	14	0.000047	0.0000033	0.023
Individual	2520	0.422570	0.0001676	1.143[a]
Individual × Side	2520	0.369590	0.0001466	2.560[a]
Measurement	532	0.030484	0.0000573	
White females				
Side	14	0.000254	0.0000181	0.097
Individual	1624	0.306240	0.0001885	1.143[a]
Individual × Side	2520	0.30606	0.0001466	2.560[a]
Measurement	532	0.030484	0.0000573	
Black males				
Side	14	0.00014	0.00001	0.052
Individual	1652	0.30609	0.0001852	0.9719
Individual × Side	1652	0.31767	0.0001905	3.325[a]
Measurement	532	0.030484	0.0000573	
Black females				
Side	14	0.00021	0.000015	0.9009
Individual	1512	0.28998	0.0001917	1.155[a]
Individual × Side	1512	0.25418	0.000168	2.934[a]
Measurement	532	0.030484	0.0000573	

Note:
[a] $p < 0.001$.

sum of squares (MS), and F statistic are provided for each group. Note that Side was significant for White females and Black females and males, indicating the presence of directional asymmetry. As expected, individual variation was significant in all four subgroups. The presence of fluctuating asymmetry (the interaction term between Side and Individual) was also significant for all groups, ranging from the highest among White females to the lowest among Black females.

The Presence and Types of Shape Asymmetry

As with the analysis of size, a two-way MANOVA model with repeated measures was used to test the main effects of Side and Individual and their interaction on total shape variation (Table 3). No directional asymmetry (Side) was present. Individual variation was significant among all groups, except Black males. Black females exhibited the largest amount of individual variation ($F = 11.51$, $p < 0.001$). Finally, all groups exhibited significant levels of fluctuating shape asymmetry.

Table 4. Test for secular change in craniofacial fluctuating size asymmetry, which is regressed onto year of birth for pooled data and by subgroup, showing polynomial regressions

	df	R^2	MS	MSE	F
Total sample term					
Linear	1	0.012	1.02	2.25	0.17
Quadratic	2	0.012	0.42	2.28	0.08
White males term					
Linear	1	0.133	10.77	2.32	2.01
Quadratic	2	0.152	6.13	2.38	1.08
White females term					
Linear	1	0.072	4.06	2.01	1.00
Quadratic	2	0.093	2.66	2.07	0.62
Black males term					
Linear	1	0.081	4.58	1.99	1.15
Quadratic	2	0.081	2.290	2.08	0.53
Black females term					
Linear	1	0.157	13.99	2.50	2.23
Quadratic	2	0.176	7.89	2.58	1.18

Secular Change of Fluctuating Asymmetry Assessed through Polynomial Regression

To explore the secular trends of fluctuating asymmetry for the data as a whole and within each group, polynomial regression was performed on the fluctuating asymmetry of size and birth cohort (the midpoint of each decade of birth) (Table 4). Since no directional asymmetry was noted for White males, the absolute right minus left difference was used as the fluctuating asymmetry score. For the other three subgroups, where directional asymmetry was observed, the directional component (the signed difference between the mean left and right sides) was subtracted so that only fluctuating asymmetry was tested. No significant overall association was detected for size asymmetry and decade of birth.

To further test the relationship between secular patterns in shape asymmetry within each group, polynomial regression was used to test the relationship between fluctuating shape asymmetry (R − L distance) on birth cohort (Table 5). Interestingly, Black females show the only significant linear ($F = 11.92$, $p = 0.005$) and quadratic ($F = 5.48$, $p = 0.024$) relationship between fluctuating shape asymmetry and the decade of birth. The bivariate Pearson's correlation between year of birth and shape asymmetry was 0.669

Table 5. Test for secular change in craniofacial fluctuating shape asymmetry, which is regressed onto year of birth for pooled data and by subgroup, showing polynomial regressions

	df	R^2	MS	MSE	F
Total sample term					
Linear	1	0.072	1648.47	13.31	1.09
Quadratic	2	0.247	3.77	1.32	2.14
White males term					
Linear	1	0.000	1.78	2.13	0.00
Quadratic	2	0.052	1.08	1.79	0.33
White females term					
Linear	1	0.002	0.13	2.09	0.03
Quadratic	2	0.290	7.64	1.84	2.25
Black males term					
Linear	1	0.051	9.44	3.66	0.70
Quadratic	2	0.055	5.06	3.80	0.35
Black females term					
Linear	1	0.520	3540.66	17.23	11.92[a]
Quadratic	2	0.523	1780.46	18.02	5.48[b]

Notes:
[a] $p < 0.001$.
[b] $p < 0.05$.

($p = 0.009$) for the total sample. While significant levels of fluctuating shape asymmetry were present among Black males, White males, and White females, no secular association of fluctuating asymmetry and birth year was present for either linear or quadratic terms.

Patterns in the Secular Trends in Fluctuating Asymmetry

The mean values of size and shape asymmetry were plotted by birth decade for White males (Figure 3) and females (Figure 4), who exhibit similar patterns. The level of fluctuating shape asymmetry appears to remain relatively constant over time, whereas there is a slight trend (though nonsignificant association) for fluctuating size asymmetry to increase over time.

The mean values of size and shape asymmetry were plotted by decade for Black males (Figure 5) and illustrate a similar pattern. Only Black females (Figure 6) exhibit a statistically significant increase in shape asymmetry over time.

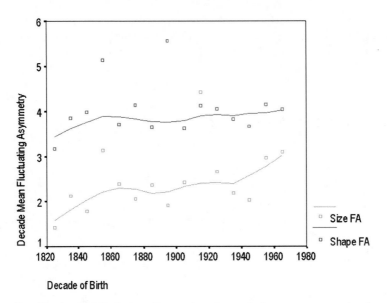

Figure 3. Secular trend of mean fluctuating size and shape asymmetry by decade of birth for White males, with fitted Lowess line.

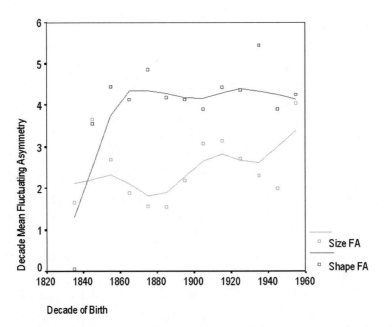

Figure 4. Secular trend of mean fluctuating size and shape asymmetry by decade of birth for White females, with fitted Lowess line.

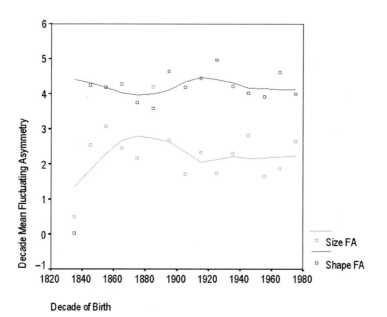

Figure 5.   Secular trend of mean fluctuating size and shape asymmetry by decade of birth for Black males, with fitted Lowess line.

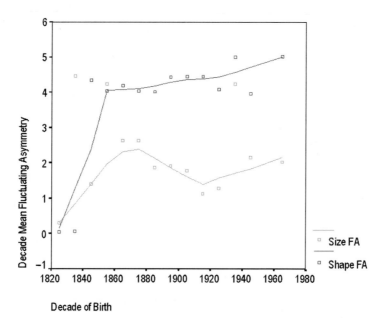

Figure 6.   Secular trend of mean fluctuating size and shape asymmetry by decade of birth for Black females, with fitted Lowess line.

DISCUSSION

The purpose of this chapter was to explore a 200-year period of American craniofacial form to determine if morphological change and increased variability is accompanied by changes in the amount or type of asymmetry observed using geometric morphometrics. Both fluctuating and directional asymmetry in the size of craniofacial form are present, with the exception that no directional component is found among White males. Overall, high levels of individual variation are significant among all groups but the highest levels are among African Americans. In contrast, no directional component is found for shape asymmetry in either racial or sex group. Individual variability in shape is significant in all subgroups, except for Black males, yet Black females rank the highest for this value. Further, fluctuating shape asymmetry is present in all groups.

To assess the trends of fluctuating shape and size asymmetry over time, these associations were investigated through polynomial regression. We find that the only significant association of fluctuating shape or size asymmetry and the birth-year cohort is for shape for Black females. This finding suggests facial morphology among this group is becoming less symmetrical over time and may reflect a decline in developmental stability and increasing levels of individual variation. Yet, only about half of the variation observed for Black females can be explained by decade of birth ($r^2 = 0.523$). From the scatter plots, we observe similar, nonlinear patterns in the secular trends of fluctuating asymmetry among all of the groups; shape asymmetry remains relatively constant while size asymmetry fluctuates with a slight increase, although significant associations among three of these groups cannot be substantiated at this time.

Due to the overall tendency toward low r^2 values, a cursory attempt was made to determine whether those individuals with the highest levels of fluctuating asymmetry, particularly during the early 19th century, shared any common life history factors. The cause of death, age at death, geographic location of birth, and even the collection in which the skeletal remains are housed were compared. To date, the only patterns observed are those for birth year.

Economic historians use the secular change of biological variables as a reflection of changing socioeconomic conditions. This is analogous to many studies of developmental stability in anthropology. However, interpreting the causal mechanisms of developmental instability is challenging given the various genetic and environmental components which may, under given conditions, result in asymmetry. In the case of American craniofacial morphology, several influential components have changed. Environmental variants known to result in high levels of asymmetry, such as nutritional deficiencies, infectious disease, and parasitic load have markedly declined in America over the time period in which we are interested. Consequently, we would expect the developmental stability of Americans to increase. Instead, the pattern observed is that developmental stability remains constant, except for Black females who show a decline. Of course, improving environmental conditions does not have a linear relationship with time. Various individuals used in this study had been subject to wide-ranging environmental and material conditions including Slavery during the 19th century, followed by the American Civil War and the Reconstruction Period in the southern United States, and later, the Great Depression. Such fluctuations throughout history likely account for the nonlinear relationship between year of birth and asymmetry. Though this study is preliminary, it suggests some interesting patterns in cranial asymmetry. We look forward to seeing if the observed patterns continue and/or become better resolved as new data are added from modern cases.

ACKNOWLEDGMENTS

We would like to thank the following people for their editorial comments, for providing access to skeletal collections, and for sharing of data: Michael Boland, Dave Hunt, Lyle Konigsberg, Dennis Slice, Paul Sledzik, Danny Wescott, Greg Berg, and Kate Spradley. Funding for this project was provided by the William M. Bass Endowment through the Forensic Anthropology Center at the University of Tennessee.

REFERENCES

Auffray, J., Ailbert, P., Renaud, A., Orth, A., and Bonhomme, F., 1996, Fluctuating asymmetry in *Mus musculus* subspecific hybridization: Traditional and Procrustes comparative approach, in: *Advances in Morphometrics*, L. F. Marcus, M. Corti, A. Loy, G. J. P. Naylor, and D. E. Slice, eds., Plenum Press, New York.

Bookstein, F. L., 1991, *Morphometric Tools for Landmark Data*, University of Cambridge Press, London.

Farkas, L. G., and Cheung, G., 1981, Facial asymmetry in healthy North American Caucasians: An anthropometrical study, *Angle Orthod.* 51:70–77.

Gower, J. C., 1985, Generalized Procrustes analysis, *Pschometrika* 40:33–50.

Hershkovitz, I., Livshits, G., Moskona, D., Arensburg, B., and Kobyliansky, E., 1993, Variables affecting dental fluctuating asymmetry in human isolates, *AJPA* 91:349–365.

Howells, W. W., 1973, *Cranial Variation in Man: A Study by Multivariate Analysis of Patterns of Difference among Recent Human Populations*, Papers of the Peabody Museum No. 67, Peabody Museum, Harvard Museum, Cambridge, MA.

Jantz, R. L. and Brehme, H., 1993, Directional and fluctuating asymmetry in the palmar interdigital ridge-counts, *Anthrop. Anz.* 51:59–67.

Jantz, R. L. and Meadows Jantz, L., 2000, Secular change in craniofacial morphology, *Am. J. Hum. Biol.* 12:327–338.

Klingenberg, C. P., Badyaev, A. V., Sowry, S. M., and Beckwith, N. J., 2001, Inferring developmental modularity from morphological integration: Analysis of individual variation and asymmetry in bumblebee wings, *Am. Nat.* 157(1):11–23.

Klingenberg, C. P. and McIntyre, G. S., 1998, Geometric morphometrics of developmental instability: Analyzing patterns of fluctuating asymmetry with Procrustes methods, *Evolution* 52(5):1361–1375.

Leamy, L. J., Routman, E. J., and Cheverud, J. M., 2002, An epistatic genetic basis for fluctuating asymmetry of mandible size in mice, *Evolution* 56(3):642–653.

Mardia, K. V., Bookstein, F. L., and Moreton, I. J., 2000, Statistical assessment of bilateral symmetry of shapes, *Biometrika* 87(2):285–300.

Palmer, A. R. and Strobeck, C., 1986, Fluctuating asymmetry: Measurement, analysis, Patterns, *Annu. Rev. Ecol. Syst.* 17:391–421.

Palmer, A. R. and Strobeck, C., 1992, Fluctuating asymmetry as a measure of developmental stability: implications of non-normal distributions and power of statistical tests, *Acta Zool. Fenn.* 191:57–72.

Peck, S., Peck, L., and Kataja, M., 1991, Skeletal asymmetry in esthetically pleasing faces, *Angle Orthod.* 61:43–47.

Richtsmeier, J., 1987, A comparative study of normal, Crouzon and Apert craniofacial morphology using finite-element scaling analysis, *AJPA* 74:473–493.

Rutherford, S. L., 2000, From genotype to phenotype: Buffering mechanisms and the storage of genetic Information, *Bioessays* 22:1095–1105.

Sciulli, P. W., Doyle, W. J., Kelley, C., Siegel, P., and Siegel, M. I., 1979, The interaction of stressors in the induction of increased levels of fluctuating asymmetry in the laboratory rat, *AJPA* 50:279–284.

Siegel, M. I., Doyle, W. J., and Kelley, C., 1977, Heat stress, fluctuating asymmetry and prenatal selection in the laboratory rat, *AJPA* 46:121–126.

Slice, D., 2001, Landmark coordinates aligned by Procrustes analysis do not lie in Kendall's shape space, *Sys. Biol.* 50(1):141–149.

Slice, D. C., 1998, *Morpheus* et al.: *Software for Morphometric Research*, Revision 01-31-00. Department of Evolution and Ecology, State University of New York, Stony Brook, NY.

Slice, D., Bookstein, F. L., Marcus, L. F., and Rohlf, F., 1996, A glossary for geometric morphometrics, in: *Advances in Morphometrics*, L. F. Marcus, M. Corti, A. Loy, G. J. P. Naylor, and D. E. Slice, eds., Plenum Press, New York, pp. 531–551.

SPSS, Inc., 1998, SYSTAT Statistics, SPSS, Inc., Chicago.

Waddington, C. H., 1957, *The Strategy of Genes*, MacMillan Co, New York.

CHAPTER TWELVE

The Morphological Integration of the Hominoid Skull: A Partial Least Squares and PC Analysis with Implications for European Middle Pleistocene Mandibular Variation

Markus Bastir, Antonio Rosas, and H. David Sheets

INTRODUCTION

An important question of comparative anatomy in paleoanthropology is that of similarity of variation patterns between closely related species on the one hand and between extant and fossil species on the other (Ackermann, 2002; Bastir, 2004). Recently, mandibular variation in hominids and modern humans

Markus Bastir and Antonio Rosas • Museo Nacional de Ciencias Naturales, Department of Paleobiology, CSIC, Madrid, Spain. **H. David Sheets** • Canisius College, Department of Physics, Buffalo, USA and SUNY, State University of New York, Buffalo, Department of Geology, USA.

Modern Morphometrics in Physical Anthropology, edited by Dennis E. Slice.
Kluwer Academic/Plenum Publishers, New York, 2005.

has been discussed based on the hypothesis of the existence of a general hominoid craniofacial variation pattern (Rosas et al., 2002a). These authors used landmark-based geometric morphometry to identify principal components of shape variation in modern humans and chimpanzees. It was hypothesized that the first relative warp describes similar patterns of shape variation in both hominoid species. What in modern humans appeared to reflect brachy- and dolichofacial growth patterns, in chimpanzees was associated with shape changes related to "airo-" or "klinorhynchy" (Bastir, 2004; Hofer, 1952; Rosas et al., 2002a). The hypothesis was based on the visual observation that in both groups a rotation of the viscerocranium and the neurocranium seemed to produce these patterns. A similarity of processes leading to this pattern of variation was assumed.

In the present study, this hypothesis is analyzed in more detail, and the following questions are addressed: Is the observed similarity between the first relative warps of humans and chimpanzees an overall similarity or is it localized? Which features show similarities in variation? Which show differences? Do variation patterns of modern humans and chimpanzees differ from a statistical point of view? What are the possible implications for mandibular variation alone? How does this variation relate to mandibles of fossil hominids, such as Neandertals or the mandibles of the hominids from Atapuerca?

The mandibular sample of the Atapuerca Sima de los Huesos (AT-SH)-site provides a unique opportunity in order to investigate intrapopulational variation in the morphology of fossil hominids (Rosas, 1995, 1997). In the present study, the morphological variation of the adult AT-SH mandibular sample ($n = 15$) and 14 Neandertals (Table 1) is investigated by geometric morphometry with the aim of evaluating morphological variation of mandibles within its craniofacial context.

The Relationship between Cranial and Mandibular Variation

Morphological variation is generally analyzed within two major conceptions of form, that is, structure and function. Coordinated variation between various parts within a functioning whole is usually conceived as morphological integration (Klingenberg et al., 2001; Lieberman et al., 2000a; Olson and Miller, 1958). In the case of the mandible and the cranium, the coordinated variation of both parts of the skull can be analyzed at different levels

Table 1. List of the fossil mandibles

	Atapuerca SH sample	Neandertals
1	AT-3880 (F)	Tabun 1
2	AT-3888 (F)	Tabun 2
3	Individual 1 (F)	Krapina E
4	Individual 3 (F)	Krapina H
5	Individual 6 (F)	Krapina J
6	Individual 23 (F)	La Chapelle 1
7	Individual 4 (F)	Aubesier 11
8	Individual 31 (F)	Monte Circeo 3
9	Individual 19 (F)	La Ferassie 1
10	Individual 26 (F)	La Quina 9
11	Individual 15 (F)	Saint Cesaire
12	Individual 12 (M)	Zafarrayah
13	Individual 22 (M)	Regourdou 1
14	Individual 27 (M)	Amud
15	Individual 7 (M)	

(Cheverud, 1996). For example, functional integration would characterize the relationship between the cranium and the mandible with respect to mastication and respiration, whereas growth and development of the basicranium and the mandible is related to developmental integration and structural morphogenetic determination (Bastir et al., 2004).

General architectonic features of the craniofacial system structurally influence growth and development. This relationship is established by the key position of the cranial base in establishing the craniofacial growth field and perimeter (Enlow and Hans, 1996). For instance, there is a tendency for a longer, narrower, and less flexed basicranium to be associated with an anteroposteriorly and vertically elongated facial pattern (Enlow and Hans, 1996; Lieberman et al., 2000b). This facial pattern is typical for dolichofacial morphologies. The brachyfacial pattern is related to a shorter and wider cranial base and a vertically shorter face vice versa. The mandibular shape is also closely related to the facial pattern (Bhat and Enlow, 1985). In dolichocephalic basicrania, the mandible tends structurally toward retrusion, in brachycephalic toward protrusion (Enlow and Hans, 1996).

Besides architectural features, such as mentioned above, further factors, that is, sexual dimorphism and allometry, are relevant for modern human craniofacial variation (Enlow and Hans, 1996; Rosas and Bastir, 2002). For variation in fossil hominids, allometry and sexual dimorphism have been studied

(Rosas, 1997; Rosas and Bastir, 2004; Rosas et al., 2002b). The reflection of facial patterns in the mandible and the analysis of possible architectural features on nonhuman primates and fossil hominids, however, have rarely been investigated. The present study aims to establish gross morphological relationships between facial patterns and variation in mandibular morphology.

Morphological Integration

Morphological integration describes coordinated morphological variation of components of a functioning whole (Klingenberg et al., 2001; Liebermann et al., 2000a; Olson and Miller, 1958). The systemic and coordinated covariation patterns of facial, mandibular, and neurocranial components of the skull (Bastir, 2004) are the basis for the assumption of integration at a functional and/or developmental level (Cheverud, 1996). It has been suggested that studies of morphological integration should proceed by quantifying, evaluating, and comparing patterns and degrees of covariation (Chernoff and Magwene, 1999). In the present study, these steps are realized by partial least squares analysis (PLS) and permutations of the corresponding vector correlations (Marcus, 1993; Rohlf and Corti, 2000; Sheets, 2001).

Design of the Study

The design of the study is depicted in Figure 1. The first analytical step is the analysis of the first relative warps of modern humans and chimpanzees (Figure 1a) to compare interspecific shape variation in the neurocranium, the nasomaxillary complex, the mandibular ramus, and corpus.

Then a PLS analysis is applied in order to identify patterns of mandibular and cranial covariation and to determine the degree of cranio-mandibular morphological integration (Figure 1b). Finally, interspecific comparison of the relative warps of modern human and fossil mandibles is used to outline similarities and dissimilarities of morphological variation and its possible relationship to facial variation revealed by the first two analyses (Figure 1c).

MATERIALS AND METHODS

The modern samples consisted of 104 adult human individuals of known age and sex (University of Coimbra) and 48 adult chimpanzees of known sex

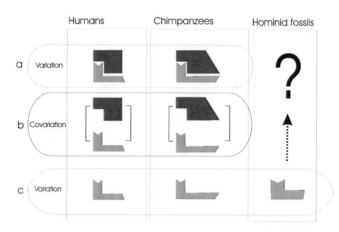

Figure 1. Design of the study. (a) Patterns of interspecific variation are evaluated by relative warps; (b) correlated singular warps are used for interspecific exploration of the morphological integration and covariation of cranio-mandibular components; (c) relative warps of humans and fossil hominid mandibles predict cranial shape.

(NHM, London). The 15 AT-SH mandibles were digitized at the MNCN, Madrid. The Neandertal mandibular sample was digitized on casts, except Tabun 2, which is housed at the NHM, London. The fossil sample is listed in Table 1.

The analysis of the whole craniofacial sample was done using 29 3D-landmarks, while the mandibular sample consisted of 11 landmarks. Landmark definitions, the 3D–2D transformation, and the treatment of missing data necessary in the TPS analysis are described elsewhere (Rosas and Bastir, 2002). In the AT-SH mandibles, missing data were replaced in shape space by the sex-specific mean coordinates (Rosas et al., 2002b). The two new specimens (AT-3880, AT-3888) were considered female. The sex attribution in the Neandertal sample was not clear, and, thus, the missing landmarks were replaced by the grand mean values of the corresponding point.

Geometric Morphometrics

In the present study, we used geometric morphometry based on the analysis of landmark coordinates to study patterns of morphological variation. At the core of geometric morphometric methods is the separation of two components of form, that is, size and shape (Bookstein, 1991). Shape is the residual geometric information remaining once differences due to location, scale, and rotational effects are removed (Kendall, 1977). Partial Procrustes superimposition techniques (Dryden and Mardia, 1998) minimize the offset between homolog landmarks, and size is obtained as a scaling factor termed "centroid size" (Bookstein, 1991; Rohlf and Slice, 1990).

The thin-plate spline (TPS) decomposition method (Bookstein, 1991; Rohlf, 1996; Rohlf et al., 1996) is used to produce Partial Warp and Uniform Component scores for further analysis, as these scores represent all information about the shape of the specimens with the same number of variables as degrees of freedom in the shape measurement. In the present study, the Procrustes mean shape of all specimens is used as the reference form for the TPS decomposition. Overall variation in shape is small enough that common statistical procedures may be used to analyze shape data in the Euclidean linear tangent space to the hyperhemispherical Generalized Procrustes Space (Slice, 2001) produced by using a GLS Partial Procrustes Superimposition (Dryden and Mardia, 1998) at a unit centroid size of 1. This approach of utilizing an orthogonal projection from the Procrustes hyperhemisphere is thought to yield the preferred linearization of distances in Kendall's shape space (Bookstein, 1991, 1996; Dryden and Mardia, 1998; Rohlf et al., 1996; Slice, 2001).

Relative Warps

Relative warps are the principal components of shape variables, such as Procrustes residuals or partial warp scores and reflect the major patterns of shape variation within a group (Bookstein, 1996; Rohlf and Slice, 1990; Rohlf et al., 1996). To compare principal component axes, a bootstrap test (Efron and Tibshirani, 1993) is used to determine if the observed angle between the first principal component axes (or relative warp axes) of different groups is statistically significant, using the null-hypothesis that the observed angle between two axes could have arisen from a random sampling from a single, homogeneous group. This procedure is similar to that used to determine the significance of the angle between growth vectors based on regression models (Webster et al., 2001; Zelditch et al., 2000). To carry out this test, the angle between the two principal component axes is first determined using an approach presented by Houle and Mezey (2002), or by using the dot product of the two vectors describing the pattern of variation along the principal component axes. To determine the range of angles possible within a single group, a pair of bootstrap sets is formed by randomly drawing with replacement (Efron and Tibshirani, 1993) from the single group. The angle between the principal component axes of the paired bootstrap sets is then determined. This is repeated for both groups used in the comparison for a large number of bootstrap sets, and the observed angle between the groups is judged statistically significant if it exceeds the 95% confidence interval of angles obtained within each group via the bootstrap procedure. This calculation was carried out using the SpaceAngle program (Sheets, 2001).

Partial Least Squares

Partial Least Squares analysis helps to find correlated pairs of linear combinations (singular vectors) between two sets (or blocks) of variables (Bookstein, 1991; Bookstein et al., 2003; Rohlf and Corti, 2000). The singular vectors are constructed in the form of new, paired (one per block) "latent" variables (also called singular warps) that account for as much as possible of the covariation between the two original sets of variables. In a similar sense to the Principal Component Analysis (PCA), the singular value decomposition (SVD) describes the data in terms of scores of each specimen along the singular axes, singular values (similar to eigenvalues), and loadings (singular vectors, similar to eigenvectors). However, SVD is applied with a different goal, that is, to maximize low-dimensional representation of between-block covariance structure (SVD) vs maximizing low-dimensional representation total sample covariance (PCA). The singular vectors express the maximal covariance between both the variables within their set (or block) and with the variables of the other set (Rohlf and Corti, 2000). The amount of covariance explained by the paired singular vectors (the SVD-axes), the correlation "r" of the scores of specimens along the singular axes of the two blocks, and permutation tests allow assessment of the statistical significance of the observed singular values and correlations. When two groups are considered, the patterns of covariance can be compared statistically by computing the angle between the SVD-axes and testing the observed angle between the two groups against the distribution of angles produced by random resampling within a single group, using the same procedure as discussed for relative warp axes, and implemented in the PLSAngle program (Sheets, 2001).

RESULTS

Similarity and Dissimilarity of Skull Shape Variation

Figure 2a shows the principal patterns of shape variation in the human skull sample. This pattern polarizes morphologies characterized by large faces with increased anterior height and reduced posterior height (Figure 2b). Thereby,

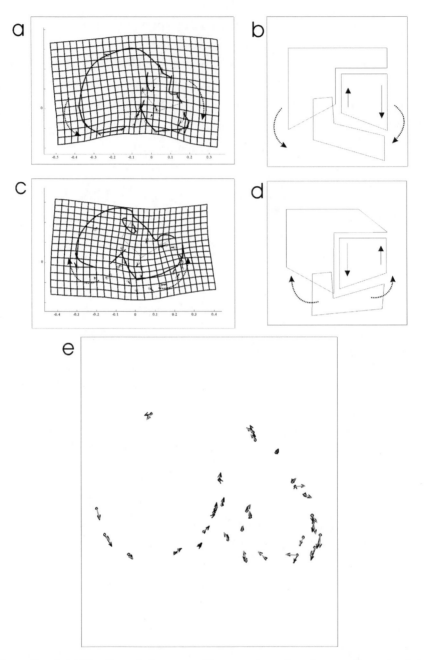

Figure 2. (a,b) The first relative warp of humans shows how increased anterior facial height and reduced posterior facial height and a downward rotation of the neurocranium contribute to a dolichofacial pattern. (c, d) The first relative warp of chimpanzees shows how the opposite shape changes produces a brachyfacial pattern. (e) The superimposed first relative warps of a grand mean consensus shape (gray vectors, chimpanzees; black vectors humans).

the mandible becomes "opened." The angle between corpus and ramus and the symphyseal height increases. A flattened nasoglabellar profile and a posterior–inferior rotation characterize shape changes at the neurocranium. These variation patterns are reflected in the superior-wards shift of the TPS grids at the posterior face, while the neurocranial and anterior facial areas are lowered. The face appears to be rotated against the neurocranium. Such morphologies are related to a dolichofacial growth pattern. The other extreme of this eigenvector is characterized by the opposite morphologies; short anterior faces, high posterior faces, rectangular mandibles, and up- and forward-rotated neurocrania. Figure 2c shows the first principal component of chimpanzee shape variation. It is a similar pattern of change. One extreme reflects morphologies characterized by short anterior faces, increased height of the posterior face, an increasingly rectangular mandible and an up- and forward-rotated neurocranium (Figure 2d). The TPS grid is lowered at the posterior face. Along the first relative warp of chimpanzees, both parts of the skull, the viscerocranium and the neurocranium, are either rotated superiorly or inferiorly against the neurocranium. Figure 2e shows a superimposition of the first relative warps of humans and chimpanzees on a grand-mean consensus shape of all human and chimpanzee specimens, which should help operationally to localize similarities and dissimilarities of shape variation. It turns out that the striking similarity of principal components of overall shape variation, as reflected by the TPS grids, can now be better localized. The neurocranium and the cranial base components share a very similar pattern of shape variation. In many cases, the magnitude, as well as the angle of landmark displacement vectors, do not differ too much between the species. Also, the posterior face appears to share similar shape variation patterns in humans and chimpanzees. Differences are located mainly at the anterior face. In humans, the anterior face varies superior–inferiorly, while the chimpanzees show horizontal vectors of variation. The quantitative data reflects this situation. The angle of the first relative warp between humans and chimpanzees is 39.54° (95% C.I.: 33.3°–54°) after 4,900 permutations. The angle range within humans is 27° and within chimpanzees 38.4°, which is close to, but smaller than, the between-group angle.

The Morphological Integration of the Skull

The correlated covariation of the cranium and the mandible is evaluated by partial least squares analysis. Shape covariation is depicted in Figure 3. Comparison

274 Markus Bastir et al.

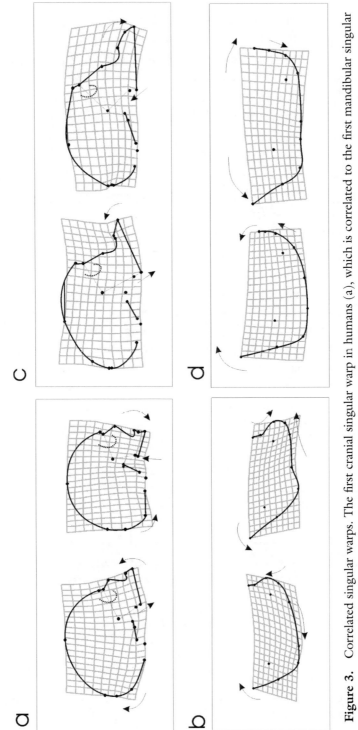

Figure 3. Correlated singular warps. The first cranial singular warp in humans (a), which is correlated to the first mandibular singular warp (b). (c, d) The first cranial and mandibular singular warps of chimpanzees.

of the SVD-axes shows that the patterns of morphological covariation in the cranio-mandibular complex analyzed separately are not the same in the two species. The quantitative data show that when the SVD-axis correlations are taken as a measurement for integration, the chimpanzees are more integrated than the humans.

The correlation coefficient between the first pair of singular warps in chimpanzees is 0.73 (with a 95% range: 0.69–0.87) and significantly higher ($p < 0.05$) than in humans ($r = 0.53$; 0.46–0.65). A bootstrap estimation of the 95% confidence interval of the difference in the correlation between blocks (a difference of 0.20 in r with a 95% C.I. of 0.08–0.35 on the difference) between chimps and humans excluded zero, leading to the assertion that the correlation was statistically significantly higher in chimps. Shape covariation patterns are also significantly different. The between-group angle of the first singular warps is 88.3° in crania. It is larger than found in either group alone (humans, 83.9°; chimpanzees, 54.6°). The same situation is found in the mandibular singular warp. The between group angle is 85.8°, that is, larger than the within-group angle ranges (humans, 83.2°; chimpanzees, 56.4°). However, the TPS grids of the singular warps show that some patterns of shape changes are shared. In Figure 3, the correlative shape changes are depicted, and both the cranium and the mandible have some aspects in common. In both species, the first cranial singular warp polarizes the relationship of anterior and posterior facial height (Figures 3a and c). This relationship contributes to the dolichofacial or brachyfacial pattern and is accompanied by similar changes in the angulation of the ramus with respect to the corpus and by changes in symphyseal height (singular warp 1 of the mandible, Figures 3b and d). The geometric relationship between correlated (relative warps) and isolated (singular warps) shape variation is shown in Figure 4. The human relative and singular warps (Figures 4a and b), as displacement vectors at each landmark, are more divergent than the corresponding warps of the chimpanzees (Figures 4c and d). This divergence is especially obvious and stronger in the human neurocranial area than in the face. The chimpanzee vectors show a tighter overall correspondence of landmark displacements.

Interspecific Mandibular Shape Variation

The relative warps of the hominid mandibles are shown in Figure 5. Statistically, these relative warps are not significantly different from each other. The between-group angle is always smaller than the within-group angle. The angle between

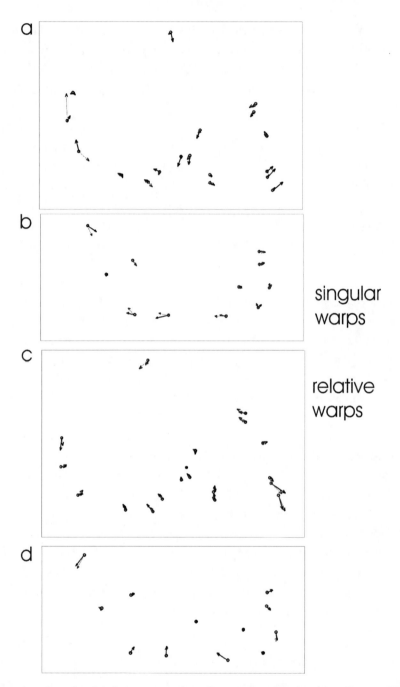

Figure 4. Correlated (relative warps, dotted gray vectors) and separated patterns (singular warps, solid black vectors) of shape variation of humans (a, b) and chimpanzees (c, d).

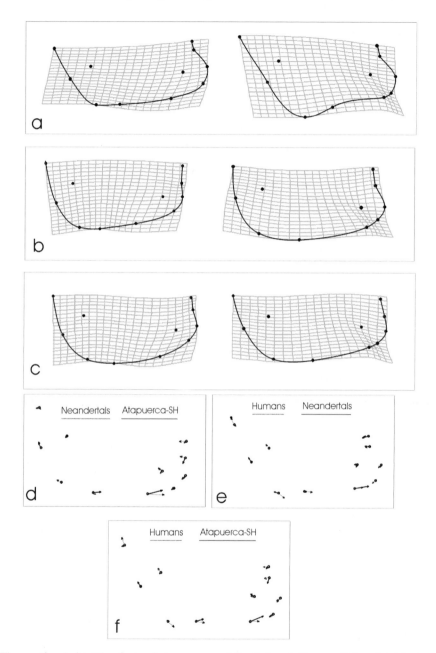

Figure 5. (a,b) The first relative warps of fossil hominid mandibles. (a) Modern humans, (b) Neandertals, and (c) Atapuerca-SH. Each of the grid pairs reflect the morphology related to the brachyfacial (left grid) and the dolichofacial (right grid) pattern. The superimposition of the first relative warps of (d) Neandertals and Atapuerca-SH specimens, (e) modern humans and Neandertals, and (f) modern humans and Atapuerca-SH specimens are displayed.

Neandertals (within-range: 83.9°) and AT-SH hominids (within-range: 87.8°) is 41.6° (95% C.I.: 35.1–87.7). The angle between humans (within-range: 86.5°) and Neandertals (within-range: 84.5°) is 65.4 (95% C.I.: 34.2–88.9). And the angle between humans (within-range: 86.5) and the AT-SH hominids (within-range: 87.6) is 55° (95% C.I.: 33.3–88.7). When ramus, corpus, and symphysis are compared, it can be seen in Figure 5 that while variation is similar at the corpus and at the symphysis, the mandibular ramus behaves differently in modern humans (Figure 5a) on the one hand and Neandertals (Figure 5b) and the AT-SH individuals (Figure 5c) on the other. These figures show a brachy-facial morphology on the left side and the dolichofacial patterns on the right side. The pairwise comparisons of Figures 5d, e, f show that in all mandibles attributed to the dolichofacial pattern, the corpus is extended anteriorly and the symphysis projects by a pronounced mandibular depression. In modern humans, the dolichofacial pattern is further characterized at the ramus by an increased height and at the gonial area by an increased angulation.

DISCUSSION

In comparative anatomy, the correct evaluation of morphological variation is of crucial importance (Ackermann, 2002; Bastir, 2004). Taxonomic hypotheses in paleontological research depend, in many cases, directly on theories that are established on the analysis of variation of shape data (Collard and Wood, 2000; Deane and Begun, 2002). In the present study, geometric morphometric applications are used to analyze patterns of variation and covariation of shape data in humans, chimpanzees, and fossil hominids. We address the question of whether variation in fossil hominid mandibles may be related to similar morpho-genetic principles to those in modern humans. We test the null-hypothesis that modern humans and chimpanzees share a common pattern of shape variation (Rosas et al., 2002a) and interpret these findings in the context of variation in European Mid-Pleistocene fossils.

The interspecific comparison of the relative warps in humans and chim-panzees shows that most of the patterns of shape variation are similar. Most of the cranio-mandibular components, the neurocranium, the basicranium, and the posterior face showed similar directions of vectors (Figure 2). Alto-gether, the compared shape changes are related to dolichofacial or brachyfacial growth patterns long known in modern humans (Enlow, 1968; Enlow and Hans, 1996).

It is interesting to note that similar patterns of shape variation to those shown in Figures 2 and 3 have been identified in a different context by Zollikofer and Ponce de León (2002), Rosas et al. (2002a), and also by Bookstein et al. (2003). In those studies, the different position of the face with respect to the braincase and/or the cranial base was found to be independent from ontogenetic (Zollikofer and Ponce de León, 2002; Bookstein et al., 2003) and phylogenetic (Bookstein et al., 2003) changes in cranial morphology of humans and hominids. It was morphologically related to pro- or retrognathism by Bookstein et al. (2003) and to early ontogenetic positional effects by Zollikofer and Ponce de León (2002).

The independence from ontogeny and phylogeny could explain why in the present chapter, variation of facial position appears as first relative and/or singular warp. In this study, only adults are investigated, partialling out thus to some degree the strong morphogenetic effect of ontogenetic or phylogenetic allometry (Bastir and Rosas, 2004), which are probably reflected in the first relative and the first singular warp of Zollikofer and Ponce de León (2002) and Bookstein et al. (2003).

The discussed shape variation describes the relationship between anterior and posterior facial height, which appears geometrically as facial rotation against the neurocranium, either inferiorly (increased anterior facial height or dolichofacial pattern, Figure 2b) or superiorly (increased posterior facial height or brachyfacial pattern, Figure 2d). Such a relationship between anterior and posterior facial heights is also observed in chimpanzees. The prognathic architecture of the chimpanzees, however, leads to the effect that increased anterior facial height leads to a more projecting face. Morphologically, this is a different effect.

Similar shape variation in primatology has been termed airo- or klinorhynchy (Bastir, 2004; Hofer, 1952), referring to elevation or declination of the jaws with respect to the basicranium. Since the present data only documents exocranial basicranial orientation, a direct relationship of within-group variation and systematic characters cannot be made. Further investigation of the interspecific relationship between exo- and endobasicranial shape variation is necessary.

However, the architectonic difference between chimpanzees and humans as expressed by their pro- and orthognathism seems to account for the tendency toward statistical difference in shape variation. The angle between the first relative warp axes of humans and chimpanzees is only slightly larger than what could be produced by the chimp sample alone. This corresponds to what the vectors indicate at each landmark, when both relative warps are superimposed onto a grand mean consensus (Sheets, 2001) (Figure 2). It becomes apparent that prognathism clearly influences patterns of morphological variation (Figure 2e). The modern human orthognathism is opposed to the chimpanzee prognathism. Increased orthognathism on the other hand is an evolutionary tendency in hominids. Since most of the craniofacial variation is similar except where evolutionary change has taken place, it is most parsimonious to assume that the nature of variability is similar. But due to structural reasons, actual morphological variation is different. We hypothesize that if humans were still prognathic (like australopiths), they probably might have varied in the same way as chimpanzees.

The Morphological Integration of the Cranio-Mandibular System

Quantitative data indicate that singular warps are differently correlated in humans and chimpanzees but show also that the chimpanzee skull is significantly more strongly integrated than that of modern humans. Common changes in geometry are related to the relationship of anterior and posterior facial height, and closing or opening of the mandibular angle (Figure 3). This seems to be further evidence that dolicho- and brachyfacial patterns are important constituents of cranio-mandibular variation in hominoids (Bastir, 2004; Enlow and Hans, 1996).

The consistency between the cranial relative warps and cranial singular warps with respect to the mandible in chimpanzees (Figure 4b) and the lack of such consistency in humans (Figure 4a) shows that much more of the overall cranial variability in chimpanzees is associated with covariation in the mandible. This suggests that the chimpanzee skull is a much more integrated unit than that of humans, in which the occipital region departs strongly in the comparison of cranial relative warps to cranial singular warps. The human cranial singular warps (Figure 4a) are very similar to overall skull variation (Figure 2). These findings indicate that in humans, predictive value of mandibular shape variation is related more to the face than to overall craniofacial variation. A reduced predictive capacity is indicated by significantly decreased SVD-axis correlations (morphological integration) in the human cranio-mandibular system.

When human mandibular relative warps (Figure 5a) are compared to those of the fossil mandibles (Figures 5b, c), permutations of the first relative warps show that there are similar patterns of shape changes. This similarity is expressed at the inferior basal border and in the symphyseal shape variation patterns. We related such shape variation to patterns of dolicho- and brachyfacial growth and use the human patterns as a standard for the diagnosis of the fossils.

Figure 5a shows the human brachyfacial (left grid) and the human dolichofacial pattern (right grid) indicated by the position of the inferior basal border and the shape of the mandibular depression. A similar pattern of these features is repeated in the fossil mandibles (Figures 5b, c). But there are also differences. While in dolichofacial humans the corpus is rotated downward with respect to the ramus, increasing the angle and leading to a stronger formation of the preangular notch (Björk, 1969), no such change is observed in the "dolichofacial" fossils. We relate this peculiarity in humans again to evolutionary changes in facial architecture and reduced prognathism. The increased verticality of the human face requires different morphogenetic adjustments than the increased horizontality in the Neandertal and Atapuerca mandibles. A different basic architecture in the hominids may be reflected in variation in facial projection. The morphogenetic adjustment of the mandible is probably executed in an antero-posterior direction. In orthognathic humans, the mandibles must adjust mainly in a supero-inferior sense, leading to a more downward flexed mandible (*sensu* Björk, 1969).

Consequently, the "dolichofacial" fossil mandibles should be associated with crania of increased anterior facial lengths and more horizontal orientation—thus more prognathic—than the fossil crania of brachyfacial mandibles.

The present findings underline the systematic and morphogenetic importance of the comparative study of variation. More interspecific recent and fossil cranio-mandibular data in a broader comparative framework are necessary to better understand the morphological integration of the hominoid and hominid cranio-mandibular complex.

ACKNOWLEDGMENTS

We thank Dennis Slice for inviting us to contribute to this volume. We are grateful to Eugenia Cunha, Chris Stringer, and Paula Jenkins for accessing the collections; Miriam L. Zelditch for discussion and the Atapuerca Research Team. We also thank Callum Ross for his helpful comments on the manuscript. This research was partly founded by the NHM-London (SYS-Resources) and the Project No. BXX2000-1258-CO3-01, MCYT. MB is supported by a MCYT-grant.

REFERENCES

Ackermann, R. R., 2002, Patterns of covariation in the hominoid craniofacial skeleton: Implications for paleoanthropological models. *J. Hum. Evol.* 42:167–187.

Bastir, M., 2004, A geometric morphometric analysis of integrative morphology and variation in human skulls with implications for the Atapuerca-SH hominids and the evolution of Neandertals. Structural and systemic factors of morphology in the hominid craniofacial system, Doctoral Dissertation, Autonoma University of Madrid.

Bastir, M. and Rosas, A., 2004, Facial heights: Implications of postnatal ontogeny and facial orientation for skull morphology in humans and chimpanzees, *Am. J. Phy. Anthropol.* 123S:60–61.

Bastir, M., Rosas, A., and Kuroe, K., 2004, Petrosal orientation and mandibular ramus breadth: Evidence of a developmental integrated petroso-mandibular unit, *Am. J. Phys. Anthropol.* 123:340–350.

Bhat, M. and Enlow, D. H., 1985, Facial variations related to headform type, *Angle Orthod.* 55:269–280.

Björk, A., 1969, Prediction of mandibular growth. *Am. J. Orthod.* 55:585–599.

Bookstein, F. L., 1991, *Morphometric Tools for Landmark Data*, Cambridge University Press, Cambridge.

Bookstein, F. L., 1996, Combining the Tools of Geometric Morphometrics, in: *Advances in Morphometrics*, L. F. Marcus, ed., Plenum Press, New York, pp. 131–151.

Bookstein, F. L., Gunz, P., Mitteroecker, P., Prossinger, H., Schaefer, K., and Seidler, H., 2003, Cranial integration in *Homo*: Singular warps analysis of the midsagittal plane in ontogeny and evolution. *J. Hum. Evol.* 44:167–187.

Collard, M. and Wood, B., 2000, How reliable are human phylogenetic hypotheses? *Proc. Natl. Acad. Sci. USA* 97:5003–5006.

Chernoff, B. and Magwene, P. M., 1999, Afterword, in: *Morphological Integration*, E. C. Olson and P. L. Miller, eds., University of Chicago, Chicago, pp. 319–353.

Cheverud, J. M., 1996, Developmental integration and the evolution of pleiotropy, *Am. Zool.* 36:44–50.

Deane, A. and Begun, D., 2002, Please don't throw the baby out with the bath water: Skeletal characters in cladistic analyses of hominoid evolution, *Am. J. Phys. Anthropol.* S115:61.

Dryden, I. L. and Mardia, K. V., 1998, *Statistical Shape Analysis*, Wile, Chichester.

Efron, B. and Tibshirani, R. J., 1993, *An Introduction to the Bootstrap*, Chapman and Hall, New York.

Enlow, D. H., 1968, *Handbook of Facial Growth*, W.B. Saunders Company, Philadelphia, London, Toronto.

Enlow, D. H. and Hans, M. G., 1996, *Essentials of Facial Growth*, W.B. Saunders Company, Philadelphia, London, New York.

Hofer, H., 1952, Der Gestaltwandel des Schädels der Säugetiere und der Vögel, mit besonderer Berücksichtigung der Knickungstypen und der Schädelbasis, *Ver. Anat. Ges.* 99:102–126.

Kendall, D. G., 1977, The diffusion of shape, *Ad. Appl. Probabil.* 9:428–430.

Klingenberg, C. P., Badyaev, A. V., Sowry, S. M., and Beckwith, N. J., 2001, Inferring developmental modularity from morphological integration: Analysis of individual variation and asymmetry in bumblebee wings, *Am. Nat.* 157:11–23.

Lieberman, D. E., Ross, C., and Ravosa, M. J., 2000a, The primate cranial base: ontogeny, function, and integration, *Yrbk. Phys. Anthropol.* 43:117–169.

Lieberman, D. E., Pearson, O. M., and Mowbray, K. M., 2000b, Basicranial influence on overall cranial shape, *J. Hum. Evol.* 38:291–315.

Marcus, L. F., 1993, Some aspects of multivariate statistics for morphometrics, in: *Contributions to Morphometrics*, L. F. Marcus, E. Bello, and A. Valdecasas-Garcia, eds., Museo Nacional de Ciencias Naturales, Consejo Superior De Investigaciones Cientificas, Madrid, pp. 95–130.

Houle, D. and Mezey, J., 2002, Personal communication.

Olson, E. C. and Miller, R. L., 1958, *Morphological Integration*, The University of Chicago, Chicago.

Rohlf, F. J. and Slice, D., 1990, Extensions of the procrustes method for the optimal superimposition of landmarks, *Syst. Zool.* 39:40–59.

Rohlf, F. J., 1996, Morphometric Spaces, Shape Components and the Effects of Linear Transformations, in: *Advances in Morphometrics*, L. F. Marcus, ed., Plenum Press, New York, pp. 117–128.

Rohlf, F. J., Loy, A., and Corti, M., 1996, Morphometric analysis of old world Talpidae (Mammalia, Insectivora) using partial warp scores. *Syst. Biol.* 45:344–362.

Rohlf, F. J. and Corti, M., 2000, The use of two-block partial least-squares to study covariation in shape, *Syst. Zool.* 49:740–753.

Rosas, A., 1995, Seventeen new mandibular specimens from the Atapuerca/Ibeas Middle Pleistocene hominids sample (1985–1992), *J. Hum. Evol.* 28:533–559.

Rosas, A., 1997, A gradient of size and shape for the Atapuerca sample and Middle Pleistocene hominid variability, *J. Hum. Evol.* 33:319–331.

Rosas, A. and Bastir, M., 2002, Thin-Plate spline analysis of allometry and sexual dimorphism in the human craniofacial complex. *Am. J. Phys. Anthropol.* 117:236–245.

Rosas, A., Bastir, M., and Martinez-Maza, C., 2002a, Morphological integration and predictive value of the mandible in the craniofacial system of hominids: A test with the Atapuerca SH mandibular sample, *Coll. Antropol.* 26:171–172.

Rosas, A., Bastir, M., Martínez Maza, C., and Bermúdez de Castro, J. M., 2002b, Sexual dimorphism in the Atapuerca-SH hominids. The evidence from the mandibles, *J. Hum. Evol.* 42:451–474.

Rosas, A. and Bastir, M., 2004, Geometric morphometric analysis of allometric variation in the mandibular morphology from the hominids of Atapuerca, Sima de los Huesos site. *Anat. Rec.*, 278A:551–560.

Sheets, H. D., 2001, IMP, Integrated Morphometric Package: Sheets, H. David, Canisius College, Buffalo, NY. http:\\www.canisius.edu\~sheets\morphsoft.html.

Slice, D. E., 2001, Landmark coordinates aligned by Procrustes analysis do not lie in Kendall's shape space, *Syst. Biol.* 50:141–149.

Webster, M., Sheets, H. D., and Hughes, N. C., 2001, Allometric patterning in trilobite ontogeny: Testing for heterochrony in Nephrolenellus, in: *Beyond Heterochrony*, M. L. Zelditch, ed., New York. Wiley Liss, pp. 105–144.

Zelditch, M. L., Sheets, H. D., and Fink, W. L., 2000. Spatiotemporal reorganization of growth rates in the evolution of ontogeny, *Evolution* 54:1363–1371

Zollikofer, C. P. E., and Ponce de León M. S., 2002, Visualizing patterns of craniofacial shape variation in *Homo sapiens, Proc. Biol. Sci.* 269:801–807.

CHAPTER THIRTEEN

A Geometric Morphometric Analysis of Late Pleistocene Human Metacarpal 1 Base Shape

Wesley Allan Niewoehner

INTRODUCTION

This chapter describes the methodology I use to quantify some of the joint structures of Late Pleistocene[1] and recent human hands. Neandertal remains are important because they are the largest, most complete sample of archaic humans, and some regions of their skeletons contrast significantly with the skeletons of more recent humans. Neandertal hand remains have a suite of features, including indications of hypertrophied hand musculature and increased mechanical advantages across many joints, that makes them unique among Late Pleistocene humans. My own research focuses on the hand because

---

[1] The Late Pleistocene refers to the part of the Pleistocene epoch beginning 130,000 years ago and ending with the onset of the Holocene epoch 10,000 years ago. Neandertals, the approximately 100,000-year-old near-modern humans from Skhul and Qafzeh, and the modern humans of the European Upper Paleolithic all existed during this time-period.

Wesley Allan Niewoehner • Department of Anthropology, California State University—San Bernardino, 5500 University Parkway, San Bernardino, CA 92407.

Modern Morphometrics in Physical Anthropology, edited by Dennis E. Slice.
Kluwer Academic/Plenum Publishers, New York, 2005.

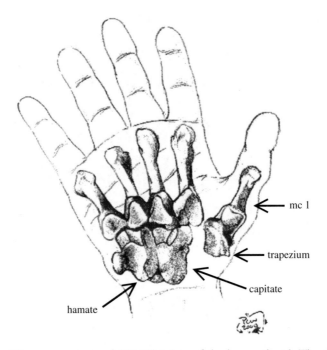

Figure 1. The carpometacarpal (CMC) joints of the human hand. The CMC joints comprise the distal carpal row (trapezium, trapezoid [not shown], capitate, hamate) and their articulations with the metacarpal bases. The metacarpal 1 articulates with the trapezium and forms a functionally important complex at the base of the thumb. The joints are not articulated in order to show their details.

Neandertal and recent human CMC^2 joints (Figure 1) have functionally significant differences in both shape and orientation. The anatomy of this region reflects (in part) the habitual levels and directions of force transmitted through the hand. The combined archaeological and anatomical evidence (Churchill, 1994; Dennell, 1983; Niewoehner, 2000, 2001) indicates there have been significant shifts in manipulative behavior during the Late Pleistocene of Europe. The functional analysis of the hand, in conjunction with behavioral inferences from the archaeological record, will continue to reveal the nature and timing of these behavioral shifts.

---

[2] Carpometacarpal; referring to the articulations between the distal carpals (trapezium, trapezoid, capitate, hamate) and the metacarpals.

MATERIALS

I collected landmark data on most of the available Neandertal, Skhul/Qafzeh, EUP[3], and LUP[4] nondeformed and nonpathological first metacarpal joint surfaces (Table 1). I also collected data on three Holocene human subsamples

Table 1.   Sample composition

Sample	Side
Recent human males ($N = 20$)[a]	11 Right , 9 Left
Recent human females ($N = 17$)[b]	11 Right, 6 Left
Neandertal ($N = 8$)	
Amud 1	Right
La Ferrassie 1	Left
La Ferrassie 2	Right
Kebara 2	Left
La Chapelle-aux-Saints	Right
Le Régourdou 1[c]	Left
Shanidar 4[c]	Left
Spy 2	Right
EUP ($N = 5$)	
Grotte des Enfants 4	Right
Paglicci	Left
Parabita	Right
Abri Pataud 163	Right
Abri Pataud 230	Left
LUP ($N = 7$)	
Arene Candide *principe*	Right
Arene Candide *tombe* 3	Right
Arene Candide *tombe* 5	Right
Barma Grande 3	Right
Bruniqel 24	Right
Ohalo 2	Right
Vado all' Arancio	Right
Skhul/Qafzeh ($N = 2$)	
Qafzeh 9	Right
Skhul 5	Right

Notes:
[a] 7 North American Urban, 5 Amerindian, 8 Mistihalj.
[b] 5 North American Urban, 5 Amerindian, 7 Mistihalj.
[c] Cast.

[3] Early Upper Paleolithic; referring to modern human remains in Europe dated to greater than 20,000 years ago.
[4] Late Upper Paleolithic; referring to Late Pleistocene modern human remains in Europe dated to less than 20,000 years ago.

that were selected to maximize between-sample differences in articular size, population activity level, and indicators of hand muscularity. When both sides were present and undamaged, lefts or rights were alternately selected during data collection. If both sides were present, but one was damaged, the best preserved side was used. The North American Urban subsample, part of an autopsied skeletal collection, and the Amerindian subsample, individuals from various Pueblo IV sites (1250–1600 A.D.) in the central Rio Grande Valley of New Mexico, are both curated at the University of New Mexico's Maxwell Museum. The Mistihalj subsample, curated at the Peabody Museum, Harvard University, is from a Yugoslavian Medieval cemetery. These three subsamples are pooled to form the recent human male and female samples used in the analysis (Table 1).

METHODS

$3D^5$ landmark coordinates, rather than caliper-derived measurements, are used for a geometric morphometric analysis in which biological shapes are ultimately visualized using computer software to determine how principal components and shapes are related. The methodology described here uses the first metacarpal base-shape as an example, but the same analytical methods have been applied to the entire CMC region (Niewoehner, 2000, 2001).

The first step, a Prucrustes superimposition, is followed by a principal components analysis of the registered landmark coordinates, and finally, by a canonical discriminant function of principal components scores. The use of landmark data requires special attention to the process of data collection since landmarks are points that must have the same meaning and same locations between specimens (Bookstein, 1991). These can be anatomical landmarks such as the intersection of sutures on a skull, the intersection of insect wing veins, or the cusps on teeth—all of which can be named and are readily identifiable on all specimens.

Creating Landmarks

Unfortunately, joint surfaces have few easily identifiable natural landmarks and fossils cannot be marked with artificial landmarks because they might be damaged. These problems were addressed by using a set of specially prepared slides and a slide projector to project the image of a 10×10 grid onto the joint surface.

---

[5] Three-dimensional.

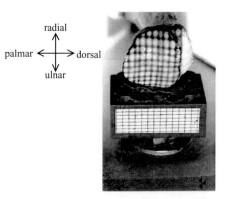

Figure 2. Protocol for projecting the grid onto the metacarpal 1 base. The projected grid is oriented with the gridlines parallel to the base dorsopalmar and radioulnar axes. The grid is scaled and proportioned to cover the maximum dorsopalmar height and radioulnar width of the metacarpal base. Note that the grid appears distorted due to the angle of the photograph.

The gridline intersections are easily identifiable, an essential requirement for digitizing; yet, they are also landmarks since each point on the grid on one specimen is the same point on the grid on another specimen. I also devised a protocol for projecting the gridlines onto the joint surface to ensure both that the grid covered the entire joint surface (by using an appropriately proportioned and scaled grid) and that the grid was oriented to the joint's dorsopalmar and radioulnar axes on all specimens to maintain the aforementioned landmark data requirements (Figure 2).

Data Collection

Actually touching the specimen with a digitizing arm, such as a Microscribe, results in movement of the specimen and alters the original grid position, so I used photogrammetry (the extraction of 3D information from two-dimensional [2D] images) because it does not require physical contact with the joint surface. After the grid is projected onto the joint surface, the specimen is photographed from three or more angles with a 35-mm camera having a 1 : 1 90-mm macro lens. The film negatives are scanned and imported into Photomodeler (EOS, 1993), a PC software photogrammetry program. The landmarks are manually digitized and the computer software calculates the 3D coordinates of each digitized landmark to within 0.1 mm.[6]

[6] More details are available in Niewoehner (2000).

Procrustes Superimposition

The data set is reduced to include only those landmarks in common to all specimens because some specimens have missing landmarks due to slight damage. This results in an array of 65 landmarks (Figure 3). The 3D landmark coordinates from all specimens are exported from Photomodeler into a spreadsheet program (Microsoft Excel) and formatted for use in the Morphologika computer program (O'Higgins and Jones, 1998). The landmark coordinates are initially registered through a generalized Procrustes analysis that removes translational and rotational differences between forms and then scales them to a common size to achieve a least-squares fit (Goodall, 1991; Gower, 1975; Rohlf and Slice, 1990). The problem here is that different patterns of form variation may be associated with different sets of reference points chosen as the basis for registration. This is especially true when comparing radically different shapes; however, when shapes are more similar to each other the choice of reference points has little effect on perceived patterns of shape variation (Dryden and

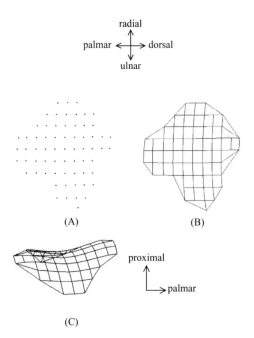

Figure 3. Sample mean metacarpal 1 base shape. (A) The 65 landmarks used in the Procrustes analysis to produce the grand sample mean shape (proximal view). (B) The wire frame of the same landmarks (proximal view). (C) The radial view of the wire frame.

Mardia, 1998). Since there are no a priori reasons for selecting a particular fixed subset of reference landmarks, configurations were registered using the best fit of all landmarks to each other.

Principal Components Analysis

The Procrustes mean shape provides a suitable pooled sample mean shape, but it is also of great interest to describe shape variability. A principal components analysis of the sample covariance matrix in Procrustes tangent space coordinates is considered the most effective means of analyzing the primary modes of variation in shape. Procrustes tangent space should not be confused with Kendall's shape space since generalized Procrustes analysis residuals lie in a hyperhemispherical space (Slice, 2001). The usefulness of principal components analysis in shape analysis is the same as in multivariate analysis because in both cases it can be used to reduce the dimensionality of the problem to a few variables (Dryden and Mardia, 1998).

The interpretation of shape variation along a principal component is accomplished solely by visualization of shape changes. The Morphologika software accomplishes this by reconstructing hypothetical forms along the principal component of interest. The mean landmark coordinates are added to the product of the principal component score of the hypothetical specimen and the eigenvectors for the principal component of interest to produce the new landmark coordinates. This morphing of the mean shape is viewed interactively by choosing a principal component and using a slider to change the principal component score. The wire frames presented in the results are the morphed Procrustes mean shape on the extremes of the principal component scores, and none of the morphed shapes discussed later are beyond the lowest and highest scores of the real specimens on the principal component being investigated. The polarity (negative or positive) of the principal components is entirely arbitrary; it only indicates direction and is a by-product of the type of analysis used here.

Canonical Discriminant Function

Each specimen's principal component scores are saved in another spreadsheet file and analyzed with a canonical discriminant function in the SAS statistical software package (SAS, 1989) to test the null hypothesis of between-sample

Table 2. Percent sample variance principal components 1 through 8

Principal component	Total variance (%)
1	26.4
2	20.1
3	13.3
4	8.9
5	6.8
6	3.9
7	3.7
8	2.5
Total	85.6

shape equivalence. The five classes consist of Neandertals, EUP, LUP, and recent human males and recent human females.[7]

The average eigenvalue of all the principal components is calculated, and only those principal components with eigenvalues equal to or greater than the average eigenvalue are retained for the analysis (Jobsen, 1992). Thus, only the first 8 of the 59 eigenvalues from the principal component analysis are retained, and these principal component scores (representing 85.6% of total sample variance [Table 2]) are the variates for the canonical discriminant function. Pooled within-group covariance matrices are used and only those canonical axes that display significant between-class variance ($p \leq 0.05$) are discussed in the results. Because five groups are included in the analysis, there are only four possible canonical axes.

Individuals were resubstituted back into the discriminant function and reclassified. The use of the same data to both construct and evaluate the discriminant functions gives optimistically biased nonerror rates (Dillon and Goldstein, 1984). Consequently, the reclassification results are used as a heuristic tool to help evaluate the degree of between-sample overlap in the discriminant function plot discussed in the results. Skhul V and Qafzeh 9 are not included in the discriminant function criteria; they are substituted without an assigned class, classified according to the discriminant function, and then forced into one of the preexisting classes. This helps to determine the degree to which they

---

[7] The North American Urban, Mistihalj, and Amerindian samples from Table 1 are pooled to obtain the recent male and recent female samples.

morphologically resemble one of the other classes and is not meant to infer any phylogenetic relationships; it only is an exercise in determining shape, and ultimately, functional similarities.[8]

Size and Shape

One additional concern is the degree to which size contributes to significant between-sample shape differences, because the Procrustes superimposition fixes only isometric shape differences (Dryden and Mardia, 1993, 1998). Tradition-ally, the study of allometry involves the fitting of linear or nonlinear regression equations between size and/or shape measures (Sprent, 1972). In a geometric morphometric analysis, one can use regressions of a shape coordinate variable on a size variable (Dryden and Mardia, 1998). The shape variable here is defined as a specimen's canonical score and size is defined as centroid size.[9] There are no significant residual size/shape correlations ($p \leq 0.05$) affecting the results.

RESULTS

Canonical Discriminant Function

Only the first canonical axis has significant between-sample variance (Table 3: p[F] axis $1 = 0.001$; p[F] axis $2 = 0.74$), and accounts for 80% of the total between-sample variance. The considerable overlap of the recent male and female samples (Figure 4) indicates their morphological similarity. Neandertals are clustered near the positive end of the first canonical axis with the other fossil specimens (EUP and LUP) falling into the overlap between the two extremes of shape variation. Skhul V and Qafzeh 9 fall within the range of recent human variation along the first canonical axis, but are also clearly within the LUP scatter.

Qafzeh 9 and Skhul V are both classified as LUP. After resubstitution, five Neandertals (63% of the sample) are correctly reclassified (Table 4). Spy 2 and La Ferrassie 1 are incorrectly reclassified as EUP, and one, Le Régourdou, as LUP. No Neandertals are reclassified as recent humans. All five EUP specimens are reclassified correctly, as is most of the LUP sample (58%; four specimens).

---

[8] The implications of the analysis of the Skhul/Qafzeh specimens are further considered in Niewoehner (2001).

[9] Centroid size is determined in the Procrustes superimposition, and is the square root of the summed squared Euclidean distances from each landmark to the centroid.

Table 3. Canonical axes percent variance and p values

Canonical axis	% variance	$P(F)$
1	80.0	0.001
2	13.4	0.738
3	6.1	0.933
4	0.5	0.996

Table 4. Canonical discriminant function reclassification results

	As Neandertal	As EUP	As LUP	As recent male	As recent female
Neandertal ($N = 8$)	5/63.0%	2/25.0%	1/12.0%	0/0.0%	0/0.0%
EUP ($N = 5$)	0/0.0%	5/100.0%	0/0.0%	0/0.0%	0/0.0%
LUP ($N = 7$)	1/14.0%	1/14.0%	4/58.0%	1/14.0%	0/0.0%
Recent male ($N = 20$)	1/5.0%	1/5.0%	2/10.0%	11/55.0%	5/25.0%
Recent female ($N = 17$)	0/0.0%	0/0.0%	4/23.5.0%	4/23.5.0%	9/53.0%

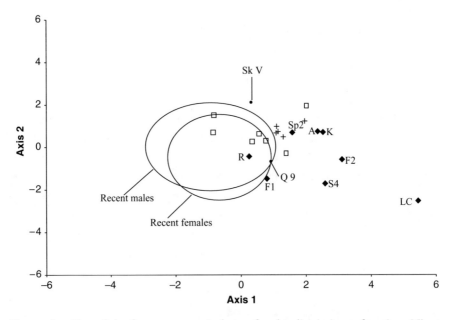

Figure 4. Plot of the first two canonical axes for the discriminant function. Ninety percent confidence ellipses encircle the recent male and female groups for the sake of clarity. Neandertals = ♦ (A, Amud 1; LC, La Chapelle-aux-Saints; F1, La Ferrassie 1; F2, La Ferrassie 2; K, Kebara 1; R, Le Régourdou; S4, Shanidar 4; Sp2, Spy 2); EUP = +; LUP = □; Qafzeh9 (Q9) and Skhul V (Sk V) are indicated.

Arene Candide *tombe* 3 and Arene Candide *principe* are misclassified as EUP and Neandertal, respectively, and Ohalo is misclassified as a recent human male.

Most of the recent humans are reclassified as recent humans (78% of the combined sample), although four recent females (23.5%) are reclassified as males and five males (25%) are reclassified as females. This is not a surprising result given their overlapping ranges of variation (Figure 4). One recent human male is reclassified as an EUP specimen and six recent humans (four females and two males) are reclassified as LUP specimens, while only one recent human (a male) is reclassified as a Neandertal.

Principal Components

The correlations between canonical coefficient scores and the principal component scores listed in Table 5 indicate that the classes are separated primarily by the linear combination of principal components two, three, and six. Shape changes along principal component two involve alterations in the relative dorsopalmar concavity-convexity of the articular surface (Figure 5). The extremes in articular shape vary from the condyloid surface of La Chapelle-aux-Saints that lacks any development of the palmar beak to the extremely concave surface of most of the recent human specimens that have well-developed palmar beaks. The third principal component is associated with a slight radial torsioning of the joint surface along the dorsopalmar axis as the palmar beak reduces in prominence (Figure 6); specimens with high positive scores have increased torsioning

Table 5. Canonical axis 1 coefficients and correlations for the discriminant function

Principal component	Canonical coefficient	Correlation between canonical score and principal component
1	0.377	0.17^a
2	−0.997	-0.73^b
3	0.972	0.53^b
4	0.699	0.19^a
5	−0.400	-0.13^a
6	0.804	0.26^b
7	0.647	0.18^a
8	0.120	0.03^a

Notes:
[a] $(p > 0.05)$.
[b] $(p \leq 0.05)$.

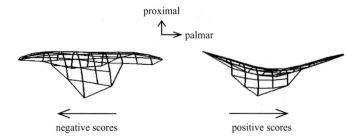

Figure 5. Metacarpal 1 base shape changes along the second principal component (ulnar view). Specimens with high positive principal component scores have dorsopalmarly convex bases. Those with high negative scores have a well-developed palmar beak, and hence, have dorsopalmarly concave bases.

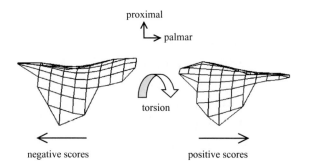

Figure 6. Metacarpal 1 base shape changes along the third principal component (ulnar view). Specimens with high positive principal component scores have a radial torsioning, or twisting, of the joint surface along the dorsopalmar axis and reduced dorsopalmar concavity.

and reduced palmar beaks. The sixth principal component represents changes in the radioulnar symmetry of the articular surface (Figure 7). Specimens with high positive scores have more symmetrically convex bases in the radioulnar direction, while those with high negative scores have flatter radial sides with a distinct ulnar shoulder.

Given the previously described shape contrasts, Neandertal first metacarpal bases fall to the positive end of the discriminant function because they have the following combination of traits: dorsopalmarly flat bases that lack palmar beak development that tend to be torsioned and more symmetrically convex in the radioulnar direction. Clearly, there is not a distinct break in morphologies between the recent human and fossil samples; rather there is a continuum of morphological variation; the Neandertals representing one extreme, the recent

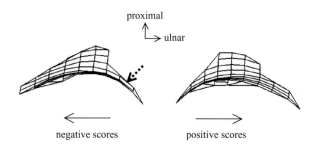

proximal

ulnar

negative scores   positive scores

Figure 7. Metacarpal 1 base shape changes associated with the sixth principal component (dorsal view). Specimens with high positive scores have more symmetrically convex bases in the radioulnar direction. Those with high negative scores have flatter radial sides with a distinct ulnar shoulder (stippled arrow).

humans representing the other. Qafzeh 9 and Skhul V, with their moderate palmar beak development, fit best with the LUP sample. The functional and behavioral interpretation of these shape differences is a separate issue that is discussed in Niewoehner (2000, 2001).

CONCLUSIONS

A crucial step in the interpretation of multivariate results is relating the principal components back to the biological shapes originally measured. An additional worry is that linear measurements cannot always capture every important aspect of shape variability. Unfortunately, the task of principal components interpretation becomes substantially more difficult when complex shapes, such as joint surfaces, are considered. On the other hand, the entire joint surface-shape was captured with 3D landmark coordinates and, as demonstrated here, the principal components results were used in a conventional multivariate analysis to test the null-hypothesis of between-sample shape equivalence, while the visualization of shape changes associated with the principal components removed any doubt of the nature of shape variation separating the Late Pleistocene samples from each other and from the recent human samples. Functional inferences generated from this type of geometric morphometric analysis of the CMC joints may lead to a better understanding of the evolution of human manipulative behavior during the Late Pleistocene (Niewoehner, 2000, 2001).

REFERENCES

Bookstein, F. L., 1991, *Morphometric Tools for Landmark Data: Geometry and Biology*, Cambridge University Press, Cambridge.

Churchill, S. E., 1994, *Human Upper Body Evolution in the Eurasian Later Pleistocene*, University of New Mexico, Albuquerque.

Dennell, R., 1983, *European Economic Prehistory: A New Approach*, Academic Press, New York.

Dillon, W. R. and Goldstein, M., 1984, *Multivariate Analysis: Methods and Applications*, Wiley and Sons, New York.

Dryden, I. L. and Mardia, K. V., 1993, Multivariate shape analysis, *Sankya* 55(A):460–480.

Dryden, I. L. and Mardia, K. V., 1998, *Statistical Shape Analysis*, John Wiley and Sons, London.

EOS Systems, 1993, Photomodeler Pro Computer Software, ver. 3.1. EOS Systems, Inc., Vancouver.

Goodall, C. R., 1991, Procrustes methods and the statistical analysis of shape, *J. R. Stat. Soc. Br.* 53:284–340.

Gower, J. C., 1975, Generalized Procrustes analysis, *Psychometrika* 40:33–50.

Jobsen, J. D., 1992, *Applied Multivariate Data Analysis*, Springer-Verlag, New York.

Niewoehner, W. A., 2000, *The Functional Anatomy of Late Pleistocene and Recent Human Carpometacarpal and Metacarpophalangeal Articulations*, University of New Mexico, Albuquerque.

Niewoehner, W. A., 2001, Behavioral inferences from the Skhul/Qafzeh early modern human hand remains, *PNAS* 98(6):2979–2984.

O'Higgins, P., and Jones, N., 1998, Morphometrika, Tools for Shape Analysis Computer Software, University College, London.

Rohlf, F. and Slice, D. E., 1990, Extension of the Procrustes method for the optimal superimposition of landmarks, *Syst. Zool.* 39:40–59.

SAS, 1989, SAS Computer Software, Carey, North Carolina.

Slice, D. E., 2001, Landmark coordinates aligned by Procrustes analysis do not lie in Kendall's shape space, *Syst. Biol.* 50:141–149.

Sprent, P., 1972, The mathematics of size and shape, *Biometrics* 28:23–37.

CHAPTER FOURTEEN

A Geometric Morphometric Assessment of the Relationship between Scapular Variation and Locomotion in African Apes

Andrea B. Taylor and Dennis E. Slice

INTRODUCTION

Investigators have long sought to establish the functional correlates of scapular variation in primates and other mammals (Anapol, 1983; Corruccini and Ciochon, 1976; Fleagle, 1976; Inouye and Shea, 1997; Inouye and Taylor, 2000; Larson, 1995; Leamy and Atchley, 1984; Oxnard, 1963, 1968, 1977; Roberts, 1974; Schultz, 1930, 1934; Shea, 1986; Smith et al., 1990; Stern and Susman, 1983; Swiderski, 1993; Taylor, 1992, 1997; Taylor and Siegel, 1995). In primates, morphological variation in scapular form has been frequently

Andrea B. Taylor • Doctor of Physical Therapy Division, Department of Community and Family Medicine, Duke University School of Medicine, and Department of Biological Anthropology and Anatomy, Duke University Medical Center, Box 3907, Durham, NC 27710. **Dennis E. Slice** • Institute for Anthropology, University of Vienna, Althanstrasse 14, A-1091 Vienna, Austria.

Modern Morphometrics in Physical Anthropology, edited by Dennis E. Slice. Kluwer Academic/Plenum Publishers, New York, 2005.

associated with the degree to which animals use their forelimbs in overhead
suspensory behaviors involving tensile forces, as compared to animals that
recruit both forelimbs and hindlimbs in behaviors primarily involving compressive forces (e.g., Oxnard, 1963, 1968). Napier and Napier's (1967) gross
locomotor classification largely reflects this dichotomy, as do more fine-grained
categorizations (Hunt et al., 1996).

A number of theoretical biomechanical models (Roberts, 1974; Schultz,
1930), some bolstered by electromyographical (EMG) data (Larson and Stern,
1986, 1989; Larson et al., 1991), have been advanced to explain the relationship between scapular form and locomotor variation. For example, the
ratio of scapular length to scapular breadth (the scapular index; Schultz, 1930)
(Figure 1) has been functionally linked to degree of overhead forelimb suspensory behavior (Coolidge, 1933; Roberts, 1974; Schultz, 1930; Shea, 1986).
Roberts (1974) suggested that among hominoids, the high scapular index
observed in hylobatids reflects their "slender and leverlike" scapulae, which
provide for greater range of forelimb motion. Thus, high index values, reflecting relatively longer scapulae (longer parallel to the scapular spine, and narrower
perpendicular to the spine) (Schultz, 1930), are typically found in accomplished
brachiators like gibbons and siamangs (Figure 1).

Among the large-bodied hominoids, a relatively well-developed supraspinous
fossa, and by implication, supraspinatus muscle, has been linked to relatively
large, heavy, powerful limbs, and the need for glenohumeral joint stabilization
during humeral elevation (Larson and Stern, 1986; Roberts, 1974). Roberts

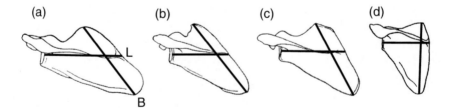

Figure 1.   Scapulae of: (a) a gibbon (*Hylobates agilis*); (b) a chimpanzee (*Pan troglodytes*); (c) gorilla (*Gorilla gorilla gorilla*); and (d) modern human. All are male scapulae,
reduced to the same height and orientation (adapted from Schultz, 1930; see also Shea,
1986). The scapular index (Schultz, 1930) is based on the generalized mammalian condition, where the longest dimension is defined parallel to the scapular spine (scapular
length, L) and scapular breadth (B) is roughly orthogonal to scapular length. Note that
in the modern human, scapular length is shorter than scapular breadth.

(1974) observes that the area of the supraspinous fossa tends to increase, relative to the length of the scapula, as the propensity for quadrumanous climbing and arm-swinging increases. Among hominoids, *Gorilla* and *Pan* possess relatively broad supraspinous fossae (Roberts, 1974; Schultz, 1930). The relative proportions of the supraspinous and infraspinous fossae have been similarly shown to vary among hominoids (Schultz, 1930). *Pongo* has the relatively broadest infraspinous fossa (Schultz, 1930), which Roberts (1974) associated with their quadrumanous climbing behavior. Gorillas, however, have relatively broader infraspinous fossae compared to *Pan*, which may reflect the former's need for greater shoulder joint stabilization (Roberts, 1974). Finally, among hominoids, gorillas (but also humans) have been shown to have the relatively longest scapular spines (Roberts, 1974). In primates, a well-developed scapular spine and acromion process have been functionally linked to both arm-swinging and vertical climbing behaviors because these features are presumed to improve the mechanical leverage for the trapezius and deltoid muscles (Roberts, 1974; Takahashi, 1990).

One of us (Taylor, 1997) has previously evaluated some of these biomechanical models in an ontogenetic, allometric comparison of scapular form in two geographically isolated and behaviorally differentiated populations of *Gorilla*. Compared to the eastern mountain gorilla (*Gorilla gorilla beringei*), the west African lowland gorilla (*Gorilla gorilla gorilla*) was predicted to exhibit features consistent with their purported greater degree of arboreality and associated higher frequencies of climbing and suspensory behaviors (Doran, 1996; Remis, 1994, 1995, 1998). Using conventional linear dimensions obtained with digital calipers, Taylor (1997) observed few differences between *G.g. gorilla* and *G.g. beringei* that could be convincingly linked to degree of suspensory behavior. Specifically, mountain gorillas displayed relatively longer scapular spines and shorter scapulae as compared to western lowland gorillas at comparable lengths of the superior borders of the scapulae. However, at comparable body weight estimates (derived from dry skeletal weight), mountain gorillas displayed relatively shorter scapular spines and superior borders. Taylor (1997) suggested that differences in scapular length relative to scapular size (but not skeletal weight) may be related to variation in thoracic shape and scapular positioning. Only the relatively longer spine in western lowland gorillas fits the prediction linking a well-developed scapular spine with increasing frequency of suspensory and vertical climbing behaviors. This lack of a clear link between scapular form and function in African apes is consistent with previous studies by Shea (1986) for *Pan*, Inouye and Shea (1997) for hominoids, Taylor (1997) for *Gorilla*, and Inouye and Taylor (2000) for the African apes.

As evidenced throughout this volume, landmark-based geometric morphometrics (GM) has become an important component of anthropological shape studies. Traditional measures, such as linear distances, indices, or angles, can be computed from the coordinates of the landmarks by which they are defined. Furthermore, the information for all possible such measures on a set of landmarks is contained within the coordinates of those landmarks, and the analysis of only a subset of these variables represents an intentional disregard of at least some of the information available (see Chapter 1). This conservation of geometric information that characterizes GM also makes possible graphical visualizations of shape differences and variation that are difficult, if not impossible, to achieve with distance or angular data (e.g., Bookstein, 1991).

An important initial step in GM is generalized Procrustes analysis (GPA) (Gower, 1975), wherein sets of landmark coordinates obtained from individual specimens are superimposed onto an iteratively computed mean configuration to achieve registration within a common coordinate system (see Chapter 1). The residuals from the grand mean are then subjected to multivariate statistical analyses. This approach has proven to be powerful and robust compared to extensions of the analysis of angular or distance-based variables designed to address the loss of information in most traditional analyses (Rohlf, 2000a, b, 2003).

The retention of geometric information, statistical performance, and the ability to generate informative and intuitive graphical displays of results make GPA a highly desirable tool for investigations of shape differences and shape variation when appropriate landmarks are available. Even in biomechanical studies in which linear distances or angles can have direct, meaningful relationships to specific mechanical models, GM, in general, and GPA, in particular, may allow for the identification of comparable and/or additional features of shape variation that could similarly be related to model components.

In this chapter, we investigate this possibility by using GM methods to analyze scapular shape variation in three African ape species that reflect an axis of increasing arboreality and arboreal-specific behaviors such as suspensory activities (Doran, 1996, 1997; Remis, 1995, 1998; Tuttle and Watts, 1985). Common chimpanzees (*Pan troglodytes*) and Virunga mountain gorillas (*Gorilla gorilla beringei; sensu* Groves, 1967; cf. Groves, 2001) represent the extreme ends of the locomotor distribution. Chimpanzees display the greatest propensity for suspensory behaviors, while mountain gorillas are essentially dedicated terrestrial, quadrupedal knuckle-walkers; as adults, Virunga mountain gorillas spend approximately 96% of their locomotor behavior in quadrupedal activities and approximately 91% of their time is spent locomoting on the ground (data averaged for males and females; Doran, 1997). *G.g. gorilla*, the western lowland gorilla, is intermediate in this regard, engaging in more suspensory activities than Virunga mountain gorillas, and more quadrupedal climbing and scrambling behaviors than either chimpanzees (with the exception of male *P.t. verus* from Täi) or mountain gorillas (Doran, 1996).

As noted by previous investigators (e.g., Doran, 1996; Remis, 1994), integrating behavioral data on locomotion across multiple studies can be problematic. This is primarily because some data were collected on unhabituated animals, which means that such observations will tend to be biased toward activities in an arboreal setting. We agree with Doran (1996) when she states that "there are adequate qualitative data to rank confidently all the African apes on their degree of arboreality," and we rely on this qualitative ranking as the framework for evaluating scapular variation as a function of arboreal climbing and suspensory behaviors amongst these African ape taxa. Nevertheless, the potential for biased estimates of arboreal activities should be kept in mind when interpreting our results.

Finally, as in previous studies of form–function relationships (Inouye and Taylor, 2000; Taylor, 2002), we follow the conservative approach of accepting that a particular regional difference in form is functionally linked to a locomotor behavior only if that shape differs consistently in the predicted direction for comparisons within *Gorilla*, and between *P. troglodytes* and *Gorilla*. We test predictions derived from various biomechanical models and compare our results to those previously obtained using traditional distance measurements.

MATERIALS AND METHODS

Most of the material for this study derives from the collection of the National Museum of Natural History in Washington, DC, USA. The specimens are all adults as judged by a combination of dental eruption and basilar suture fusion, and include right or left scapulae from west African lowland gorillas, *G.g. gorilla* (5 males, 2 females), east African mountain gorillas, *G.g. beringei* (8 males, 5 females), and common chimpanzees, *P. troglodytes* (7 males, 10 females). The west African subspecies (*P.t. verus*) is not included in this

Table 1.   Sample composition

	Total N	Males	Female	Sex ratio (m/f)
Gorilla gorilla gorilla	17	8	9	0.89
Gorilla gorilla beringei	17	8	9	0.89
Pan troglodytes	17	7	10	0.70

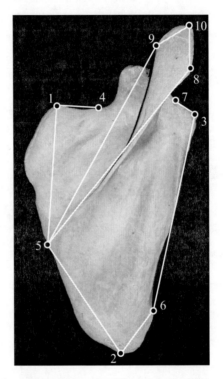

Figure 2.   Landmarks used in this study. White lines indicate graphical links used to aid visualization of the scapula in plots. Numbering system is the same as that of Table 2. See text and Table 2 for details.

study. We supplemented these samples with data randomly drawn from Taylor's (1992) extensive collection of images of gorilla scapulae to achieve equal sample sizes and approximately equal sex ratios: 3 males and 7 females of *G.g. gorilla* and 4 females of *G.g. beringei*. The relatively limited sample sizes used in this study required pooling of the sexes. The final sample composition is shown in Table 1.

This study was done using 10 landmarks (Figure 2 and Table 2). These landmarks were chosen because they were presumed to be representative of

Table 2. Landmark numbers (see Figure 2) and definitions

Number	Name/definition
1	Superior angle of the scapula
2	Inferior angle of the scapula
3	Most inferior point on glenoid fossa
4	Base of superior notch of scapula
5	Intersection of scapular spine and medial border
6	Distal end of the lateral expansion of the subscapular fossa
7	Midpoint of the glenoid fossa
8	Lower intersection of scapular spine and acromion process
9	Upper intersection of scapular spine and acromion process
10	Tip of scapular spine

scapular shape (Roberts, 1974; Schultz, 1930; Shea, 1986; Taylor, 1997) and, at least, operationally homologous. Scapulae were oriented with the scapular blade parallel to a flat surface and photographed using either a digital (Olympus D-520 Zoom for the NMNH specimens) or an analog (Nikon FM2 in the case of the archival images) camera (see Taylor [1992, 1997] for additional details).

Coordinates of the landmarks were recorded using tpsDig (Rohlf, 2001). GPA was carried out in Morpheus et al. (Slice, 1998) with the option "allow-reflections" set to "ON" to appropriately reflect scapulae as needed. Mean coordinates from separate group GPAs were substituted for the small number of missing (obscured by the spine) landmarks (landmark 7 in one *G.g. gorilla* and one *G.g. beringei*, and landmark 4 in one *P. troglodytes*). Randomization tests for shape differences were carried out in Morpheus et al. after super-imposition of the entire sample onto its GPA-estimated mean configuration. In these tests, the total, within-group, and between-group cross-product matrices were computed for the entire superimposed data set. The trace of the between-group matrix was compared to the same value computed for 999 random relabelings of group membership. The reported *p*-values are the proportion of samples (original plus randomizations) with a between-group trace equal to or greater than that of the original data. Centroid size (CS) (see Chapter 1), the "scale $(1/r)$" parameter saved from the superimpos-ition in Morpheus et al., was similarly tested. Principal components (PC) analysis and general plotting were done using the R statistical computing package (http://www.r-project.org). Thin-plate spline plots were produced in Morpheus et al. Illustrations of results are arbitrarily presented as right scapulae.

RESULTS

Centroid size in the two gorilla subspecies is bimodally distributed due to the admixture of sexes, while those of the chimpanzees are somewhat leptokurtic as judged from Q–Q plots and histograms. Randomization tests showed that the samples of lowland (*G.g. gorilla* min/mean/max CS mm = 242.0/287.0/341.7) and mountain (*G.g. beringei* = 234.5/296.6/356.8) gorillas do not differ significantly from each other in this measure of size (p = 0.452). Chimpanzees (*P. troglodytes* = 186.0/207.0/224.1), on the other hand, differed significantly in size from both gorilla subspecies ($p_{G.g.g.}$ = 0.001, $p_{G.g.b.}$ = 0.001). Not surprisingly, chimpanzees do not overlap in centroid size with either of the gorilla subspecies.

Figure 3 shows the results of the GPA of the entire sample. The pattern of residuals suggests no obvious within-group structure (though some degree of

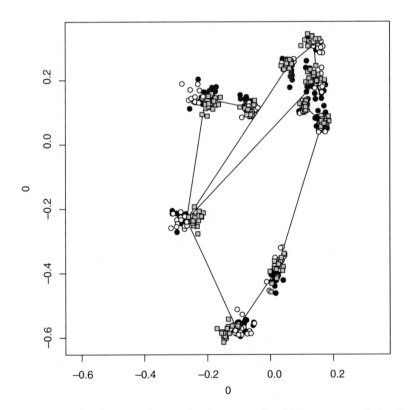

Figure 3. Scapular data superimposed using generalized Procrustes analysis. Open circles are *G.g. gorilla*, solid circles are *G.g. beringei*, and gray squares are *P. troglodytes*.

sexual dimorphism is likely present; see Chapter 1), but it does suggest some between-group differences. Randomization tests confirm significant differences between groups at the limit of the power of the test ($p = 0.001$). Furthermore, differences between every pair of groups are equally significant by the same test (Table 3). All differences are significant, but there is a suggestive pattern in the proportion of variation associated with group membership. Between-group variation accounted for nearly one third of the total sample variation between each of the gorilla samples and the chimpanzees. Group membership, however, accounted for less than 20% of the differences in Procrustes residuals between the two gorilla groups. These results suggest a greater degree of resemblance of scapular shape between gorilla subspecies than between gorillas and chimpanzees (Table 3).

A principal components analysis (PCA) of the total sample variation of the Procrustes residuals found about 76% of total shape variation to be associated with the first four PCs—33, 26, 11, and 7%, respectively. Projections of the data onto these components are shown in Figure 4. Only the first two show any apparent separation between the three groups, with the first distinguishing the chimpanzees from the gorillas and the second partially separating the two gorilla subspecies. The samples largely overlap on PC3 and 4, as they do on all higher PCs.

A major benefit of the GM methods is realized in Figure 5, which shows the differences between the three groups in mean shape found to be significant by the randomization tests. Differences are shown without magnification as thin-plate splines mapping the mean landmark locations of *G.g. gorilla* onto the mean of *G.g. beringei* (Figure 5a) and a similar mapping of the mean of the *P. troglodytes* sample onto the mean configurations of *G.g. gorilla* (Figure 5b) and *G.g. beringei* (Figure 5c).

Table 3.   *P*-values (below diagonal) and proportion of total variance accounted for by group membership (above diagonal) from pairwise randomization testing of group differences in scapular shape. All *p*-values are at the limit of test resolution, $p = 0.001$

	G.g. beringei	G.g. gorilla	P. troglodytes
G.g. beringei	—	0.18	0.31
G.g. gorilla	0.001	—	0.34
P. troglodytes	0.001	0.001	—

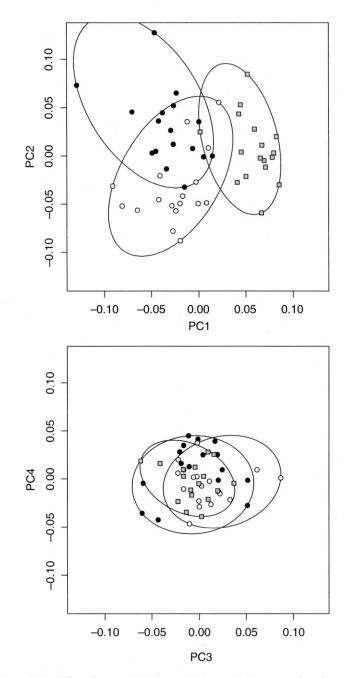

Figure 4. Plots of first four principal components (PC) scores for the generalized Procrustes analysis (GPA) superimposed data. Open circles are *G.g. gorilla*, solid circles are *G.g. beringei*, and gray squares are *P. troglodytes*. Ellipses are ellipsoidal hulls—minimum-area ellipses just containing all of the members of a particular group.

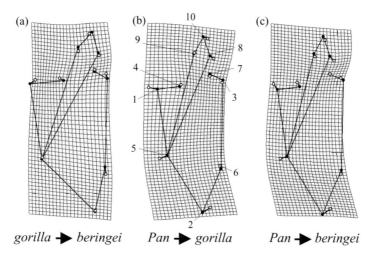

Figure 5. Thin-plate splines showing deformations of group means: (a) *G.g. gorilla* to *G.g. beringei*; (b) *P. troglodytes* to *G.g. gorilla*; and (c) *P. troglodytes* to *G.g. beringei*. The splines show the planar deformation mapping the landmarks of the reference specimen (solid circles with lines to aid visualization) onto those of the target specimen (open circles).

The mapping of *gorilla* to *beringei* indicates that gorillas share similar scapular (landmarks 5 and 7) and infraspinous fossa (landmarks 5 and 2) lengths (Figure 5a). Spine length (landmarks 5 and 10) appears to be slightly shorter in *beringei*, and there are minor shifts in the positioning of the acromial and glenoid fossa landmarks. The two regions that show the greatest degree of difference are the superior border (landmarks 1 and 4) and the axillary border (landmarks 3 and 6), which reflects, to some extent, the length of the lateral expansion of the subscapular fossa (Larson, 1995). The reduction in distance between landmarks 1 and 4, noted by the compression in the spline grids, suggests that *beringei* has relatively shorter superior scapular borders. The region between landmarks 3 and 6 appears expanded in *beringei* compared to *gorilla*, suggesting a relative elongation of the lateral expansion of the subscapular fossa in *beringei*.

The degree of difference, as reflected in the thin-plate splines, is greater between *P. troglodytes* and gorillas than between gorilla subspecies, consistent with the PCA results (Figures 4 and 5b–c). There are a number of similarities in the mappings between *P. troglodytes* and the two gorilla subspecies. For example, both mappings show a lateral displacement of landmark 8, indicating a relatively wider acromion process in both *G.g. gorilla* and *G.g. beringei* as compared to *P. troglodytes*. Both mappings show a medial displacement of land-mark 1, suggesting a relative elongation of the superior border (landmarks 1 and 4), though this elongation is greater in *gorilla* than *beringei* because of the lateral displacement of landmark 4 in the former. Landmark 5 is displaced medially and landmark 2 displaced superolaterally in both gorilla subspecies. These displacements correspond to a relatively (a) longer scapula (landmarks 5 and 7); (b) longer scapular spine (landmarks 5 and 10); and (c) mediolat-erally expanded (landmarks 5 and 6) and vertically compressed (landmarks 6 and 2) inferior region of the infraspinous fossa in gorillas. The combined medial displacements of landmarks 1 and 5 additionally suggest a relative elongation of the supraspinous fossa in *gorilla*, and to a lesser extent, in *beringei*.

Differences in the mappings between *P. troglodytes* and the two gorilla sub-species involve landmarks 3, 4, 6, 7, and 9 (Figures 5b–c). For example, the axillary border (landmarks 3 and 6) is slightly shorter in *gorilla*, yet slightly longer in *beringei*, as compared to *P. troglodytes*. In the *P. troglodytes* to *beringei* mapping, landmarks 1 and 4 are shifted superiorly as well as medially. By con-trast, in *gorilla*, landmark 1 is shifted laterally and landmark 4 is shifted medially, with little superior displacement of either.

DISCUSSION

Functional Correlates of Scapular Variation

To summarize our morphological findings, there are differences in scapular size and shape between *P. troglodytes* and the two gorilla subspecies, but shape alone differs between gorilla subspecies. Degree of shape difference is greater between *P. troglodytes* and gorillas than between gorilla subspecies. The differences in scapular size (as reflected by CS) between *P. troglodytes* and *Gorilla* suggest that some of the variation in scapular shape may be allometric; this has been shown to be the case in previous ontogenetic, allometric studies of scapular shape in the African apes using different measures of size (Inouye, 2003; Inouye and Taylor, 2000; Shea, 1986; Taylor, 1997).

Compared to *P. troglodytes*, gorillas exhibit to varying degrees relative increases in lengths of the superior scapular border, scapula, and spine, a mediolateral expansion and superoinferior compression of the inferior portion of the infraspinous fossa, and a relatively wider acromion process (Figures 5b–c). *G.g. beringei* has a relatively shorter superior border, an elongated axial border, and slightly reduced scapular spine compared to *G.g. gorilla* (Figure 5a). Apart from those differences, scapular shape in gorillas is quite similar.

Overall, GM methods provide results that are consistent with previous studies of scapular variation in the African apes (Inouye, 2003; Inouye and Shea, 1997; Inouye and Taylor, 2000; Shea, 1986; Taylor, 1997). When scaled for CS, we observe some differences in scapular shape, both between *P. troglodytes* and the gorillas, and between gorilla subspecies. There is, however, no apparent systematic or consistent pattern of differentiation across taxa that fits the predictions associated with frequency of suspensory or vertical climbing behaviors (compare *Pan → gorilla* and *gorilla → beringei* mappings in Figure 5).

For example, a well-developed scapular spine and acromion process have been hypothesized to be mechanically advantageous in both arm-swinging and vertical climbing behaviors (Roberts, 1974; Takahashi, 1990). *G.g. beringei* appears to have a marginally shorter scapular spine as compared to *gorilla*, accompanied by slight changes in acromial width (see also Taylor, 1997). However, contrary to expectations, gorillas are shown to have relatively longer scapular spines and wider acromial processes as compared to *Pan*.

If a well-developed spine is functionally linked to arm-swinging behavior, then we would expect chimpanzees to exhibit the relatively longest, and mountain gorillas the relatively shortest, scapular spines. Conversely, if a well-developed spine is functionally linked to vertical climbing behavior, western lowland gorillas should have the relatively longest, and mountain gorillas the relatively shortest, scapular spines. Neither of these expectations is borne out by the data. It is worth pointing out that although the two gorilla subspecies exhibit relatively longer spines as compared to *P. troglodytes*, this difference is largely reflected in the position of the spine as it meets the vertebral (medial) border of the scapula (landmark 5) and not as an extension of the spine relative to the glenoid fossa. It seems clear from Roberts (1974) that enhanced leverage of the deltoid and trapezius muscles is functionally linked to elongation of the spine relative to the glenoid fossa. Thus, differences in spine length observed in this study may not reflect the mechanical advantages hypothesized previously.

It has been previously argued (Coolidge, 1933; Roberts, 1974) that a relatively longer and narrower scapula (parallel and perpendicular to the scapular spine, respectively) is functionally linked to greater frequency of suspensory behaviors. Thus, chimpanzees, which engage in higher frequencies of overhead suspensory behaviors than either subspecies of gorilla, should have the relatively longest and narrowest scapulae. Our results show that, on the contrary, gorillas have relatively longer scapulae (landmarks 5 and 7) and vertically compressed infraspinous fossae (landmarks 2 and 6) as compared to *P. troglodytes* (Figures 5b–c). While *P. troglodytes* exhibits the expected relatively narrower supraspinous fossa as compared to gorillas (landmarks 1, 5, and 7; Figures 5b–c), this difference is not observed in comparisons between gorilla subspecies, which runs contrary to expectations based on a higher frequency of suspension during feeding (Remis, 1994) and locomotion (Doran, 1996; Remis, 1998) in western lowland gorillas.

One last example of how scapular variation deviates from predictions based on locomotor behavior involves differences along the axillary border (landmarks 3 and 6). Elongation between landmarks 3 and 6 may reflect a relative increase in the length of the lateral expansion of the subscapular fossa (in this study, this dimension is measured from the inferior glenoid to the distal end of the lateral expansion of the subscapular fossa; cf. Larson, 1995). EMG studies (Larson and Stern, 1986) have demonstrated that in chimpanzees, the subscapularis muscle is important during the support (or "pull-up") phase of vertical climbing, and length of the lateral expansion of the subscapular fossa has been hypothesized (Larson, 1995) to reflect the relative importance of climbing during locomotion. Western lowland gorillas (*G.g. gorilla*) incorporate considerably higher frequencies of vertical climbing as compared to Virunga mountain gorillas (Doran, 1996). Thus, if vertical climbing is a behavior that is functionally linked to a longer lateral expansion of the subscapular fossa, we would predict that *G.g. gorilla* would show a relative expansion between landmarks 3 and 6. On the contrary, the thin-plate splines indicate that *beringei* has a relatively longer lateral expansion than *gorilla* (Figure 5a).

We note that findings from this study do not preclude the possibility that scapular morphology is functionally linked to locomotor behavior. We also point out that sample sizes were relatively small, and we did not test all of the variables that have been hypothesized to confer mechanical advantages related to locomotor specialization (see Larson, 1995 for example). That said, we emphasize that no single morphology examined here differed in the predicted direction in any set of pairwise comparisons among taxa.

Linear Dimensions vs Coordinates

The application of GM to evaluate the locomotor correlates of scapular variation, both here and elsewhere (e.g., Taylor and Siegel, 1995), produces results that are compatible with those previously obtained through bivariate and multivariate analyses of traditional linear distances and angles. These findings indicate that GM, specifically GPA, is at least as robust as more conventional methods of analysis in identifying regional shape variation. One appeal of the Procrustes method is that it retains all of the geometric information contained within the coordinate locations throughout an analysis, thereby producing biologically intuitive graphic representations that display the spatial organization of changes in shape in an integrated fashion. These visual descriptions of entire structures contribute to our interpretations of change in shape in ways that differ from plots of ratios. In this way, the two approaches are mutually informative.

There are a number of additional theoretical advantages of coordinate-based Procrustes methods over analyses of limited subsets of distances and angles based on the same set of landmarks. Recent simulation studies (Rohlf, 2000a, b, 2003), for example, have shown that for simple models of triangles and isotropic error, the Procrustes methods are better behaved (do not introduce artifactual structure into the data), have better statistical power, and have less systematic bias in mean shape estimates than extensions of the analysis of traditional measures designed to address their inherent shortcomings, such as the use of geometrically sufficient angles or distances or EDMA (see Chapter 1).

In this study, these additional theoretical advantages do not appear to be fully realized, since GPA and bivariate and multivariate analyses of conventional linear dimensions produce comparable results. In other words, no newly recognized differences in shape are recovered that were obscured or disregarded (intentionally or otherwise) in previous studies of linear distances, nor are systematic patterns of scapular variation revealed that map predictably to locomotor differences. Investigations of biomechanical models of form and function, however, may not inherently favor Procrustes methods of analysis of landmarks over bivariate and multivariate analyses of distance measurements. Biomechanical models of form and function frequently involve estimates of bony regions that do not lend themselves to recognizably homologous landmarks, and/or estimates of moment arms, lever arms, and structural changes in bony morphology that are hypothesized to provide resistance to internal and external loads. These types of analyses often explicitly isolate distance, angle, or areal measurements that are considered to be of primary biological importance. In such cases, GM methods, like other multivariate methods of analysis, provide an important first-step in identifying areas of shape variation. These areas can then be further dissected using measurements appropriate to questions being addressed, to determine whether such differences fit *a priori* expectations of patterns and degrees of difference based on theoretical biomechanics, comparative studies, and/or other criteria.

The unsatisfactory relationship between scapular form and locomotor behavior in the African apes has been emphasized by others (e.g., Jungers and Susman, 1984). Some may suggest that behavioral differences between gorilla subspecies, or even between gorillas and chimpanzees, are too subtle to reflect functional differences related to locomotion. Others (e.g., Larson, 1995) have noted that hominoids as a group probably retain morphological traits of the shoulder and forelimb that are inherited from a common ancestor, but which still confer a mechanical advantage. Alternatively, the "resultant or average biomechanical situation" (Oxnard, 1979) may be so overwhelmingly similar as to obscure subtle differences, particularly if certain behaviors or biological functions involving the generation of large forces have a disproportionate influence on the total morphological pattern over those functions that involve relatively small forces (Oxnard, 1972). It seems reasonable, however, to expect that scapular variables that have clear-cut functional links to locomotor behavior would systematically and predictably distinguish between arm-hanging, suspensory chimpanzees and the specialized terrestrial knuckle-walking Virunga mountain gorillas. Others have argued, and we agree, that our confidence that a particular feature of the scapula is indeed correlated with a particular locomotor behavior is greater when the relationship can be demonstrated in all populations that bear that particular trait (Bock, 1979; Kay and Cartmill, 1977; Kay and Covert, 1984) at both micro- and macroevolutionary levels (Arnold, 1983).

In applying "modern" morphometric methods to biological questions of form and function, it may be worthwhile to consider the cautionary advice offered a quarter century ago by Oxnard (1972), whose mathematical methods and attempts to quantify Thompson's (1942) Cartesian coordinate grids provide some of the foundation for current GM methods. 'Oxnard states, "however careful are the techniques of data collection, and however complicated the methods of analysis, new insights into morphology cannot be expected if the biological basis is not understood or is too simplistic'" (p. 308). The continued development and testing of biomechanical models, more than the application of different methods of analysis, could have the greatest impact in advancing our understanding of scapular form and biomechanics in both fossil and living primates.

CONCLUSIONS

In this study, we carried out a GM evaluation of the functional correlates of scapular variation in the African apes using GPA applied to landmark data and compared our findings with previous results based on linear measurements obtained using digital calipers. Scapular size and shape vary between *Pan* and gorillas, while gorilla subspecies differ only in scapular shape. Shape variation is significant in comparisons between all three groups, with the greatest degree of difference observed between *Pan* and the two gorilla subspecies. There is, however, no systematic pattern of differentiation that maps predictably to differences in frequency of suspensory or arboreal climbing behaviors. We conclude that the relationship between scapular morphology and locomotor behavior, at least in the African apes, is not a compelling one. Our findings generally confirm patterns of variation between gorillas, and between *Pan* and gorilla subspecies, observed by previous investigators using conventional distance measurements, and demonstrate the mutually informative nature of GM and other methods of analysis when used to address questions of biological variation.

ACKNOWLEDGMENTS

We thank F. James Rohlf, Patricia Vinyard, and one anonymous reviewer for comments on earlier versions of this chapter. A. Taylor thanks S. Inouye, B. Shea, and C. Vinyard for ongoing, critical discussions of this work. Work by DES was supported, in part, by the Austrian Ministry of Culture, Science, and Education and the Austrian Council for Science and Technology (grant numbers: GZ 200.049/3-VI/I 2001 and GZ 200.093/1-VI/I 2004 to Horst Seidler) and Dr. Edward G. Hill and members of the Winston-Salem community. Work by ABT was supported, in part, by the National Science Foundation (BNS-9016522), a Pennsylvania State System of Higher Education Faculty Research Grant, the Wenner-Gren Foundation for Anthropological Research (GR. 5323), and Sigma-Xi Grant-in-Aid-of Research.

REFERENCES

Anapol, F., 1983, Scapula of *Apidium phiomense*: A small anthropoid from the Oligocene of Egypt, *Folia Primatol.* 40:11–31.
Arnold, S. J., 1983, Morphology, performance and fitness, *Am. Zool.* 23:347–361.

Bock, W. J., 1979, The synthetic explanation of macroevolutionary change—a reductionistic approach, *Bulletin of the Carnegie Museum of Natural History* 13:20–69.

Bookstein, F. L., 1991, *Morphometric Tools for Landmark Data: Geometry and Biology*, Cambridge University Press, Cambridge.

Coolidge, H. J., 1933, *Pan paniscus*: Pygmy chimpanzee from south of the Congo River, *Am. J. Phys. Anthropol.* 18:1–57.

Corruccini, R. S. and Ciochon, R. L., 1976, Morphometric affinities of the human shoulder, *Am. J. Phys. Anthropol.* 45:19–37.

Doran, D. M., 1996, Comparative positional behavior of the African apes, in: *Great Ape Societies*, W. McGrew, L. Marchant, and T. Nishida, eds., Cambridge University Press, Cambridge, pp. 213–224.

Doran, D. M., 1997, Ontogeny of locomotion in mountain gorillas and chimpanzees, *J. Hum. Evol.* 32:323–344.

Fleagle, J. G., 1976, Locomotor behavior and skeletal anatomy of sympatric Malaysian leaf-monkeys (*Presbytis obscura* and *Presbytis melalophos*), *Yearb. Phys. Anthropol.* 20:440–453.

Gower, J. C., 1975, Generalized Procrustes analysis, *Psychometrika* 40:33–51.

Groves, C. P., 1967, Ecology and taxonomy of the gorilla, *Nature* 213:890–893.

Groves, C. P., 2001, *Primate Taxonomy*, Smithsonian Institution Press, Washington.

Hunt, K. D., Cant, J. G. H., Gebo, D. L., Rose, M. D., Walker, S. E., and Youlatos, D., 1996, Standardized descriptions of primate locomotor and postural modes, *Primates* 37:363–386.

Inouye, S. E., 2003, Intraspecific and ontogenetic variation in the forelimb morphology of *Gorilla*, in: *Gorilla Biology: A Multidisciplinary Perspective*, A. B. Taylor and M. L. Goldsmith, eds., Cambridge University Press, Cambridge, pp. 194–235.

Inouye, S. E. and Shea, B. T., 1997, What's your angle? Size-correction and bar-glenoid orientation in "Lucy" (A.L. 288-1), *Int. J. Primatol.* 18:629–650.

Inouye, S. E. and Taylor, A. B., 2000, Ontogenetic variation in scapular form in African apes, *Am. J. Phys. Anthropol. (Suppl.)* 30:185.

Jungers, W. L. and Susman, R. L., 1984, Body size and skeletal allometry in African apes, in: *The Pygmy Chimpanzee: Evolutionary Biology and Behavior*, R. L. Susman, ed., New York, Plenum Press, pp. 131–177.

Kay, R. F. and Cartmill, M., 1977, Cranial morphology and adaptations of *Palaechthon nacimienti* and other Paromomyidae (Plesiadapoidea, ?Primates), with a description of a new genus and species, *J. Hum. Evol.* 6:19–53.

Kay, R. F. and Covert, H. H., 1984, Anatomy and behavior of extinct primates, in: *Food Acquisition and Processing in Primates*, D. J. Chivers, B. A. Wood, and A. Bilsborough, eds., Plenum Press, New York, pp. 467–508.

Larson, S. G., 1995, New characters for the functional interpretation of primate scapulae and proximal humeri, *Am. J. Phys. Anthropol.* 98:13–35.

Larson, S. G. and Stern, Jr., J. T., 1986, EMG of scapulohumeral muscles in the chimpanzee during reaching and "arboreal" locomotion, *Am. J. Anat.* 176:171–190.

Larson, S. G. and Stern, Jr., J. T., 1989, The role of supraspinatus in the quadrupedal locomotion of vervets (*Cercopithecus aethiops*): Implications for interpretation of humeral morphology, *Am. J. Phys. Anthropol.* 79:369–377.

Larson, S. G., Stern, Jr., J. T., and Jungers, W. L., 1991, EMG of serratus anterior and trapezius in the chimpanzee: Scapular rotators revisited, *Am. J. Phy. Anthropol.* 85:71–84.

Leamy, L. and Atchley, W. R., 1984, Morphometric integration in the rat (*Rattus* sp.) scapula, *J. Zool.* 202:43–56.

Napier, J. R. and Napier, P. H., 1967, *A Handbook of Living Primates,* Academic Press, New York.

Oxnard, C. E., 1963, Locomotor adaptations of the primate forelimb, *Symp. Zool. Soc.* London 10:165–182.

Oxnard, C. E., 1968, The architecture of the shoulder in some mammals, *J. Morphol.* 126:249–290.

Oxnard, C. E., 1972, Functional morphology of primates: Some mathematical and physical methods, in: *The Functional and Evolutionary Biology of Primates,* R. H. Tuttle, ed., Aldine-Atherton, Chicago, pp. 305–336.

Oxnard, C. E., 1977, Morphometric affinities of the human shoulder, *Am. J. Phys. Anthropol.* 46:367–374.

Oxnard, C. E., 1979, Some methodological factors in studying the morphological–behavioral interface, in: *Environment, Behavior, and Morphology: Dynamic Interactions in Primates,* M. E. Morbeck, H. Preuschoft, and N. Gomberg, eds., Gustav Fischer, New York, pp. 183–227.

Remis, M. J., 1994, *Feeding Ecology and Positional Behavior of Lowland Gorillas in the Central African Republic,* Ph.D. dissertation, Yale University, New Haven, Connecticut.

Remis, M. J., 1995, Effects of body size and social context on the arboreal activities of lowland gorillas in the Central African Republic, *Am. J. Phys. Anthropol.* 97:413–434.

Remis, M. J., 1998, The gorilla paradox: The effects of habitat and body size on the positional behavior of lowland and mountain gorillas, in: *Primate Locomotion,* E. Strasser, J. G. H. Fleagle, A. Rosenberger, and H. McHenry, eds., Plenum Press, New York, pp. 95–106.

Roberts, D., 1974, Structure and function of the primate scapula, in: *Primate Locomotion,* F. A. Jenkins, ed., Academic Press, New York, pp. 171–200.

Rohlf, F. J., 2000a, Statistical power comparisons among alternative morphometric methods, *Am. J. Phys. Anthropol.* 111:463–478.

Rohlf, F. J., 2000b, On the use of shape spaces to compare morphometric methods, *Hystrix* 11:9–25.

Rohlf, F. J., 2001, tpsDig version 1.31, Department of Ecology and Evolution, State University of New York at Stony Brook, New York.

Rohlf, F. J., 2003, Bias and error in estimates of mean shape in morphometrics, *J. Hum. Evol.* 44:665–683.

Schultz, A. H., 1930, The skeleton of the trunk and limbs of higher primates, *Hum. Biol.* 2:303–438.

Schultz, A. H., 1934, Some distinguishing characters of the mountain gorilla, *J. Mammal.* 15:51–61.

Shea, B. T., 1986, Scapula form and locomotion in chimpanzee evolution, *Am. J. Phys. Anthropol.* 70:475–488.

Slice, D. E., 1998, Morpheus et al.: Software for morphometric research. Recent beta revision, Department of Ecology and Evolution, State University of New York at Stony Brook, New York.

Smith, T. D., Mooney, M. P., and Taylor, A. B., 1990, Variation in cetacean scapular shape as a function of locomotor adaptations to salt and fresh water environments. A tensor biometric analysis, *Proc. 18th Annu. Meet. Intl. Mar. Animal Trainers Assoc.* 4:88–94.

Stern, J. and Susman, R., 1983, The locomotor anatomy of *Australopithecus afarensis*, *Am. J. Phys. Anthropol.* 60:279–317.

Swiderski, D. L., 1993, Morphological evolution of the scapula in tree squirrels, chipmunks, and ground squirrels (Sciuridae): An analysis using thin-plate splines, *Evolution* 47:1854–1873.

Takahashi, L. K., 1990, Morphological basis of arm-swinging: Multivariate analyses of the forelimbs of *Hylobates* and *Ateles, Folia Primatol.* 54:70–85.

Taylor, A. B., 1992, A morphometric study of the scapula in *Gorilla* (*Gorilla gorilla gorilla* and *G.g. beringei*), Ph.D. dissertation, University of Pittsburgh, Pittsburgh, Pennsylvania.

Taylor, A. B., 1997, Scapula form and biomechanics in gorillas, *J. Hum. Evol.* 33:529–553.

Taylor, A. B., 2002, Masticatory form and function in the African apes, *Am. J. Phys. Anthropol.* 117:133–156.

Taylor, A. B. and Siegel, M. I., 1995, Modeling differences in biological shape in two species of *Peromyscus, J. Mammal.* 76:828–842.

Thompson, D., 1942, *On Growth and Form*, The Macmillan Company, New York.

Tuttle, R. H. and Watts, D. P., 1985, The positional behavior and adaptive complexes of Pan (Gorilla), in: *Primate Morphophysiology, Locomotor Analyses and Human Bipedalism*, S. Kondao, ed., University of Tokyo Press, Tokyo, pp. 261–288.

CHAPTER FIFTEEN

Functional Shape Variation in the Cercopithecine Masticatory Complex

Michelle Singleton

INTRODUCTION

The study of cranial shape variation is central to physical anthropology, and the prevalence of cranial allometry in cercopithecine monkeys (Cercopithecidae Gray, 1821) has made them favorite subjects for investigating links between cranial development, function, and form (Antón, 1996; Bouvier, 1986a; Hylander, 1979, 1985; Leigh et al., 2003; Lucas, 1981, 1982; Ravosa, 1990; Ravosa and Profant, 2000; Ravosa and Shea, 1994; Shea, 1992; Vinyard and Ravosa, 1998). By distinguishing shape differences due primarily to body size from those with specific functional or phylogenetic significance (Gould, 1975; Shea, 1983a, 1985), such studies both elucidate modern primate adaptations and improve interpretations of cranial shape variation in the primate fossil record. The cercopithecine tribe Papionini (Table 1)—a monophyletic taxon comprising macaques (genus *Macaca*), mangabeys (*Cercocebus* and *Lophocebus*), mandrills and drills (*Mandrillus*), and baboons (*Papio* and *Theropithecus*) (Delson, 1975a, b; Hill, 1974; Kuhn, 1967; Strasser and

Michelle Singleton • Department of Anatomy, Midwestern University, Downers Grove, IL 60515.

Modern Morphometrics in Physical Anthropology, edited by Dennis E. Slice.
Kluwer Academic/Plenum Publishers, New York, 2005.

Table 1. Taxonomy of species sampled

Family Cercopithecidae Gray, 1821
  Subfamily Colobinae: Leaf monkeys
    Colobus
  Subfamily Cercopithecinae: Cheek-pouch monkeys
    Tribe Cercopithecini: Guenons and related taxa
      Allenopithecus, Miopithecus, Cercopithecus, Erythrocebus
    Tribe Papionini
      Subtribe Macacina (predominantly Asian)
      Macaca
      Subtribe Papionina (predominantly African)
        Cercocebus, Lophocebus, Mandrillus, Papio, Theropithecus

Note: Cercopithecid cladistic relationships are shown in Figure 1.
Source: Cercopithecid taxonomy after Szalay and Delson (1979).

Delson, 1987; Szalay and Delson, 1979)—has received particular attention from researchers interested in allometry as a source of cranial homoplasy (Collard and Wood, 2000; Fleagle and McGraw, 2001; Leigh et al., 2003; Lockwood and Fleagle, 1999). The mangabeys, small- to medium-sized monkeys with moderately prognathic faces, were historically considered sister taxa (Figure 1a) to the exclusion of *Papio* and *Mandrillus*, large-bodied taxa with long faces and enlarged canines (Hill, 1974; Kuhn, 1967; Strasser and Delson, 1987; Szalay and Delson, 1979; Thorington and Groves, 1970). But molecular phylogenetic analyses (Figure 1b) have consistently rejected mangabey monophyly, instead linking *Cercocebus* and *Lophocebus* with *Mandrillus* and *Papio*, respectively (Barnicot and Hewett-Emmett, 1972; Cronin and Sarich, 1976; Disotell, 1994; Disotell et al., 1992; Dutrillaux et al., 1979, 1982; Harris, 2000; Harris and Disotell, 1998; Hewett-Emmett et al., 1976; Page et al., 1999; Van Der Kuyl et al., 1995). Given the marked size differences between the newly recognized papionin sister taxa (Delson et al., 2000), simple ontogenetic scaling—shape difference arising from truncation or extension of shared ancestral ontogenetic trajectories (Gould, 1966; Shea, 1983b, 1985)—was considered the most likely cause of papionin facial homoplasy (Harris, 2000; Lockwood and Fleagle, 1999; Ravosa and Profant, 2000; Shah and Leigh, 1995). But comparative studies have revealed a complex interplay of interspecific size allometry (Singleton, 2002), shared ontogenetic scaling patterns (Collard and O'Higgins, 2001), and allometric dissociations (Leigh et al., 2003; Shah and Leigh, 1995) contributing to the pervasive, nonhomologous similarities between like-sized members of the disparate papionin clades.

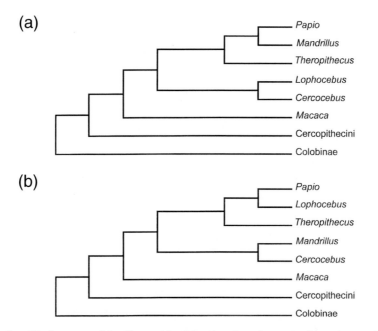

Figure 1. Cladograms of the Cercopithecidae showing alternative hypotheses of papionin relationships. Traditional phylogenies (a) identified mangabeys—*Cercocebus* and *Lophocebus*—as sister taxa (Delson and Dean, 1993; Strasser and Delson, 1987; among others); molecular phylogenies (b) reject mangabey monophyly (Disotell, 1994; Harris, 2000; Harris and Disotell, 1998).

While these studies have increased our understanding of papionin facial allometries and clarified the ontogenetic basis of the resultant cranial shape similarities, the polarity and adaptive significance of these trends are still contested (Harris, 2002; Ravosa and Profant, 2000). The majority of workers favor the view that long faces are derived in papionins (Benefit and McCrossin, 1991, 1993; Collard and O'Higgins, 2002; Cronin and Sarich, 1976; Delson, 1975a, b; Disotell, 1994, 1996; Harris, 2000, 2002; Jolly, 1970; Strasser and Delson, 1987), but others maintain that facial prognathism is the basal African papionin condition (Groves, 1978; Kingdon, 1997). Whereas ontogenetic studies incorporating a broader range of cercopithecine taxa should help to establish the polarity of papionin facial growth patterns, consideration of the biomechanical consequences of differing facial allometries is needed to ascertain their broader evolutionary significance. Deviations from ancestral ontogenetic patterns, such as observed in papionins (Collard and O'Higgins, 2001; Leigh et al., 2003; Shah and Leigh, 1995), frequently indicate selection for altered size–shape relationships that have functional and adaptive implications beyond simple size change (Gould, 1975; Shea, 1983b, 1985). Specifically, transposition of allometric trajectories often reflects the maintenance of geometric similarity or biomechanical equivalence as species evolve into new size ranges (Gould, 1971; Ravosa, 1992; Shea, 1983b, 1995; Smith, 1993; Vinyard and Ravosa, 1998). Conversely, selection for new or enhanced functional capacities may cause dissociation of scaling trajectories between closely related taxa (Ravosa, 1990; Shea, 1985) or convergence of ontogenetic allometries and homoplastic similarity in distantly related forms (Demes et al., 1986; Ravosa, 1992).

The mechanical constraints of mastication are a fundamental determinant of maxillofacial form (Sakka, 1985), and diet is a major influence upon cercopithecine facial scaling (Antón, 1996; Hylander, 1977, 1979; Ravosa, 1990; Shea, 1983b; Vinyard and Ravosa, 1998). Conveniently, the African papionin clade (subtribe Papionina) encompasses two distinct ecomorphs: small-bodied and relatively arboreal mangabeys, which are heavily reliant on hard fruits, nuts, and seeds; and large-bodied, terrestrial forms (*Papio, Mandrillus*) characterized by extreme facial prognathism, greatly enlarged canines, and diets incorporating a variety of resistant foods. Thus, it is highly plausible that documented allometric shifts among papionins (Collard and O'Higgins, 2001; Leigh et al., 2003; Shah and Leigh, 1995) result from selection for specific functional capacities, that is, functional convergence. Confirmation that facial allometries in *Papio* and *Mandrillus* (or *Cercocebus* and *Lophocebus*) produce novel size–shape associations (i.e., forms) with similar biomechanical properties would support a hypothesis of functional convergence, clarify extant papionin adaptations, and potentially shed light on the selective pressures to which ancestral papionins were subject.

Scaling of the primate facial complex has traditionally been studied using linear regression of interlandmark distances analyzed within the framework of the bivariate allometric model (Gould, 1966; Huxley, 1932). The multivariate generalization of allometry (Jolicoeur, 1963) and Euclidean distance matrix analysis (EDMA)(Richtsmeier and Lele, 1993) have also been used to investigate size-related patterns of shape variation. More recent studies (Collard and O'Higgins, 2001; O'Higgins and Jones, 1998; Penin et al., 2002; Singleton, 2002; Vidarsdottir et al., 2002) have combined classic allometric models with landmark-based geometric morphometric analysis (Bookstein, 1996; Dryden and Mardia, 1998) to investigate interspecific and ontogenetic allometries in a range of anthropoid primates. This geometric approach has yet to be applied to comparative analysis of functional allometries. Yet, it would seem to be ideally suited to investigations of functional scaling. Whereas traditional size allometry deals with scaling of individual variables relative to size and biomechanical allometry is narrowly concerned with scaling relationships among variables relative to *a priori* biomechanical models (Smith, 1993), neither approach adequately describes how the functional geometry of a configuration changes with changing size. By contrast, allometric analysis of geometrically derived shape variables quantifies differences in relative distance *and* relative position among multiple landmarks relative to size (Rohlf and Marcus, 1993), potentially bridging the gap between these previously incommensurate allometric categories.

The cercopithecine masticatory apparatus is an ideal test case for this proposition. Certain aspects of jaw function may be modeled as simple lever systems (Greaves, 1974; Hylander, 1979; Ravosa, 1990; Spencer, 1999), and primate masticatory muscle forces are known to scale isometrically with cranial size (Cachel, 1984; Hylander, 1985). Thus, geometric shape differences affecting the relative positions of joints, muscle attachments, and bite points have predictable biomechanical consequences (Hylander, 1985; Ravosa, 1990), easily interpretable in terms of relative functional capacities. Prior studies of facial biomechanics (Antón, 1996; Bouvier, 1986a; Greaves, 1974, 1995; Hylander, 1979; Lucas, 1981, 1982; Ravosa, 1990; Smith, 1984; Vinyard and Ravosa, 1998) furnish models for the functional interpretation of shape variation and permit validation of results. Therefore, this study employs a combination of geometric morphometric and statistical analytic methods to examine patterns of shape variation in the cercopithecine masticatory complex. Its goals are to evaluate the efficacy of geometric morphometrics for functional allometric analysis; to document the phylogenetic distribution of adult masticatory scaling patterns within Cercopithecinae; to interpret the functional consequences of differing adult facial allometries in light of established biomechanical models; and to explore the potential adaptive significance of papionin facial homoplasy.

MATERIALS AND METHODS

The study sample comprised 450 adult individuals representing all commonly recognized cercopithecine genera and two colobine outgroups (see Table 2).

Table 2. Within-species correlations

Taxon	n		PC1		PC2	
	Female	Male	SIZE	MLVR	SIZE	FL
Colobus angolensis cottoni	12	22	0.69	0.57	0.56[b]	0.60[b]
Colobus guereza kikuyuensis	10	20	0.66	0.50[b]	0.44[a]	0.47[a]
Cercopithecus ascanius ngamiensis	19	8	0.76	0.63	0.81	0.81
Cercopithecus mona	23	20	0.63	0.53	0.65	0.67
Allenopithecus nigroviridis	3	3	0.89[a]	0.94[a]	0.94[b]	0.91[a]
Erythrocebus patas	4	9	0.89	0.80	0.84	0.88
Miopithecus ougouensis	15	6	0.53[a]	0.34[NS]	0.67	0.56[a]
Cercocebus galeritus agilis	9	7	0.77	0.58	0.86	0.89
Cercocebus torquatus torquatus	11	18	0.75	0.65	0.73	0.73
Lophocebus albigena johnstoni	15	24	0.82	0.62	0.62	0.61
Macaca fascicularis	17	26	0.87	0.67	0.67	0.64
Mandrillus leucophaeus	8	19	0.88	0.76	0.85	0.86
Mandrillus sphinx	8	13	0.87	0.81	0.80	0.81
Papio hamadryas anubis	21	42	0.90	0.82	0.71	0.74
Theropithecus gelada	13	25	0.78	0.63	0.68	0.68

Notes: Within-species Pearson product moment correlations for principal shape components with log centroid size (SIZE), log masseter lever arm length (MLVR), and log facial length (FL). All correlations are significant at $p < 0.001$ except as indicated.
NS, not significant.
[a] $p < 0.05$.
[b] $p < 0.01$.

Following published protocols (Frost et al., 2003; Singleton, 2002), 45 three-dimensional craniofacial landmarks were recorded; missing data points were estimated by reflection (Singleton, 2002). Linear interlandmark distances (see below) were computed from raw coordinates and log-transformed. Specimens were scaled to unit centroid size and aligned via generalized Procrustes analysis (GPA) using *tpsSmall* vs 1.19 (Rohlf, 1998). Where shape variation about the Procrustes mean is sufficiently small, Procrustes-aligned coordinates may be used in lieu of Kendall tangent-space coordinates as the basis of parametric statistical analysis (Dryden and Mardia, 1998; Rohlf, 1999; Slice, 2001). The least squares regression of Procrustes distances against Euclidean (tangent-space) distances (slope = 0.9962, $r = 0.9999$) computed by *tpsSmall* vs 1.19 (Rohlf, 1998) supports this assumption, and Procrustes-aligned coordinates were accepted as a reasonable approximation of the orthogonal tangent space projection.

A subset of five landmarks—Postglenion (PGL), Zygomaxillare Inferior (ZMI), Distal M3 (DM3), M1–M2 Contact (M12), and Prosthion (PRO) (Figure 2; see Frost et al., 2003 for landmark definitions)—was selected to

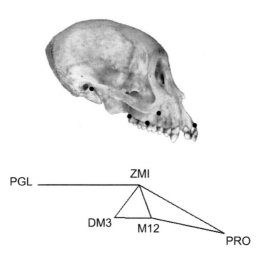

Figure 2. Top: Lateral view of representative cercopithecine skull (female *Lophocebus albigena*) showing masticatory landmarks employed in this study. Bottom: Wireframe representation of the sample Procrustes mean configuration for the five masticatory landmarks (right side only). PGL = Postglenion, ZMI = Zygomaxillare Inferior, DM3 = Distal M3 at alveolar margin, M12 = M1−M2 contact at alveolar margin, PRO = Prosthion.

capture specific functional aspects of the masticatory system including: (a) the relative positions of the temporomandibular joint (TMJ) and zygomatic root; (b) the length and relative position of the palate; and (c) the positions of maxillary bite points. These landmarks were chosen to correspond to endpoints of linear measures employed in prior studies of the primate masticatory system (e.g., Ravosa, 1990). By so doing, it was hoped to validate the results of geometric analysis against documented patterns of masticatory scaling and to evaluate the utility of geometric analysis for elucidating patterns of functional variation.

To compensate for the lack of statistical independence among landmarks due to morphological integration and the constraints of translation, rotation, and scaling imposed by GPA (Dryden and Mardia, 1998; Rohlf, 1999), principal components analysis (PCA) was conducted on the covariance matrix of Procrustes-aligned coordinates for the masticatory landmarks. This procedure ordinates specimens relative to mutually orthogonal axes of shape variation and yields a small number of uncorrelated summary shape variables, namely the principal component (PC) scores (Dryden and Mardia, 1998). The signs (positive or negative) of these scores are arbitrary and analysis-specific; thus, it is their relative (rather than absolute) values that signify differences in shape among specimens. It should also be emphasized that, because GPA eliminates only the effects of scale (i.e., absolute size), principal shape components incorporate both size-correlated (allometric) and size-independent (residual) variation. Patterns of shape variation summarized by selected shape components were explored using the *Morphologika* morphometrics package (O'Higgins and Jones, 1999) to generate wireframe representations of variation along corresponding shape axes. Bivariate scatterplots of shape component (PC) scores against independent size variables were used to identify potential allometric relationships, the strengths of which were assessed by correlation analyses within species and tribes.

The choice of independent variables for cranial allometry has been a subject of some debate (Bouvier, 1986a, b; Hylander, 1985; Ravosa, 1990; Smith, 1993). Ontogenetic and functional allometries are most commonly examined relative to measures of cranial or facial length; for biomechanical allometry, the choice of independent variable is determined by the mechanical relationships under consideration (Bouvier, 1986a, b; Hylander, 1985; Ravosa, 1990; Smith, 1993). For Procrustes-based analyses, the appropriate size metric is centroid size, a Mosimann variable (Mosimann and Malley, 1979) approximately uncorrelated with all shape variables assuming isotropic landmark error (Bookstein, 1996; Slice et al., 1996); thus, log centroid size was adopted as the sole estimate of cranial size. As previously noted, scaling relationships for the principal shape components are expected to subsume both biomechanical allometry—scaling relationships *among* landmarks (Smith, 1993)—and size allometry—scaling relationships for *individual* landmarks. To test this supposition, scaling of shape components against biomechanically appropriate variables—masseter lever arm length (MLVR, distance from PGL to ZMI) and facial length (FL, distance from PGL to PRO)—was examined to confirm certain functional interpretations.

Between-species differences in adult intraspecific scaling were tested by least squares regression and analysis of covariance (ANCOVA). Species-by-covariate interaction effects were tested to rule out heterogeneity of slopes, and *a posteriori* comparisons of estimated marginal means—species means adjusted for covariate effects—were performed to identify significant differences in regression elevations. This approach was adopted in preference to interspecific regression of species means because it both permits pairwise significance testing of between-species scaling differences and, more importantly, avoids *a priori* functional or taxonomic grouping of species.

RESULTS

Functional Shape Variation

The first 8 of 15 PCs accounted for 95% of total shape variance. Although specimens are scaled to unit centroid size, the 1st Principal Component (66% of total shape variance) appears to ordinate individuals on the basis of size, separating small- from large-bodied species and females from conspecific males (Figure 3). Visualization of shape variation along the 1st Principal Component axis (Figure 3) indicates that animals with more negative scores possess relatively short palates, which are retracted relative to ZMI, such that molar bite points lie posterior to the zygomatic. ZMI, which marks the anteriormost extent of the masseter origin, lies well anterior to the TMJ. In biomechanical terms, more negative scores correspond to relatively long masseter lever arms and relatively short dental load arms. This favorable input- to output-ratio results in increased mechanical advantage (MA) for the masseter and increased molar and incisal bite force magnitudes relative to cranial size (Hylander, 1985; Ravosa, 1990). By contrast, animals with more positive scores are characterized by more posteriorly positioned zygomatics, anteriorly positioned bite points with the M12 point anterior to ZMI, and relatively long palates. This configuration results in less favorable input- to output-ratios, reduced masseter MA, and decreased relative bite forces, particularly in the incisal region (Hylander, 1985; Ravosa, 1990; Spencer, 1999).

The 2nd Principal Component (10% of total shape variance) summarizes variation in the height of the TMJ independent of facial length (Figure 3). Negative scores signify decreased vertical separation between the TMJ and alveolar margin relative to facial length; positive scores, increased relative TMJ height. *Miopithecus* and papionin species fall toward the negative end of the axis; *Colobus, Theropithecus*, and the remaining cercopithecins at its positive end. Females exhibit more negative scores than conspecific males. Variation in TMJ height relative to facial length is an important determinant of jaw gape, that is, maximum mandibular opening (Greaves, 1974; Herring and Herring, 1974; Lucas, 1981, 1982). With all other factors held constant, decreased relative TMJ height results in increased vertical mandibular displacement and greater gape (Greaves, 1974; Lucas, 1982; Ravosa, 1990; Smith, 1984). It may also increase the vertical component of bite force, especially in the incisor region (Ravosa, 1990). Thus, relatively negative scores on the 2nd Principal shape component signify enhanced gape and potentially more forceful incisal

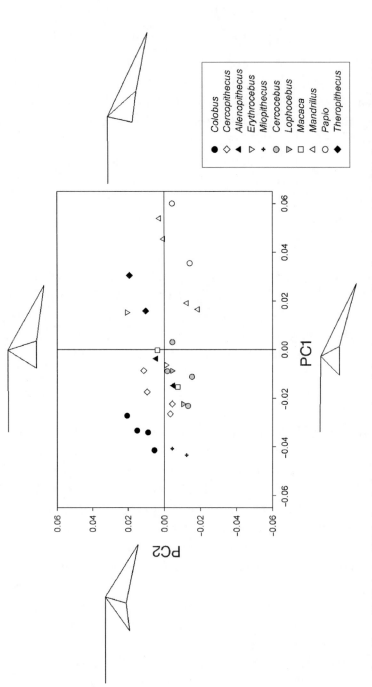

Figure 3. Plot of PC1 versus PC2. Symbols represent male and female mean values for each species. Wireframes represent the extremes of shape variation along each shape component axis.

biting, whereas positive scores indicate decreased relative gape and reduced vertical bite force components.

Functional Allometry—Principal Component 1

The 1st Principal Component is strongly linearly correlated with log centroid size (Figure 4a) within the majority of cercopithecine species sampled (Table 2) and within the two cercopithecine tribes (Table 3). The ANCOVA of PC1 by species with log centroid size as the covariate is highly significant ($F = 371.91$, $p < 0.001$, $r^2 = 0.96$). Homogeneity of slopes is confirmed, but ANOVA of estimated marginal means (linearly independent contrasts) finds significant differences ($F = 21.8, p < 0.001$) among regression elevations. *Colobus*, mangabeys, and *Cercopithecus mona* have negative adjusted mean values, in contrast with all remaining cercopithecins and papionins, which have positive means (Table 4). There are relatively few significant pairwise differences in elevations among species either within or between groups (Table 5). However, the two colobine species are significantly different from all cercopithecines except *Miopithecus* and *C. galeritus*. Among papionins, mangabey species are significantly different from *Macaca*, *Mandrillus leucophaeus*, and *Papio* but not *M. sphinx* or *Theropithecus*.

Scaling of PC1 against MLVR (not shown) resembles that for centroid size, although full-sample, within-species, and within-tribe correlations are generally weaker (Tables 2, 3). The ANCOVA of PC1 by species with MLVR as the covariate is highly significant statistically ($F = 233.1, p < 0.001, r^2 = 0.94$), homogeneity of slopes is confirmed, and the test of independent contrasts is again significant ($F = 70.2, p < 0.001$). Adjusted mean values (Table 4) separate colobines from cercopithecines and mangabeys from all other papionins, *Erythrocebus* and *Allenopithecus*. However, adjusted means for *Miopithecus* and the two *Cercopithecus* species are more similar to mangabeys, and taxonomic patterning of between-species comparisons is less clear than for centroid size (Table 5).

Scaling of the 1st Principal Component, summarizing shape variation tied to MA of the masseter muscle, is consistent with known patterns of cercopithecine facial allometry (Bookstein, 1985; Cheverud and Richtsmeier, 1986; Cochard, 1985; Collard and O'Higgins, 2001; Freedman, 1962, 1963; Leigh et al., 2003; McNamara et al., 1976; Profant and Shea, 1994; Ravosa and Profant, 2000; Ravosa and Shea, 1994; Singleton, 2002; Swindler and Sirianni, 1973;

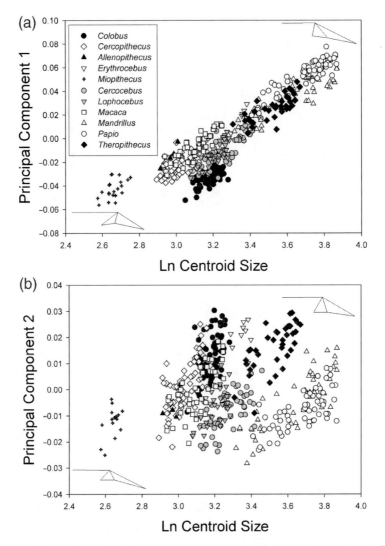

Figure 4. Plots of principal shape components against log centroid size. Wireframes represent extremes of shape variation for each component. (a) Intraspecific scaling of PC1 relative to size is largely uniform among cercopithecines, but negative displacement of allometric lines in mangabeys reflects facial retraction resulting in increased mechanical advantage and enhanced relative bite forces. (b) Scaling of PC2 differs between cercopithecine tribes. Negative displacement of papioninan allometric lines implies decreased relative temporomandibular joint height and increased relative jaw gape; further dissociations results in geometric similarity of gape among African papionins.

Table 3.   Within-tribe correlations by sex

	PC1		PC2	
	SIZE	MLVR	SIZE	FL
Cercopithecin				
Female	0.94^a	0.86^{NS}	0.98^b	0.98^b
Male	0.97^b	0.94^a	0.96^a	0.95^a
Papionin				
Female	0.93^b	0.78^a	0.04^{NS}	0.04^{NS}
Male	0.97^b	0.90^b	0.17^{NS}	0.12^{NS}

Notes: Within-tribe Pearson product moment correlations for
species mean principal component scores by sex. Abbreviations
as in Table 2.
NS not significant.
^{a} $p < 0.05$.
^{b} $p < 0.01$.

Table 4.   Estimated marginal means

Taxon	PC1		PC2	
	SIZE	MLVR	SIZE	FL
C. angolensis	−0.027	−0.039	0.016	0.016
C. guereza	−0.022	−0.030	0.024	0.024
C. ascanius	0.005	−0.004	0.028	0.026
C. mona	−0.004	−0.012	0.023	0.021
A. nigroviridis	0.012	0.005	0.022	0.021
E. patas	0.009	0.009	0.014	0.014
M. ougouensis	0.015	−0.012	0.051	0.033
M. fascicularis	0.015	0.010	0.012	0.010
C. torquatus	−0.004	−0.008	−0.010	−0.004
C. galeritus	−0.011	−0.016	−0.003	−0.003
L. albigena	−0.004	−0.011	−0.002	−0.012
M. leucophaeus	0.011	0.016	−0.024	−0.024
M. sphinx	0.002	0.014	−0.022	−0.023
P. hamadryas	0.013	0.027	−0.024	−0.024
T. gelada	0.005	0.013	0.002	0.002

Note: Species mean values adjusted for the effects of the
covariate.

Swindler et al., 1973). With size increase, relative palate length increases; relative masseter lever length decreases; and molar bite points shift forward relative to the zygomatic root, all leading to reduced relative MA and decreased relative bite forces at larger cranial sizes (Hylander, 1979; Ravosa, 1990). The significant negative displacement of the *Colobus* regression reflects a shortening and retraction of the palate, conferring increased MA and greater bite forces

Table 5. Pairwise comparisons of regression elevations for PC1

Taxon	CAC	CGK	CAN	CM	AN	EP	MO	MFA	CTT	CGA	LAJ	ML	MS	PHA	TG	MLVR
C. angolensis	—	*	*	*	*	*	NS	*	*	*	*	*	*	*	*	
C. guereza	*	—	*	*	*	*	NS	*	*	NS	*	*	*	*	*	
C. ascanius	*	*	—	NS	NS	NS	NS	NS	NS	NS	NS	*	NS	*	NS	
C. mona	*	*	NS	—	NS	*	NS	*	NS	NS	NS	*	*	*	*	
A. nigroviridis	*	*	NS	NS	—	NS	NS	NS	NS	NS	NS	NS	NS	NS	NS	
E. patas	*	*	NS	*	NS	—	NS	NS	*	*	*	NS	NS	*	NS	
M. ougouensis	NS	NS	NS	NS	NS	NS	—	NS	NS	NS	NS	NS	NS	NS	NS	
M. fascicularis	*	*	*	*	NS	NS	NS	—	*	*	*	*	*	*	*	
C. torquatus	*	*	NS	NS	NS	*	NS	*	—	NS	NS	*	*	*	*	
C. galeritus	*	NS	NS	NS	NS	*	NS	*	NS	—	NS	*	*	*	*	
L. albigena	*	*	NS	NS	NS	*	NS	*	NS	NS	—	*	*	*	*	
M. leucophaeus	*	*	NS	NS	NS	NS	NS	NS	*	NS	NS	—	NS	NS	NS	
M. sphinx	*	*	NS	*	NS	NS	NS	NS	NS	NS	*	NS	—	NS	NS	
P. hamadryas	*	*	NS	*	NS	NS	NS	NS	*	*	*	NS	NS	—	*	
T. gelada	*	*	NS	NS	NS	NS	NS	NS	NS	*	NS	NS	NS	NS	—	
SIZE																

Note: A posteriori pairwise comparisons of differences in estimated marginal means of PC1 by species controlling for the effects of log centroid size (below diagonal) and log MLVR (above diagonal). NS, not statistically significant; *, statistically significant at Bonferroni-corrected $p < 0.05$.

in comparison with similarly sized cercopithecines. Although few statistically significant differences are detected among cercopithecines, the patterning of allometric dissociations is clear. Like *Colobus*, the mangabeys exhibit negative displacements relative to other papionins and most cercopithecins (Table 4, Figure 4a), implying increased MA and bite force relative to cranial size. Scaling of PC1 relative to MLVR has identical biomechanical implications: negative displacement of mangabey allometric lines results in shortened dental load arms relative to MLVR length. The mangabey MLVR regressions differ significantly from other papionins but overlap those of *Cercopithecus* and *Miopithecus*, which also exhibit negative displacement of elevations and a concomitant increase in MA relative to most cercopithecines. Owing to differences in scaling relative to centroid size, however, mangabeys maintain this advantage at relatively larger cranial sizes.

Functional Allometry—Principal Component 2

Scaling relationships for the 2nd Principal Component (10% of total shape variance) are relatively complex. It is only weakly correlated with cranial centroid size across all individuals ($r = 0.09, p = 0.04$) but moderately correlated within most cercopithecine species (Figure 4b, Table 2). At the tribal level, PC2 is strongly correlated with size in cercopithecins but uncorrelated in papionins (Table 3). The ANCOVA of PC2 by species with centroid size as covariate is significant ($F = 51.6, p < 0.001, r^2 = 0.78$), but homogeneity of slopes is rejected ($p = 0.002$). Heterogeneity of slopes potentially invalidates pairwise comparisons; however, separate ANCOVA analyses conducted for cercopthecins and papionins, respectively, confirm homogeneity of slopes within each tribe and yield species ranks and pairwise comparison patterns almost identical to those observed for the full-sample analysis (Table 6). Estimated marginal means (Table 4) segregate African papionins (excluding *Theropithecus*) from all remaining species including *Macaca*. Among papionins, mangabey elevations are uniformly significantly different from the large-bodied taxa, with some discrepancy as to the position of *Cercocebus torquatus*. Only the position of *Macaca* is meaningfully affected by sample composition. The full-sample analysis finds it significantly different from all African papionins and indistinguishable from cercopithecins, but exploratory analyses (not shown) suggest a scaling pattern intermediate between the cercopithecin and African papionin clades.

Table 6. Pairwise comparisons of regression elevations for PC2

Taxon	CAC	CGK	CAN	CM	AN	EP	MO	MF	CTT	CGA	LAJ	ML	MS	PHA	TG	FL
C. angolensis	—	NS	NS	NS	NS	NS	NS	NS	*	*	*	*	*	*	*	
C. guereza	NS	—	NS	NS	NS	*	NS	*	*	*	*	*	*	*	*	
C. ascanius	NS	NS	—	NS	NS	NS	NS	*	*	*	NS	*	*	*	*	
C. mona	NS	NS	NS	—	NS	NS	NS	NS	*	*	*	*	*	*	*	
A. nigroviridis	NS	NS	NS	NS	—	NS	NS	NS	*	*	*	*	*	*	NS	
E. patas	NS	NS	NS	NS	NS	—	NS	NS	*	*	*	*	*	*	NS	
M. ogouensis	NS	NS	NS	NS	NS	NS	—	NS	NS	NS	NS	*	*	*	NS	
M. fascicularis	NS	*	NS	NS	NS	NS	NS	—	*	*	*	*	*	*	NS	
C. torquatus	*	*	*	*	*	*	*	*	—	*	*	*	NS	*	*	
C. galeritus	*	*	*	*	*	*	*	*	NS	—	NS	*	*	*	NS	
L. albigena	*	*	NS	*	*	*	*	*	*	NS	—	*	*	*	NS	
M. leucophaeus	*	*	*	*	*	*	*	*	*	*	*	—	NS	NS	*	
M. sphinx	*	*	*	*	*	*	*	*	*	*	*	NS	—	NS	*	
P. hamadryas	*	*	*	*	*	*	*	*	*	*	*	NS	NS	—	*	
T. gelada	*	*	*	*	NS	NS	NS	NS	*	NS	NS	*	*	*	—	
SIZE																

Note: *A posteriori* pairwise comparisons of differences in estimated marginal means of PC2 by species controlling for the effects of log centroid size (below diagonal) and log FL (above diagonal). NS, not statistically significant; *, statistically significant at Bonferroni corrected $p < 0.05$.

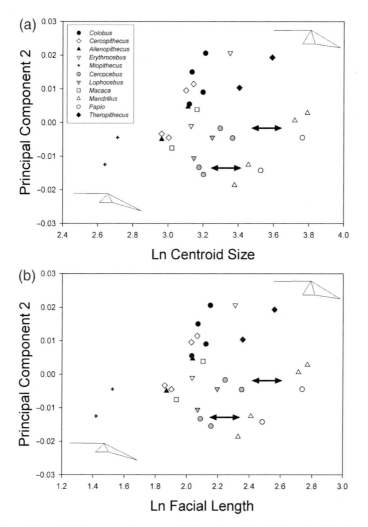

Figure 5.   Plots of PC2 against log centroid size (a) and facial length (b) indicate geometric similarity (double-headed arrows) among African papioinins irrespective of size. Symbols represent male and female mean values for each species. Wireframes represent extremes of shape variation for PC2.

As with centroid size, FL shows strong linear correlations with PC2 within and across cercopithecin species (not shown), while papionins show moderate correlations within species and absence of correlation across species (Tables 2, 3). The ANCOVA of PC2 by species with FL as covariate is significant ($F = 53.2, p < 0.001, r^2 = 0.79$), homogeneity of slopes is confirmed, and pairwise comparisons (Table 6) are extremely similar to those for centroid size.

Among African papionins, *Papio* and *Mandrillus* are significantly different from *Lophocebus* and *C. galeritus* but not *C. torquatus*, which also differs from other mangabeys. However, analysis restricted to papionins finds *C. torquatus* significantly different from all large-bodied African papionins and indistinguishable from other mangabey species.

The 2nd Principal Shape Component summarizes shape variation linked to relative gape. Intraspecific scaling of PC2 results in decreased relative gape and decreased vertical bite force components in males relative to conspecific females. A similar pattern is observed among cercopithecins, with larger taxa showing reduced gape relative to cranial size. Negative transposition of the African papionin regressions results in greater gape and potentially larger vertical bite forces than in comparably sized cercopithecins or *Macaca*. The autapomorphic facial morphology of *Theropithecus* hinders comparisons, but the negative displacement of its allometric line relative to cercopithecin species (Figure 4b) suggests that it shares this attribute. Dissociation of allometries among African papionins results in geometric similarity of relative TMJ position among species independent of size or FL (Figures 5a, b). Thus, *Papio* and *Mandrillus* maintain enhanced relative gape while dramatically increasing both FL and cranial size.

DISCUSSION

The present study is modeled upon Ravosa's (1990) comparison of masticatory scaling in colobines and cercopithecines and successfully replicates his major findings. Both analyses identify allometric dissociations between subfamilies resulting in enhanced relative MA in colobine species, and both show cercopithecines to be characterized by decreased relative MA and enhanced jaw gape. Whereas the prior study required 20 separate ANCOVA analyses (Ravosa, 1990), complicating interpretation of results and diminishing statistical power, analysis of Procrustes-based shape components required only four analyses, two of which ultimately proved redundant. In fact, scaling of principal shape components relative to centroid size and the biomechanical variables (MLVR and FL) yielded virtually identical taxonomic and functional interpretations. Thus, geometric analysis of functional shape variation does appear to provide a unified analysis of size and biomechanical allometries. Traditional linear regression analysis of metric variables will continue to be invaluable where parameter estimation and hypothesis testing relative to theoretical slopes are required. But where taxonomic differences in functional scaling and qualitative comparisons of functional shape trends are of primary interest, geometric methods offer significant advantages in terms of data reduction, analytic efficiency, internal consistency, and functional interpretation of results.

Ravosa (1990) hypothesized that differences in facial scaling between colobines and cercopithecines reflect functional tradeoffs between mechanical efficiency and gape. The present study identifies patterns of masticatory scaling indicative of similar, although less marked, functional divergences within and between cercopithecine tribes. Mangabey species exhibit facial allometries distinct from most other cercopithecines and similar in key respects to those of colobines. The facial shortening and retraction characteristic of these taxa increases the MA of the masseter muscle, resulting in greater relative bite force magnitudes, while simultaneously improving the dissipation of occlusal forces (Antón, 1996; Hylander, 1977, 1979; Ravosa, 1990). In colobines, which consume large volumes of fibrous matter requiring many chewing cycles, this configuration is seen as an adaptation for greater masticatory efficiency (Ravosa, 1990). In the case of *Cercocebus* and *Lophocebus*, both heavily reliant upon hard fruits, nuts, and seeds (Fleagle and McGraw, 1999; Kingdon, 1997; McGraw and Fleagle, 2000), maximization of bite force relative to cranial size is more likely to be the primary selective factor, and possibly key to successful exploitation of resistant foods by these small-bodied species. Although workers typically emphasize the link between facial shortening and molar biting (Antón, 1996; Du Brul, 1977; Hylander, 1979; Ravosa, 1990), recent studies have challenged the role of facial retraction in altering postcanine, but not incisal, bite forces (Greaves, 1995; Spencer, 1999). Perhaps not coincidentally, *Cercocebus* and *Lophocebus* are both characterized by enlarged incisors relative to molar size (Groves, 1978; Jablonski, 2002). Whether mangabey facial form is specifically adaptive for incisal preparation of hard fruits (Chalmers, 1968), postcanine crushing of nuts and seeds (Fleagle and McGraw, 2001), or both, is a question requiring further study.

Based upon outgroup comparisons with *Macaca* and cercopithecins (Figure 1), the shared masticatory form of mangabeys is most parsimoniously interpreted as homoplasious, the apparent result of functional convergence related to hard-object feeding. However, the intermediate position of the *Macaca* regression and apparent similarities in masticatory scaling between mangabeys, *Miopithecus* and *C. mona* combine to raise doubts concerning polarity of this trait. Because *Miopithecus* is a dwarf species characterized by decreased growth rates and greatly reduced terminal body size (Shea, 1992), it is likely that its facial geometry is the result of biomechanical scaling. To compensate for greatly reduced *absolute* masticatory muscle forces, changes in facial proportion similar to those observed for mangabeys are likely to have evolved to maintain minimum biomechanical competence (Shea, 1985, 1992; Smith, 1993). The case of *C. mona*, which differs from its congener in neither size nor dietary consistency, is less amenable to post hoc explanation (Kingdon, 1997). This discrepancy may reflect latent morphological diversity within the guenon clade or simple sampling effects, and its significance is unclear. Pending confirmatory studies incorporating additional guenon species, the hypothesis that shared mangabey facial forms are derived and homoplasious stands, albeit with a large asterisk.

Differences in the scaling of the 2nd Principal Component cause African papionins to exhibit increased mandibular gape relative both to cranial size and FL. The possession of enhanced relative gape distinguishes African papionins, including *Theropithecus,* from all cercopithecins as well as *Macaca fascicularis.* It is unknown whether African papionins are similarly distinguished from all *Macaca* species, but a finding of homogeneous facial scaling in *M. fascicularis, M. mulatta,* and *M. sylvanus* (Collard and O'Higgins, 2002)—the latter being the most basal and arguably the most primitive of the macaques (Morales and Melnick, 1998)—suggests this is likely. The intermediate position of *Macaca* between tribes contributes to an inconsistent pattern of pairwise comparisons and highlights the difficulty of drawing strong inferences of polarity for continuous traits, analyses of which are particularly susceptible to sample composition. Pending further investigation of this trait within genus *Macaca,* increased relative gape is most parsimoniously interpreted as a synapomorphy of the African papionin clade, that is, subtribe Papionina. Whereas cercopithecins exhibit decreased relative gape as cranial size increases, a dissociation of papionin elevations results in isometry among species and similar relative gape capacity irrespective of size or FL. Such displacements are commonly associated with selection for maintenance of functional equivalence as species enter new size ranges (Gould, 1971; Ravosa, 1992; Shea, 1983b, 1995; Smith, 1993; Vinyard and Ravosa, 1998), implying this aspect of papionin facial geometry has adaptive significance separate from cranial size (Gould, 1971; Shea, 1985). Assuming directional selection upon a continuous trait (relative gape), the intermediate position of mangabey regressions between the outgroups (*Macaca* and cercopithecins) and large-bodied papionins suggests that mangabeys most closely resemble the basal African papionin form. Thus, *Papio* and *Mandrillus* are inferred to have experienced homoplastic shifts in the scaling of gape leading to maintenance of enhanced gape at relatively large body sizes.

Gape is functionally correlated with both jaw length and canine size, and has been hypothesized to be the principal factor limiting canine height (Lucas, 1981, 1982; Ravosa, 1990). Gape is thus expected to be emphasized in large-bodied species with pronounced sexual size and canine dimorphism (Greaves, 1974; Lucas, 1981, 1982; Lucas et al., 1986; Plavcan and van Schaik, 1993). Species which engage canines as weapons or in threat display—that is, terrestrial species subject to predation risk and social species with high levels of intraspecific competition—are also expected to exhibit increased gape (Lucas et al., 1986; Plavcan and van Schaik, 1993). These expectations are met in *Papio*, *Mandrillus*, and *Theropithecus*—large-bodied, terrestrial monkeys with extreme sexual size and canine dimorphism and intensive mate competition (Jolly, 1970)—but are less apt for the smaller bodied and more arboreal mangabeys. An alternative explanation for increased gape lies in its masticatory significance. Enhanced gape is thought to result in increased vertical bite forces, particularly in the incisal region (Ravosa, 1990), and has been hypothesized to confer selective advantages upon primates specializing in large-diameter foods or hard foods requiring incisal preparation (Hylander, 1979; Ravosa, 1990; Smith, 1984). Certainly, mangabeys fall into this category (see above), and large-bodied African papionins are also known to consume a variety of resistant foods including hard fruits, nuts, seed pods, roots, and herbaceous matter (Kingdon, 1997). The last common ancestor of the African papionin radiation is believed to have resembled *Parapapio*, a moderately sized, terrestrial or semiterrestrial stem papioninan living in dry, mixed-cover, and open-country environments (Jablonski, 2002; Szalay and Delson, 1979). The ability to exploit resistant foods, especially nutrient-dense roots and seeds, would have been strongly selected under these conditions; thus, enhanced gape may have arisen initially as an adaptation to hard-consistency diets.

Under this scenario, subsequent allometric and functional shifts within the African papionin clade are hypothesized to reflect selection for maintenance or enhancement of specific functional capacities coincident with the origin of distinct papioninan ecomorphs. In the case of the small-bodied mangabeys, decreased *absolute* masticatory force is offset by altered masticatory scaling patterns that increase relative bite force and permit maintenance of the ancestral hard-object feeding regime at the smaller body sizes necessary for exploitation of arboreal niches. Conversely, increased body size in *Papio* and *Mandrillus* is associated with allometric displacements that effectively decouple gape from size. By increasing the efficiency of incisal biting (Ravosa, 1990), maintenance of enhanced gape may partially offset the loss of MA associated with increased FL; however, increased *absolute* masticatory forces associated with extreme body size reduce the practical impact of this effect (Ravosa, 1990). Similarly, at larger absolute cranial sizes, enhanced relative gape is no longer required to accommodate the largest food items typical to the primate diet. In the absence of clear masticatory benefits, dissociation of gape-size allometries— which in turn relaxes constraints on adult male canine size (Lucas, 1982; Lucas et al., 1986; Ravosa, 1990)—is best interpreted as a response to selection for increased canine size due to social or ecological factors (Harris, 2000, 2002; Jolly, 1970).

The preceding scenario draws upon phylogenetic, ontogenetic, biomechanical, and behavioral studies to infer the evolutionary history of papionin facial forms. The most plausible alternate scenario—that enhanced relative gape arose in the common African papionin ancestor in response to direct selection for increased canine height—is less consistent with current understanding of ancestral papioninan morphotype. Although estimated male body masses for the largest *Parapapio* species overlap those of extant *Papio* and *Mandrillus* (Delson et al., 2000), the smaller average body sizes, primitive cranial morphology, and moderate canine dimorphism of these basal papionins (Jablonski, 2002) tend to discount baboon-like canine enlargement as key to early papioninan adaptations. Still, the possibility that enhanced gape in combination with extreme canine dimorphism was the primitive African papionin condition cannot be dismissed. In this case, enhanced gape in mangabeys would be symplesiomorphic, maintained as a secondary adaptation to consumption of large-diameter foods or perhaps simply through phylogenetic inertia.

Of the inferences drawn here, the most controversial concern the polarity of mangabey masticatory allometries. The presence of shared patterns of masticatory scaling in these species mirrors previous findings that mangabeys share static and ontogenetic facial allometries distinct from those of large-bodied papionins (Collard and O'Higgins, 2001; Leigh et al., 2003; Shah and Leigh, 1995; Singleton, 2002); however, most of these studies have interpreted mangabey facial allometries as symplesiomorphic. These two sets of results are not necessarily discordant. Patterns of localized functional scaling may have been "swamped" in previous analyses concerned with larger scale patterns of craniofacial allometry. Also, minor perturbations of shared ancestral ontogenetic allometries, arising in parallel under strong functional selection, could conceivably result in novel, homoplastic adult morphologies. Given the complexities of the systems in question, both developmental studies sampling the full range of cercopithecine taxonomic and morphological diversity and more nuanced functional analyses will be required to ultimately reconcile patterns of ontogenetic and functional allometry in this group.

CONCLUSIONS AND SUMMARY

This paper demonstrates but one possible application of geometric morphometrics to the functional interpretation of allometric shape variation—an area of historical bioanthropological interest. Geometric analysis is shown here to yield qualitatively similar results to distance-based linear regression analysis while offering advantages in terms of analytic efficiency and functional interpretation. Results reveal differences in functional scaling of the masticatory complex within and between cercopithecine tribes signaling functional and adaptive divergences. Relative to cercopithecins and *Macaca*, African papionins exhibit decreased relative height of the TMJ leading to enhanced mandibular gape and increased incisal bite forces. This shift is interpreted as a papioninan synapomorphy and is linked to selection for hard-object feeding capabilities in the last common ancestor of the African clade. Further dissociations within this group are tied to the subsequent ecomorphological divergence of small- and large-bodied African papionins. Allometric dissociations in *Papio* and *Mandrillus* that maintain enhanced gape at markedly increased body size are interpreted as homoplastic and related to accommodation of enlarged male canines. Similarly, homoplastic displacements of functional shape allometries in *Cercocebus* and *Lophocebus* enhance MA and increase relative bite forces, enabling continued exploitation of resistant food items at the smaller body sizes required for (semi)arboreality. An alternative scenario positing increased body size, canine enlargement, and enhanced gape as the basal papioninan condition cannot be excluded but is considered less likely based on reconstructions of the African ancestral morphotype. The scope of this study is, by necessity, limited, its results largely qualitative, and its conclusions tentative. Future studies sampling the true range of guenon and macaque craniofacial diversity are required to firmly establish morphocline polarities for the cercopithecine masticatory complex and construct robust functional and adaptive hypotheses. As demonstrated here, geometric morphometric analysis of functional allometry should contribute substantially to this effort.

ACKNOWLEDGMENTS

I thank Dennis Slice for inviting me to contribute to this volume, and I dedicate this work to the life and legacy of Les Marcus. His patient instruction and gentle (or not so) remonstrances are daily missed by those privileged to have worked and studied with him. I wish to thank Eric Delson for his continuing support of this research and David Reddy for his ongoing technical assistance. I am grateful to Steve Frost and Tony Tosi for their substantial contributions to the NYCEP Primate Morphometrics Database, to the numerous museum curators and collections managers who made amassing this resource possible, and to Richard Thorington and Linda Gordon (NMNH) for permission to photograph specimens in their care. I thank Dennis Slice, James Rohlf, Nicholas Jones, and Paul O'Higgins, among others, for continuing to upgrade and distribute the software upon which so many of us rely. Finally, I am grateful to Dennis Slice and an anonymous reviewer, whose comments considerably improved this contribution, as well the many colleagues who have shared advice and insights including Steve Frost, Kieran McNulty, Katerina Harvati, Sandra Inouye, Brian Shea, Edgar Allin, and Jim Cheverud. Any errors of fact, interpretation, or polarity determination are, needless to say, my own. This work was conducted with the support of the National Science Foundation via the New York Consortium in Evolutionary Primatology Morphometrics Group (NSF Research & Training Grant BIR-9602234, NSF Special Program Grant ACI-9982351). This chapter is NYCEP Morphometrics Contribution No. 12.

REFERENCES

Antón, S. C., 1996, Cranial adaptation to a high attrition diet in Japanese macaques, *Intl. J. Primatol.* 17:401–427.

Barnicot, N. A. and Hewett-Emmett, D., 1972, Red cell and serum proteins of *Cercocebus, Presbytis, Colobus* and certain other species, *Folia Primatol.* 17: 442–457.

Benefit, B. R. and McCrossin, M. L., 1991, Ancestral facial morphology of Old World higher primates, *Proc. Nat. Acad. Sci. USA* 88:5261–5271.

Benefit, B. R. and McCrossin, M. L., 1993, Facial anatomy of *Victoriapithecus* and its relevance to the ancestral cranial morphology of Old World monkeys and apes, *Am. J. Phys. Anthropol.* 92:329–370.

Bookstein, F. L., 1985, Modeling differences in cranial form, with examples from primates, in: *Size and Scaling in Primate Biology*, W. L. Jungers, ed., Plenum Press, New York, pp. 207–229.

Bookstein, F. L., 1996, Combining the tools of geometric morphometrics, in: *Advances in Morphometrics*, L. F. Marcus, M. Corti, A. Loy, G. J. P. Naylor, and D. E. Slice, eds., Plenum Press, New York, pp. 131–151.

Bouvier, M., 1986a, A biomechanical analysis of mandibular scaling in Old World monkeys, *Am. J. Phys. Anthropol.* 69:473–482.

Bouvier, M., 1986b, Biomechanical scaling of mandibular dimensions in New World monkeys, *Intl. J. Primatol.* 7:551–567.

Cachel, S. M., 1984, Growth and allometry in primate masticatory muscles, *Arch. Oral Biol.* 29:287–293.

Chalmers, N. R., 1968, Group composition, ecology and daily activity of free living mangabeys in Uganda, *Folia Primatol.* 8:247–262.

Cheverud, J. M. and Richtsmeier, J. T., 1986, Finite-element scaling applied to sexual dimorphism in rhesus macaque (*Macaca mulatta*) facial growth, *Syst. Zool.* 35:381–399.

Cochard, L. R., 1985, Ontogenetic allometry of the skull and dentition of the rhesus monkey (*Macaca mulatta*), in: *Size and Scaling in Primate Biology*, W. L. Jungers, ed., Plenum Press, New York, pp. 231–255.

Collard, M. and O'Higgins, P., 2001, Ontogeny and homoplasy in the papionin monkey face, *Evol. Devel.* 3:322–331.

Collard, M. and O'Higgins, P., 2002, Why such long faces? A response to Eugene Harris, *Evol. Devel.* 4:169.

Collard, M. and Wood, B. A., 2000, How reliable are human phylogenetic hypotheses? *Proc. Nat. Acad. Sci. USA* 97:5003–5006.

Cronin, J. E. and Sarich, V. M., 1976, Molecular evidence for dual origin of mangabeys among Old World primates, *Nature* 260:700–702.

Delson, E., 1975a, Evolutionary history of the Cercopithecidae, in: *Contributions to Primatology.* Vol. 5: *Approaches to Primate Paleobiology*, F. S. Szalay, ed., Karger, Basel, pp. 167–217.

Delson, E., 1975b, Paleoecology and zoogeography of the Old World monkeys, in: *Primate Functional Morphology and Evolution*, R. Tuttle, ed., Mouton, The Hague, pp. 37–64.

Delson, E., Terranova, C. J., Jungers, W. L., Sargis, E. J., Jablonski, N. G., and Dechow, P. C., 2000, Body mass in Cercopithecidae (Primates, Mammalia): Estimation and scaling in extinct and extant taxa, *Anthropol. Pap. Am. Mus. Nat. Hist.* 83:1–159.

Demes, B., Creel, N., and Preuschoft, H., 1986, Functional significance of allometric trends in the hominoid masticatory apparatus, in: *Primate Evolution*, J. G. Else and P. C. Lee, eds., Cambridge University Press, Cambridge, pp. 229–237.

Disotell, T. R., 1994, Generic level relationships of the Papionini (Cercopithecoidea), *Am. J. Phys. Anthropol.* 94:47–57.

Disotell, T. R., 1996, The phylogeny of Old World monkeys, *Evol. Anthropol.* 5:18–24.

Disotell, T. R., Honeycutt, R. L., and Ruvolo, M., 1992, Mitochondrial DNA phylogeny of the Old World monkey tribe Papionini, *Mol. Biol. Evol.* 9:1–13.

Dryden, I. L. and Mardia, K. V., 1998, *Statistical Shape Analysis*, John Wiley, New York.

Du Brul, E. L., 1977, Early hominid feeding mechanisms, *Am. J. Phys. Anthropol.* 47:305–320.

Dutrillaux, B., Couturier, J., Muleris, M., Lombard, M., and Chauvier, G., 1982, Chromosomal phylogeny of forty-two species or subspecies of cercopithecoids (Primates, Catarrhini), *Ann. Genet.* 25:96–109.

Dutrillaux, B., Fosse, A. M., and Chauvier, G., 1979, Étude cytogénétique de six espèces ou sous-espèces de mangabeys (Papiinae [sic], Cercopithecoidea), *Ann. Genet.* 22:88–92.

Fleagle, J. G. and McGraw, W. S., 1999, Skeletal and dental morphology supports diphyletic origin of baboons and mandrills, *Proc. Nat. Acad. Sci. USA* 96:1157–1161.

Fleagle, J. G. and McGraw, W. S., 2001, Skeletal and dental morphology of African papionins: Unmasking a cryptic clade, *J. Hum. Evol.* 42:267–292.

Freedman, L., 1962, Growth of muzzle length relative to calvaria length, *Growth* 26:117–128.

Freedman, L., 1963, A biometric study of *Papio cynocephalus* skulls from northern Rhodesia and Nyasaland, *J. Mammal.* 44:24–43.

Frost, S. R., Marcus, L. F., Bookstein, F. L., Reddy, D. P., and Delson, E., 2003, Cranial allometry, phylogeography, and systematics of large-bodied papionins (Primates: Cercopithecinae) inferred from geometric morphometric analysis of landmark data, *Anat. Rec.* 275A:1048–1072.

Gould, S. J., 1966, Allometry and size in ontogeny and phylogeny, *Biol. Rev.* 41:587–640.

Gould, S. J., 1971, Geometric similarity in allometric growth: A contribution to the problem of scaling in the evolution of size, *Am. Nat.* 105:113–136.

Gould, S. J., 1975, Allometry in primates, with emphasis on scaling and the evolution of the brain, in: *Contributions to Primatology.* Vol. 5: *Approaches to Primate Paleobiology.* F. S. Szalay, ed., Karger, Basel, pp. 244–292.

Greaves, W. S., 1974, The mammalian jaw mechanism—the high glenoid cavity, *Am. Nat.* 116:432–440.

Greaves, W. S., 1995, Functional predictions from theoretical models of the skull and jaws in reptiles and mammals, in: *Functional Morphology in Vertebrate Paleontology*, J. J. Thomason, ed., Cambridge University Press, Cambridge, pp. 99–115.

Groves, C. P., 1978, Phylogenetic and population systematics of the mangabeys (Primates: Cercopithecoidea), *Primates* 19:1–34.

Harris, E. E., 2000, Molecular systematics of the Old World monkey tribe Papionini: Analysis of the total available genetic sequences, *J. Hum. Evol.* 38:235–256.

Harris, E. E., 2002, Why such long faces? Response to Collard and O'Higgins, *Evol. Devel.* 4:167–168.

Harris, E. E. and Disotell, T. R., 1998, Nuclear gene trees and the phylogenetic relationships of the mangabeys (Primates: Papionini), *Mol. Biol. Evol.* 15:892–900.

Herring, S. W. and Herring, S. E., 1974, The superficial masseter and gape in mammals, *Am. Nat.* 108:561–576.

Hewett-Emmett, D., Cook, C. N., and Barnicot, N. A., 1976, Old World monkey hemoglobins: Deciphering phylogeny from complex patterns of molecular evolution, in: *Molecular Evolution*, M. Goodman and R. E. Tashian, eds., Plenum Press, New York, pp. 257–275.

Hill, W. C. O., 1974, *Primates: Comparative Anatomy and Taxonomy.* Vol. 7: *Catarrhini, Cercopithecinae, Cercocebus, Macaca, and Cynopithecus,* Edinburgh University Press, Edinburgh.

Huxley, J. S., 1932, *Problems of Relative Growth*, Methuen, London.

Hylander, W. L., 1977, The adaptive significance of Eskimo craniofacial morphology, in: *Orofacial Growth and Development*, A. A. Dahlberg and T. M. Graber, eds., Mouton, Paris, pp. 129–170.

Hylander, W. L., 1979, The functional significance of primate mandibular form, *J. Morph.* 106:223–240.

Hylander, W. L., 1985, Mandibular function and biomechanical stress and scaling, *Am. Zool.* 25:315–330.

Jablonski, N. G., 2002, Fossil Old World monkeys: The late Neogene radiation, in: *The Primate Fossil Record*, W. C. Hartwig, ed., Cambridge University Press, Cambridge, pp. 255–299.

Jolicoeur, P., 1963, The multivariate generalization of the allometry equation, *Biometrics* 19:497–499.

Jolly, C. J., 1970, The large African monkeys as an adaptive array, in: *Old World Monkeys: Evolution, Systematics, and Behavior*, J. R. Napier and P. H. Napier, eds., Academic Press, New York, pp. 141–174.

Kingdon, J., 1997, *The Kingdon Field Guide to African Mammals*, Academic Press, San Diego.

Kuhn, H.-J., 1967, Zur systematik der Cercopithecidae, in: *Neue Ergebnisse der Primatologie*, D. Starck, R. Schneider, and H.-J. Kuhn, eds., G. Fischer, Stuttgart, pp. 25–46.

Leigh, S. R., Shah, N. F., and Buchanan, L. S., 2003, Ontogeny and phylogeny in papionin primates, *J. Hum. Evol.* 45:285–316.

Lockwood, C. A. and Fleagle, J. G., 1999, The recognition and evaluation of homoplasy in primate and human evolution, *Yearb. Phys. Anthropol.* 42:189–232.

Lucas, P. W., 1981, An analysis of canine size and jaw shape in some Old and New World non-human primates, *J. Zool.* 195:437–448.

Lucas, P. W., 1982, An analysis of the canine tooth size of Old World higher primates in relation to mandibular length and body weight, *Arch. Oral Biol.* 27:493–496.

Lucas, P. W., Corlett, R. T., and Luke, D. A., 1986, Sexual dimorphism of tooth size in anthropoids, *Hum. Evol.* 1:23–39.

McGraw, W. S., and Fleagle, J. G., 2000, Biogeography and evolution of the *Cercocebus–Mandrillus* clade, *Am. J. Phys. Anthropol. Suppl.* 30:225.

McNamara, J. A., Riolo, M. L., and Enlow, D. H., 1976, Growth of the maxillary complex in the rhesus monkey (*Macaca mulatta*), *Am. J. Phys. Anthropol.* 44:15–26.

Morales, J. C. and Melnick, D. J., 1998, Phylogenetic relationships of the macaques (Cercopithecidae: *Macaca*), as revealed by high resolution restriction site mapping of mitochondrial ribosomal genes, *J. Hum. Evol.* 34:1–23.

Mosimann, J. E. and Malley, J. D., 1979, Size and shape variables, in: *Multivariate Methods in Ecological Work*, L. Orioci, C. R. Rao and W. M. Stiteler, eds., International Co-Operative Publishing House, Fairland, MD, pp. 175–189.

O'Higgins, P. and Jones, N., 1998, Facial growth in *Cercocebus torquatus*: An application of three-dimensional geometric morphometric techniques to the study of morphological variation, *J. Anat.* 193:251–272.

O'Higgins, P. and Jones, N., 1999, *Morphologika*, University College London, London.

Page, S. L., Chiu, C., and Goodman, M., 1999, Molecular phylogeny of Old World monkeys (Cercopithecidae) as inferred from γ-globin DNA sequences, *Mol. Phyl. Evol.* 13:348–359.

Penin, X., Berge, C., and Baylac, M., 2002, Ontogenetic study of the skull in modern humans and the common chimpanzee: Neotenic hypothesis reconsidered with a tridimensional Procrustes analysis, *Am. J. Anat.* 118:50–62.

Plavcan, J. M. and van Schaik, C. P., 1993, Canine dimorphism, *Evol. Anthropol.* 2:208–214.

Profant, L. P. and Shea, B. T., 1994, Allometric basis of morphological diversity in the Cercopithecini vs Papionini tribes of Cercopithecine monkeys, *Am. J. Phys. Anthropol. Suppl.* 18:162–163.

Ravosa, M. J., 1990, Functional assessment of subfamily variation in maxillomandibular morphology among Old World monkeys, *Am. J. Phys. Anthropol.* 82:199–212.

Ravosa, M. J., 1992, Allometry and heterochrony in extant and extinct Malagasy primates, *J. Hum. Evol.* 23:197–217.

Ravosa, M. J., and Profant, L. P., 2000, Evolutionary morphology of the skull in Old World monkeys, in: *Old World Monkeys*, P. F. Whitehead and C. J. Jolly, eds., Cambridge University Press, Cambridge, pp. 237–268.

Ravosa, M. J., and Shea, B. T., 1994, Pattern in craniofacial biology: Evidence from the Old World monkeys (Cercopithecidae), *Intl. J. Primatol.* 15:801–822.

Richtsmeier, J. T. and Lele, S., 1993, A coordinate-free approach to the analysis of growth patterns: Models and theoretical considerations, *Biol. Rev.* 68:381–411.

Rohlf, F. J., 1998, *tpsSmall*. Department of Ecology and Evolution, State University of New York–Stony Brook, Stony Brook, New York.

Rohlf, F. J., 1999, Shape statistics: Procrustes superimpositions and tangent spaces, *J. Class.* 16:197–223.

Rohlf, F. J. and Marcus, L. F., 1993, A revolution in morphometrics, *Trends Ecol. Evol.* 8:129–132.

Sakka, M., 1985, Cranial morphology and masticatory adaptations, in: *Food Acquisition and Processing in Primates*, D. J. Chivers, B. A. Wood, and A. Bilsborough, eds., Plenum Press, New York, pp. 415–427.

Shah, N. F. and Leigh, S. R., 1995, Cranial ontogeny in three papionin genera, *Am. J. Phys. Anthropol. Suppl.* 20:194.

Shea, B. T., 1983a, Allometry and heterochrony in the African apes, *Am. J. Phys. Anthropol.* 62:275–289.

Shea, B. T., 1983b, Size and diet in the evolution of African ape craniodental form, *Folia Primatol.* 40:32–68.

Shea, B. T., 1985, Ontogenetic allometry and scaling: A discussion based on the growth and form of the skull in African apes, in: *Size and Scaling in Primate Biology*, W. L. Jungers, ed., Plenum Press, New York, pp. 175–205.

Shea, B. T., 1992, Ontogenetic scaling and skeletal proportions in the talapoin monkey, *J. Hum. Evol.* 23:283–307.

Shea, B. T., 1995, Dissociability of size and shape in studies of allometry and morphometrics, *Am. J. Phys. Anthropol. Suppl.* 20:194.

Singleton, M., 2002, Patterns of cranial shape variation in the Papionini (Primates: Cercopithecinae), *J. Hum. Evol.* 42:547–578.

Slice, D. E., 2001, Landmark coordinates aligned by Procrustes analysis do not lie in Kendall's shape space, *Syst. Biol.* 50:141–149.

Slice, D. E., Bookstein, F. L., Marcus, L. F., and Rohlf, F. J., 1996, Appendix I—A glossary for geometric morphometrics, in: *Advances in Morphometrics*,

L. F. Marcus, M. Corti, A. Loy, G. J. P. Naylor, and D. E. Slice, eds., Plenum Press, New York, pp. 531–551.

Smith, R. J., 1984, Comparative functional morphology of maximum mandibular opening (gape) in primates, in: *Food Acquisition and Processing in Primates*, D. J. Chivers, B. A. Wood, and A. Bilsborough, eds., Plenum, New York, pp. 231–255.

Smith, R. J., 1993, Categories of allometry: Body size versus biomechanics, *J. Hum. Evol.* 24:173–182.

Spencer, M. A., 1999, Constraints on masticatory system evolution in anthropoid primates, *Am. J. Phys. Anthropol.* 108:483–506.

Strasser, E. and Delson, E., 1987, Cladistic analysis of cercopithecid relationships, *J. Hum. Evol.* 1 6:81–99.

Swindler, D. S. and Sirianni, J. E., 1973, Palatal growth rates in *Macaca nemestrina* and *Papio cynocephalus, Am. J. Phys. Anthropol.* 38:83–92.

Swindler, D. S., Sirianni, J. E., and Tarrant, L. H., 1973, A longitudinal study of cephalofacial growth in *Papio cynocephalus* and *Macaca nemestrina* from three months to three years, in: *Symposium of the IVth International Congress of Primatology.* Vol. 3: *Craniofacial Biology of the Primates*, M. Zingeser, ed., Karger, Basel, pp. 227–240.

Szalay, F. S. and Delson, E., 1979, *Evolutionary History of the Primates*, Academic Press, New York.

Thorington, R. W. and Groves, C. P., 1970, An annotated classification of the Cercopithecoidea, in: *Old World Monkeys—Evolution, Systematics, and Behavior*, J. R. Napier and P. H. Napier, eds., Academic Press, New York, pp. 629–648.

Van Der Kuyl, A. C., Kuiken, C. L., Dekker, J. T., Perizonius, W. R. K., and Goudsmit, J., 1995, Phylogeny of African monkeys based upon mitochondrial 12S rRNA sequences, *J. Mol. Evol.* 40:173–180.

Vidarsdottir, U. S., O'Higgins, P., and Stringer, C., 2002, A geometric morphometric study of regional differences in the ontogeny of the modern human facial skeleton, *J. Anat.* 201:211–229.

Vinyard, C. J. and Ravosa, M. J., 1998, Ontogeny, function, and scaling of the mandibular symphysis in papionin primates, *J. Morph.* 235:157–175.

CHAPTER SIXTEEN

A Geometric Morphometric Assessment of the Hominoid Supraorbital Region: Affinities of the Eurasian Miocene Hominoids *Dryopithecus, Graecopithecus,* and *Sivapithecus*

Kieran P. McNulty

INTRODUCTION

Supraorbital morphology in extant hominoids is typically diagnosed using simple, descriptive labels: a rim in *Hylobates*, a costa in *Pongo*, a torus in the African apes, and superciliary arches in modern humans. Similar terms are

Kieran P. McNulty • Department of Sociology & Anthropology, Baylor University, One Bear Place #97326, Waco, TX 76798-7326.

Modern Morphometrics in Physical Anthropology, edited by Dennis E. Slice.
Kluwer Academic/Plenum Publishers, New York, 2005.

also applied to character states in fossil specimens, though often with a qualifier (e.g., "poorly developed"). While these definitions may be adequate to
delineate among extant forms, they provide no objective basis for comparison—
among either alternate morphologies or researchers. Moreover, such descriptive
diagnoses cannot meaningfully characterize variability within taxa. Thus, the
demarcation between one character state and another remains largely a matter of individual preference. For these reasons, verbal description is inadequate for distinguishing among the diverse supraorbital morphologies of
fossil apes.

This is particularly evident in the disagreement surrounding Late Miocene
hominoids from Eurasia. Based on recent discoveries of *Dryopithecus* (Begun
and Moyà Solà, 1992; Kordos, 1987; Kordos and Begun, 2001), a variety
of opinions have emerged over the shape of its supraorbital morphology: an
incipient torus, indicative of African apes and early humans (Begun, 1992;
Kordos and Begun, 2001); a costa, as in *Pongo* (Köhler et al., 2001); or,
primitive morphology of great apes (e.g., Andrews, 1992) or even catarrhines
(Benefit and McCrossin, 1995). Similar hypotheses have been put forth for the
partial cranium of *Graecopithecus*, suggesting that its supraorbital features share
affinities with pongines (e.g., Köhler et al., 2001), hominines (i.e., African
apes and humans; Begun, 1992; Benefit and McCrossin, 1995), hominins
(i.e., members of the human clade; e.g., Bonis and Koufos, 2001), or with
Gorilla (Dean and Delson, 1992). Such dependence on descriptive labels for
supraorbital character states can result in substantial differences in phylogenetic
hypotheses (compare, e.g., Begun, 1994; Benefit and McCrossin, 1995; Köhler
et al., 2001).

This study used landmark-based morphometrics to quantify morphology
and variation in the supraorbital region of extant and fossil hominoids. The
goals of this project were to (a) assess the ability of supraorbital morphology to differentiate among modern taxa, (b) determine phenetic affinities
of fossil specimens, and (c) examine the affinities of fossil morphologies
within a phylogeny of extant hominoids. The much-debated *Dryopithecus* and
Graecopithecus specimens were included here, as well as the partial face of
Sivapithecus (GSP 15000). While most authors agree that this latter fossil
shares many similarities with *Pongo*, there is disagreement as to whether
these are synapomorphies (e.g., Ward and Kimbel, 1983), symplesiomorphies
(e.g., Benefit and McCrossin, 1995), or convergently derived (see Pilbeam and
Young, 2001).

MATERIALS

Extant hominoid specimens were measured at the American Museum of Natural History, National Museum of Natural History, Museum of Comparative Zoology, Peabody Museum, Powell-Cotton Museum, Humboldt University Museum für Naturkunde, and Musée Royal de l'Afrique Centrale. Only adult, wild-shot (for non-humans) specimens were included. Extant hominoid genera were represented by the following sample sizes: *Gorilla*—70m, 44f; *Homo*— 21m, 19f; *Hylobates*—66m, 59f; *Pan*—71m, 91f; and *Pongo*—33m, 39f. Excepting *Hylobates*, all commonly recognized subspecies were sampled; such sampling was rejected for hylobatids due to the multitude of species and subspecies attributed to this genus. Of the four hylobatid subgenera (after Marshall and Sugardjito, 1986), *H. (Hylobates)* and *H. (Symphalangus)* were included here. The former is represented by three subspecies each of *H. agilis* and *H. muelleri*.

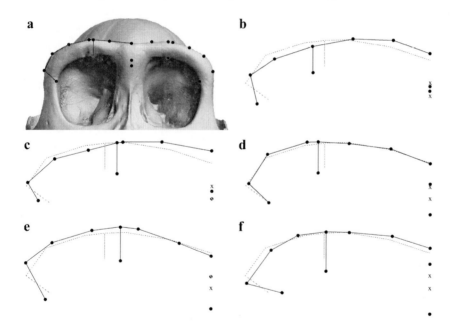

Figure 1. Variation in extant and fossil hominoid supraorbital morphology. (a) Female *Gorilla* cranium showing landmarks used in this study (courtesy of Eric Delson). (b–f) Frontal view of fossil landmark configurations depicting midline and right-side morphology. Fossil landmarks are represented by • connected with solid lines; the overall consensus configuration is shown for contrast using dotted lines and Xs. (b) *Dryopithecus*, RUD 77; (c) *Dryopithecus*, RUD 200; (d) *Dryopithecus*, CLI 18000; (e) *Sivapithecus*, GSP 15000; (f) *Graecopithecus*, XIR 1.

The Miocene Eurasian hominoid sample comprised specimens attributed to *Dryopithecus* (RUD 77, RUD 200, CLI 18000), *Graecopithecus* (XIR 1), and *Sivapithecus* (GSP 15000). Data were collected from original specimens for RUD 77, RUD 200 (Geological Institute of Hungary), and GSP 15000 (in the care of Jay Kelley, University of Illinois College of Dentistry). CLI 18000 and XIR 1 data were collected from high quality casts. Figure 1 illustrates the brow morphology of an extant ape and the landmark configurations of the fossils analyzed in this project. The laser-scan of a female *Gorilla* (Figure 1a) includes the landmarks used for analysis; fossil specimens (Figures 1b–f) are illustrated here by landmarks from the right and midline morphology only, superimposed over the consensus configuration.

METHODS

Data Collection and Processing

Three-dimensional coordinate data were collected using a Microscribe 3DX digitizer (Immersion Corp., San Jose, CA) and recorded in centimeters to four decimal places. The 10 landmarks used to quantify supraorbital morphology are listed and defined in Table 1; abbreviations given there are referred to in subsequent text and figures. To further describe this region, semilandmark data from a single space curve (Dean, 1993; Harvati, 2001; McNulty, 2003; Rohlf

Table 1. Landmark definitions, abbreviations, and intraobserver error

Landmark	Abbreviation	Definition	Side	Mean error (mm)
Frontomalare Temporale	FMT	Intersection of the frontozygomatic suture and the temporal line	RIGHT LEFT	0.23 0.66
Frontomalare Orbitale	FMO	Intersection of the frontozygomatic suture and the orbital rim	RIGHT LEFT	0.24 0.49
Mid-torus Inferior	MTI	Point on the inferior margin of the supraorbital torus at the middle of the orbit	RIGHT LEFT	0.26 0.63
Mid-torus Superior	MTS	Point on superior aspect of the torus directly above mid-torus inferior	RIGHT LEFT	0.37 0.72
Glabella	GLA	Most anterior midline point on frontal bone	MIDLINE	0.46
Nasion	NAS	Most superior point on the internasal suture	MIDLINE	0.52

and Marcus, 1993) were also included. This "line" was represented by a series of closely spaced points collected along the superior border of the supraorbital morphology, bounded by right and left frontomalare temporale. Each curve was then resampled down to nine evenly spaced semilandmarks (L1–L9) for inclusion in analyses.

Ten replicate data series were collected from a single female *Gorilla* cranium to assess the effect of intraobserver error. The mean distance of each replicate landmark to the overall landmark mean was computed to provide an average error estimate for each landmark. These results are given in Table 1. The root mean squared distance of all landmarks to their individual means, computed as the square root of the trace of the covariance matrix, was 0.2 mm.

Of the fossils included here, only RUD 77 completely preserves the relevant morphology. Therefore, missing bilateral landmarks and semi-landmarks were reconstructed by reflecting antimeres across the sagittal plane. Rather than basing such reconstructions on only these few supraorbital landmarks, this procedure was undertaken within the context of a large, comprehensive set of cranial landmarks (see McNulty, 2003 for a detailed overview and discussion of this procedure). For each fossil, mirrored configurations were created by switching the coordinates of bilateral landmark pairs and then multiplying the z-coordinates of all landmarks by -1. Subsequently, each fossil was superimposed (disallowing reflection) with its mirror configuration according to a fit of midline landmarks. Missing bilateral landmarks were then estimated from the corresponding superimposed mirror configurations.

Morphometric Methods

A generalized Procrustes analysis (GPA) performed in *Morpheus et al.* (Slice, 1998) was used to superimpose all landmark configurations (e.g., Dryden and Mardia, 1998; O'Higgins and Jones, 1998; Slice et al., 1996). This is an iterative, least-squares procedure that scales specimens to a unit size, translates them to a common origin, and rotates them to minimize the sum of squared distances across all landmarks and specimens. Because semi-landmarks have fewer degrees of freedom in which to vary, the GPA was performed on landmark coordinates only; space-curve data were transformed through the superimposition matrix of the landmarks. Other approaches to analyzing space curves—requiring different sets of assumptions—have been developed elsewhere (e.g., Bookstein et al., 1999; Dean et al., 1996; Chapters 3 and 4, this volume), but were not used here.

As GPA results in the data being mapped to a curved, non-Euclidean space (Slice, 2001), fitted specimen configurations were projected into a Euclidean space tangent to this at the sample mean (e.g., Dryden and Mardia, 1998). To test the correspondence between coordinates in these spaces, Procrustes distances were regressed against Euclidean distances using *tpsSmall* (Rohlf, 1999). A strong correlation ($r = 0.9999$, slope $= 0.9974$) indicated close unity between these spaces.

ANALYSES AND RESULTS

Phenetic Analyses

Initial phenetic analyses included both principal components analysis (PCA) and canonical discriminant analysis. The former was used both to reduce dimensionality in the dataset and to explore shape variation among specimens, particularly regarding the relationships of fossils to extant clusters. Given that group membership is reliably known for extant ape genera, discriminant analysis was used to examine shape differences among these taxa and to assign fossil specimens to extant genera. Mahalanobis distance estimates generated from the canonical analyses were also used to study hierarchical relationships among taxa.

Principal Components Analysis: A PCA was performed on the covariance matrix of the aligned coordinates. As PCA generates linear combinations of the original variables ordered sequentially to account for the greatest amount of sample variation (Slice et al., 1996), seven eigenvectors with zero variance were dropped from subsequent analyses. PCA was also used to examine the total sample variance in relatively few dimensions (Neff and Marcus, 1980). Table 2 lists the eigenvalue, proportion of variance, and cumulative variance represented by the first 12 (of 57) PCs—accounting for more than 90% of the total sample variance. An analysis of variance (ANOVA) was performed on PC scores to test for statistical significance among genera along each eigenvector; these results are also shown in Table 2. To determine the groups contributing to such differences, pairwise t-tests with a Bonferroni adjustment were performed on the least-squares adjusted means for each genus. T-test results are given in Table 3. Fossil taxa were not included in ANOVAs and t-tests. Although all principal component (PC) axes were examined during analysis, the majority of group differences were represented by the first four eigenvectors. These are discussed in detail below.

Table 2. Summary of PCA and ANOVA results for the first 12 (of 57) eigenvectors[a]

Eigenvector	PCA results			ANOVA results	
	Eigenvalue	Proportion	Cumulative	F value	Pr > F
1	0.01034	0.4168	0.4168	477.89	<0.0001
2	0.00442	0.1783	0.5950	416.76	<0.0001
3	0.00204	0.0824	0.6774	174.22	<0.0001
4	0.00134	0.0544	0.7318	72.09	<0.0001
5	0.00128	0.0518	0.7835	1.52	0.1959
6	0.00100	0.0407	0.8242	10.28	<0.0001
7	0.00057	0.0232	0.8474	3.20	0.0130
8	0.00051	0.0209	0.8683	4.31	0.0019
9	0.00035	0.0145	0.8828	0.97	0.4216
10	0.00034	0.0140	0.8968	2.33	0.0554
11	0.00028	0.0114	0.9081	0.42	0.7915
12	0.00026	0.0106	0.9188	6.37	<0.0001

Note:
[a] PCA results show statistics for the latent roots of the sample variance (eigenvalues), the proportion of variance explained by each component, and the cumulative proportion of the variance summarized by each PC and those preceding it. F-values and probabilities are given for one-way ANOVAs testing the null-hypothesis that all genera share a common mean.

Table 3. T-test results with a Bonferroni adjustment showing the probability that pairs of genera share the same mean on the first 4 (of 57) PCs

	Gorilla	Homo	Hylobates	Pan
PC 1				
Homo	0.0094	—		
Hylobates	<0.0001	<0.0001	—	
Pan	1.0000	0.0226	<0.0001	—
Pongo	<0.0001	<0.0001	<0.0001	<0.0001
PC 2				
Homo	<0.0001	—		
Hylobates	<0.0001	0.0007	—	
Pan	<0.0001	<0.0001	<0.0001	—
Pongo	<0.0001	<0.0001	<0.0001	<0.0001
PC 3				
Homo	<0.0001	—		
Hylobates	<0.0001	0.0482	—	
Pan	<0.0001	0.2183	<0.0001	—
Pongo	<0.0001	1.0000	1.0000	<0.0001
PC 4				
Homo	<0.0001	—		
Hylobates	0.0489	<0.0001	—	
Pan	1.0000	<0.0001	0.0094	—
Pongo	0.0010	<0.0001	<0.0001	0.0007

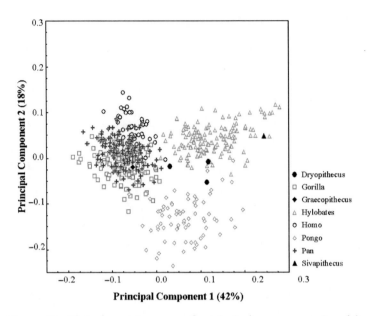

Figure 2.   Plot of specimen scores for principal components 1 and 2.

Principal Component One: The first eigenvector summarizes more than 41% of the total sample variation. T-test results indicate significant differences between all pairs of taxa except *Pan* and *Gorilla*. It is clear from a plot of PC scores (Figure 2) that the first PC axis largely separates hominines from *Pongo* and *Hylobates*. Specimens of *Dryopithecus* and *Sivapithecus* fall beyond the hominine ranges on this axis—the latter encompassed only by *Hylobates*. *Graecopithecus*, on the other hand, is situated among the African apes and humans.

Principal Component Two: This second axis summarizes nearly 18% of the total variability. T-tests were significant for all pairs, although the main distinction in this vector (see Figure 2) is between *Pongo* and the other hominoids. Though overlapping slightly with the ranges of other nonhuman primates, this axis clearly demarcates the orangutan morphology. All of the fossil specimens fall amid the hominine–hylobatid ranges, although only GSP 15000 is completely beyond the range of *Pongo*; CLI 18000 falls closest to the mean for *Pongo*.

Principal Component Three: PC 3 represents only 8% of the sample variance and does not separate taxa as well and the first two. The primary distinction is between *Gorilla* and the other apes, as shown by t-test results and a plot of the

PC scores (see Figure 3). Although t-tests indicate other significant differences between genus means, all five ranges overlap along this axis. RUD 77, RUD 200, and GSP 15000 fall closest to the mean score for *Gorilla*; conversely, CLI 18000 and XIR 1 lie near the center of the other hominoid ranges.

Principal Component Four: The fourth PC captures 5% of the sample variation. All pairwise t-tests show significant differences with the exception of *Gorilla–Pan*. Like PC 3, however, this axis does not visually separate genera. The major difference is between humans and the nonhuman apes (see Figure 3). Among fossil taxa, *Dryopithecus* and *Graecopithecus* specimens group with the latter. Alternatively, *Sivapithecus* has the most negative score—beyond even the range of the modern human specimens.

Discriminant Analysis: A canonical discriminant analysis was performed on the nonzero PCs. Unlike PCA, this procedure uses group membership data to maximize the among-group variation relative to the pooled within-group variation (Slice et al., 1996). Thus, it is useful for exploring group differences and is usually preferable to PCA when group membership is reliably known (Neff and Marcus, 1980). Three related goals were accomplished through this analysis. First, canonical axes were used to explore the effectiveness of these data in distinguishing among extant hominoid genera. Second, Mahalanobis D^2

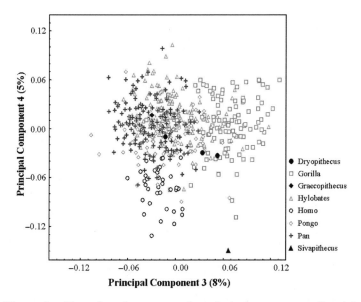

Figure 3.   Plot of specimen scores for principal components 3 and 4.

values generated from the analysis were used to compute differences between pairs of genera and to perform a cluster analysis. Finally, the discriminant functions were used to classify fossil specimens into extant genera.

To account for unequal sample sizes among genera, a randomization procedure was also employed. For this, 30 specimens were randomly chosen from each genus and subjected to discriminant analysis. This was repeated 10,000 times, recording cross-validation data, Mahalanobis D^2 values, and fossil classifications. Results of discriminant analyses are discussed below in terms of both the whole-sample analysis and the randomized, equal-sample replicates.

Differences Among Extant Genera: Four canonical axes were computed for the five extant hominoid genera. ANOVAs found highly significant ($p < 0.0001$) generic differences on each; t-tests with a Bonferroni adjustment demonstrated that all pairwise differences were also highly significant ($p < 0.0001$). A cross-validation test was performed to assess the overlap of genus ranges in the canonical space (Neff and Marcus, 1980). This procedure computed the posterior probabilities of correctly reassigning each extant specimen based on discriminant functions calculated from all other specimens. Cross-validation results from both whole-sample and equal-sample analyses are shown in Table 4. Based on the whole sample, all genera scored better than 95% reassignment except *Homo* (92%); mean values for cross-validations in the replicate series were better than 99% in all taxa. Results of ANOVAs, t-tests, and cross-validation tests all indicate that supraorbital morphology is highly robust in discriminating among hominoid genera.

Table 4.   Cross-validation results from discriminant analyses[a]

Into	Whole-sample discriminant analysis					Randomization	
	Gorilla	Homo	Hylobates	Pan	Pongo	Mean	Min.
From							
Gorilla	97.37	0.00	0.00	2.63	0.00	99.78	93.33
Homo	0.00	92.11	0.00	7.89	0.00	99.99	96.67
Hylobates	0.00	0.00	100.00	0.00	0.00	100.00	100.00
Pan	0.60	0.00	0.00	99.40	0.00	99.96	96.67
Pongo	0.00	0.00	0.00	0.00	100.00	100.00	100.00

Note:
[a] Whole-sample results show the percentage of extant specimens from genera on the left that were assigned to genera listed across the top. Randomization results list the mean and minimum percentage of correct reassignments calculated in 10,000 equal-sample replicate analyses. See text for further discussion.

Table 5.   Matrix of Mahalanobis D^2 values[a]

	Gorilla	Homo	Hylobates	Pan	Pongo
Gorilla	—	66.583	69.256	24.017	74.776
Homo	88.397	—	62.704	30.613	106.546
Hylobates	74.531	74.078	—	42.554	62.395
Pan	25.362	49.828	47.034	—	68.850
Pongo	93.851	134.824	69.980	86.322	—

Note:

[a] Scores in the lower triangle were generated from the whole-sample analysis; mean values from equal-sample replicate analyses are shown in the upper triangle.

Mahalanobis D^2:   Mahalanobis D^2 values, with a correction for bias (Marcus, 1969), were generated to estimate the distance in canonical variates space between population centroids (Neff and Marcus, 1980). Table 5 lists D^2 values based on the whole sample (lower triangle) and on average values from the equal-sample replicates (upper triangle). The two sets of numbers cannot be individually compared as they represent differently scaled canonical spaces (Neff and Marcus, 1980). Relative distances among genera ought to be comparable, however, if sample size differences did not impact results. As shown in Table 5, however, distances between *Homo* and the other taxa are significantly reduced when sample size is held equal; this is particularly evident in its distances to *Gorilla* and *Pan*. The difference between whole-sample and equal-sample D^2 values suggests that the sample size of humans used here—markedly smaller than other samples—*does* affect the outcome of these analyses and must be taken into consideration.

Mahalanobis D^2 is also the basis of the multivariate extension of pairwise t-tests, Hotelling's T^2, and follows an F distribution (e.g., Neff and Marcus, 1980). It was used here to evaluate the probability that two population centroids are statistically different across the entire canonical space. Results indicate that all pairwise groups were highly significantly different ($p < 0.0001$)—not surprising, perhaps, in a genus-level analysis.

To visualize distance relationships hierarchically, the unweighted pair-group method using arithmetic averages (UPGMA) was used to cluster genera by D^2 values. Figure 4 shows cluster diagrams based on the whole sample (Figure 4a) and on randomization means (Figure 4b). The cophenetic correlations of both trees are similar at 0.83 and 0.81, respectively. The main difference between them is the clustering of hominines (*Gorilla, Pan,* and *Homo*) in the replicate

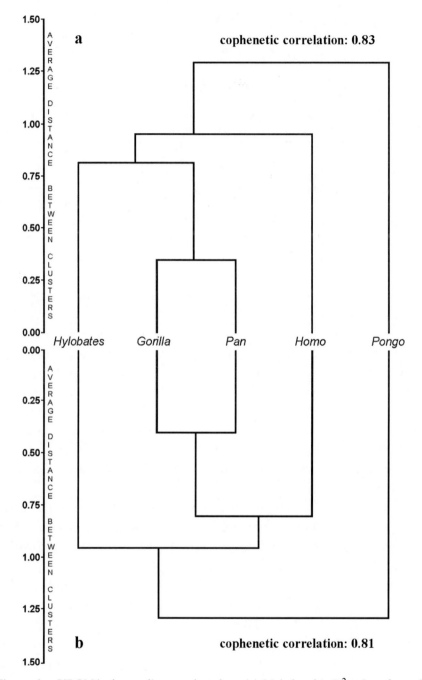

Figure 4. UPGMA cluster diagrams based on (a) Mahalanobis D^2 values from the entire sample of specimens, and (b) mean Mahalanobis D^2 values from 10,000 equal-sample replicates.

series. This again suggests that the smaller sample of *Homo* specimens affects the outcome of these analyses.

As the cophenetic coefficients demonstrate, the UPGMA amalgamation criterion distorts the distances between taxon pairs. It is useful, therefore, to consider cluster results conjointly with the D^2 matrix (see Table 5). The similarity of *Gorilla* and *Pan* is manifest in both matrix and phenogram. The next closest pair is *Pan–Hylobates* (whole-sample) or *Pan–Homo* (randomization). The latter pairing is more likely to be accurate, as it derives from analyses of equal-sample sizes (Neff and Marcus, 1980). In either case, however, their clustering with the *Gorilla–Pan* group is driven by similarities to *Pan*, rather than *Gorilla*. Indeed, *Homo* and *Hylobates* are marginally more similar to each other than either is to *Gorilla*. It is interesting to note that, in both whole-sample and equal-sample analyses, *Homo* was most similar in morphology to *Pan*; indeed, in the replicate series, the difference between *Pan* and *Homo* was only marginally greater than that between the two African apes. Thus, while evolutionary expansion of the brain has dramatically altered the outward appearance of human supraorbital morphology, the actual configuration of landmarks in this region seems largely unaltered (see also McNulty, 2003). *Pongo* is most similar to *Hylobates* in canonical space. Its overall D^2 values, however, indicate that its morphology is substantially different from those of other extant hominoids.

Fossil Assignments: Discriminant functions computed above were also used to assign fossil specimens to extant genera. Table 6 lists the posterior probabilities of grouping fossils in each genus, the percentage of assignments from replicate analyses, and the single resulting classifications. Among *Dryopithecus* specimens, RUD 200 shows strong affinities to *Gorilla*; probabilities of classifying it in other genera were negligible in both whole-sample and equal-sample analyses. RUD 77 was assigned to *Hylobates*, but with only a 92% probability from the whole sample analysis; when sample size is controlled, RUD 77 was placed within *Hylobates* only marginally more often than in *Pan*. CLI 18000 grouped among hylobatids with much stronger support ($p = 0.9997$) in the whole-sample analysis; in the randomization procedure, it grouped with *Hylobates*, *Gorilla*, and *Pan* 58%, 21%, and 15% of the time, respectively. Importantly, none of the *Dryopithecus* specimens demonstrated any affinity to *Pongo*. Results for XIR 1 were unambiguous, placing it with *Gorilla* in all analyses. Of the five fossils analyzed here, only GSP 15000 demonstrated any similarity to *Pongo*. The probability of its assignment to any other genus was

Table 6. Fossil assignments based on discriminant analyses of extant genera[a]

Specimen	*Gorilla*	*Homo*	*Hylobates*	*Pan*	*Pongo*	Assignment
RUD 200	1.0000	<0.0001	<0.0001	<0.0001	0.0000	*Gorilla*
	96.92%	0.46%	1.42%	0.21%	0.99%	
RUD 77	<0.0001	<0.0001	0.9183	0.0817	<0.0001	*Hylobates*
	13.39%	13.87%	35.32%	32.12%	5.3%	
CLI 18000	<0.0001	<0.0001	0.9997	<0.0001	<0.0001	*Hylobates*
	20.86%	6.24%	57.81%	14.57%	0.52%	
XIR 1	1.0000	<0.0001	<0.0001	<0.0001	<0.0001	*Gorilla*
	96.84%	0.36%	1.44%	0.24%	1.12%	
GSP 15000	<0.0001	<0.0001	<0.0001	<0.0001	1.0000	*Pongo*
	0.02%	6.06%	14.26%	4.17%	75.49%	

Note:
[a] The first line for each specimen lists results from the whole-sample analysis, including posterior probabilities of grouping with each extant genus and the final discriminant assignment for the fossil. The second line shows the distribution of assignments from 10,000 equal-sample replicate discriminant analyses.

highly unlikely ($p < 0.0001$) from the whole-sample results. In replicate analyses, *Sivapithecus* showed a slightly broader distribution among genera, but still grouped with *Pongo* 75% of the time.

Phylogenetic Node Discrimination

To interpret these data within an evolutionary framework, phylogenetic node discriminations were also undertaken (McNulty, 2003). The fragmentary nature of fossils has made it commonplace for some authors (e.g., Andrews, 1992) to perform cladistic analyses on extant taxa and then place fossils among the resulting branches according to their preserved features. Phylogenetic node discrimination is a morphometric analog to this approach. Given a phylogeny of extant forms (see, e.g., Figure 5), each node in the branching pattern can be treated as a two-group discrimination between taxa on the left branch and taxa on the right. Based on this series of analyses, determined by the assumed phylogeny, one can test the efficacy of extant morphology in delineating branches at each node. In addition, fossil specimens can be tested for membership along each branch. In this manner, one can examine the affinities of fossil specimens to the extant morphologies that comprise the assumed phylogenetic divisions.

Most authors (e.g., Begun et al., 1997; Collard and Wood, 2000; Pilbeam and Young, 2001; Ruvolo, 1994; but see, e.g., Schwartz, 1987) agree on the phylogenetic relationships of extant hominoids (Figure 5), although the

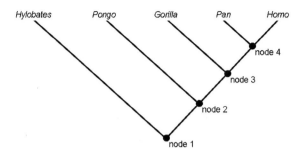

Figure 5.   Consensus phylogeny of extant hominoids. Nodes are referred to in the text.

configuration of hominines shown here is better supported from molecular rather than morphological data. Nevertheless, this consensus constitutes a reasonable hypothesis upon which to base node discriminations. Because of substantial differences in sample sizes, a randomization procedure similar to that described above was also used here. In this case, however, random samples were drawn from the two groups defined by each node, rather than from every genus. Table 7 lists the mean cross-validations and fossil assignments that resulted from node discriminations. In all cases, cross-validation scores demonstrated a clear distinction between clades. Fossil assignments are discussed below. Shape differences that correspond to these bifurcations are discussed in detail by McNulty (2003). Those features relevant to the fossils studied here, however, are described below.

Node 1 Analysis: Node 1 separates the hylobatids from the hominids. Among *Dryopithecus* fossils, RUD 200 shows strong support for placement along the hominid lineage. RUD 77 is less well supported here, grouping with hominids only 60% of the time; results for CLI 18000 were equivocal. XIR 1 has the strongest support (89%) among these fossils for clustering with the hominids. Interestingly, GSP 15000 groups most often with hylobatids at a frequency of 68%.

Node 2 Analysis: Node 2 separates the pongines from the hominines. As has been demonstrated here (e.g., Table 5; Figure 4) and elsewhere, however, the supraorbital morphology of *Pongo* is unique among extant apes, and may not represent a reasonable "outgroup" for the hominines. Therefore, node 2 analyses were also run between hylobatines and hominines. Results of the pongine–hominine discrimination demonstrate overwhelming support for grouping all three *Dryopithecus* specimens with the latter. In the hylobatine–hominine

Table 7. Results of phylogenetic node discriminations[a]

	Node 1		Node 2 (*Pongo*)		Node 2 (*Hylobates*)		Node 3		Node 4	
	Hylobatid	Hominid	Pongine	Hominine	Hylobatine	Hominine	*Gorilla*	*Pan–Homo*	*Pan*	*Homo*
% Reclassified correctly	99.98	99.82	100.00	100.00	100.00	100.00	100.00	100.00	100.00	100.00
RUD 200	16.48	83.52	4.30	95.70	17.12	82.88	57.28	42.72	65.02	34.98
RUD 77	39.39	60.61	7.34	92.66	38.46	61.54	49.54	50.46	51.39	48.61
CLI 18000	51.47	48.53	1.13	98.87	80.20	19.80	46.78	53.22	52.77	47.23
XIR 1	11.37	88.63	3.84	96.16	11.35	88.65	67.54	32.46	75.84	24.16
GSP 15000	67.84	32.16	91.92	8.08	58.75	41.25	45.43	54.57	30.08	60.92

Note:
[a] The first row lists the mean percentage of extant specimens correctly reassigned in 10,000 replicate cross-validations for each node discrimination. Subsequent rows show the percentages at which fossil specimens were assigned to branches deriving from each node. Node 2 analyses were run between both pongines and hominines, as well as hylobatines and hominines. See text for further discussion.

analysis, however, only RUD 200 retained this support: RUD 77 ground weakly with hominines and CLI 18000 grouped with *Hylobates* in 80% of the replicates. *Graecopithecus* was strongly supported as a hominine in both analyses. *Sivapithecus* was well supported among pongines (92%), but only marginally placed with hylobatines (59%) in their respective analyses.

Node 3 Analysis: Node 3 separates *Gorilla* from the *Pan–Homo* clade. All three specimens of *Dryopithecus* split among these clades nearly evenly; RUD 200 showed the most distinction, grouping with *Gorilla* in 57% of the cases. Such balanced results, however, indicate that these fossils do not share specific morphology with either clade. XIR 1 shows more differentiation, classifying with *Gorilla* 68% of the time. Like *Dryopithecus*, GSP 15000 showed no particular affinity with *Gorilla* or the *Pan–Homo* group.

Node 4 Analysis: Node 4 separates *Pan* from *Homo*. Given the results from node 3, it would be unlikely to find that these fossils placed strongly in either clade. Both RUD 77 and CLI 18000 demonstrate this, dividing evenly between *Pan* and *Homo*. RUD 200 shows more distinction, grouping with *Pan* in 65% of the replicates. XIR 1 has the strongest support here of any fossil, classifying with *Pan* 75% of the time. GSP 15000, unlike the other fossils, groups more closely with *Homo*, though at a fairly low frequency (61%).

DISCUSSION

It is clear from the above analyses that supraorbital morphology, as captured by the landmarks used here, is very robust in distinguishing among extant hominoids. This is true even in PCA, but particularly evident from the cross-validation and Hotelling's T^2 computed from discriminant analysis. Looking at Mahalanobis D^2 scores and the UPGMA cluster, these data generally support three descriptive character states typically ascribed to hominoid supraorbital morphologies: *Gorilla* and *Pan* share very similar features, with *Hylobates* and *Pongo* distinct from them and each other. A separate character state for *Homo* is not supported here, however, given its affinity to *Pan* (see also McNulty, 2003). The morphology of *Pongo* is most distinct, having the largest D^2 values in both whole-sample and randomization analyses. Beyond providing quantitative support for delineating among extant morphologies, these analyses provided statistical methods for testing the placement of fossil specimens. Figure 6 illustrates some of the shape differences associated with these fossil assignments (see discussion below).

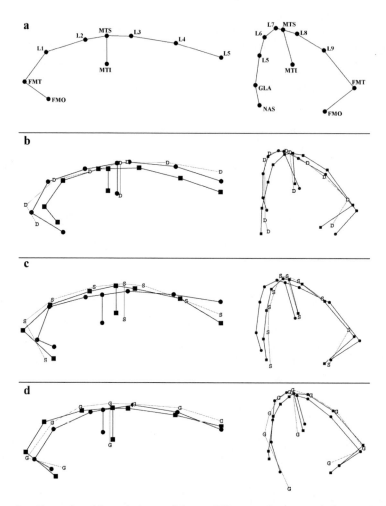

Figure 6. Frontal and lateral views of shape differences between phylogenetic nodes 1–3, with fossil specimens superimposed to illustrate the results of node discriminations. (a) Consensus configuration with landmarks (see Table 1) and semilandmarks (L1–L9) labeled; (b) node 1: ■ = hylobatid, • = hominid, **D** = RUD 200; (c) node 2: ■ = pongine, • = hominine, **S** = GSP 15000; (d) node 3: ■ = *Gorilla*, • = *Pan-Homo*, **G** = XIR 1.

Dryopithecus

The most salient conclusion emerging from analyses of *Dryopithecus* is that these three specimens are not monomorphic. In particular, results for CLI 18000 were substantively different from those of RUD 77 and RUD 200 (see Figure 1). The Spanish specimen was most similar to hylobatids in all analyses.

Results of the replicate discriminant analyses, however, suggest that this is probably not indicative of a close affinity: among extant genera, it grouped with *Hylobates* only 58% of the time. It is possible that the differences between CLI 18000 and the Hungarian specimens represent species-level or even higher taxonomic distinction (see Cameron, 1999). Indeed, these CLI 18000 results might be expected for a stem hominid (e.g., Andrews, 1992) that does not share any particular features with extant apes. Alternatively, one must also consider the possibility that such results reflect poor specimen preservation in the fossil. While this specimen was assumed to retain the entire left zygomatic process of the frontal (Begun, 1994; Moyà Solà and Köhler, 1995), it is possible that the lateral termination is a break rather than a suture. If so, morphometric data from this specimen would be incomparable to those of other specimens.

RUD 200 showed the greatest affinity to hominines. Genus-level discriminant analyses strongly linked this specimen to *Gorilla*. In node discriminations, it closely tracked African apes and humans at the first two nodes and grouped with *Gorilla*—albeit marginally—at node 3. RUD 77 also demonstrated hominine affinities, though not as strongly. The whole-sample analysis assigned this specimen to *Hylobates*; correcting for unequal sample sizes, however, placed RUD 77 in *Hylobates* and *Pan* at similar frequencies. Moreover, RUD 77 favors hominids (60%) to *Hylobates* at node 1, and hominines (61%) to *Hylobates* at node 2. These results add some support to Begun's (e.g., 1994) diagnosis of the *Dryopithecus* brow as a torus. Additional work by McNulty (2003), however, has suggested that the node 2 results shown here may reflect differences in *Dryopithecus* from the pongine and hylobatine morphologies, rather than strong affinities to the hominine form. As such, the Hungarian *Dryopithecus* brows my represent stem hominid morphology.

Figure 6b illustrates, in frontal and lateral views, the differences between hylobatids and hominids at node 1. Superimposed over these is the RUD 200 configuration. As demonstrated by discriminant analyses at node 1, this fossil is most similar to the hominid morphology. In frontal view, the *Dryopithecus* and hominid space curves are fairly consistent in both contour and scope; they differ from the hylobatid in being superiorly placed relative to the orbital rim and the fronto-malar suture, and in their reduced inferior displacement of the midline space curve (L5). Both configurations show thicker brows relative to the hylobatid, with mid-torus landmarks (MTI, MTS) medially placed. While RUD 200 bears some resemblance to the hylobatid in lateral morphology, this frontal view is somewhat misleading. Considering both frontal and lateral aspects, it is clear that the hylobatid fronto-malar suture is narrow medio-laterally and broader antero-posteriorly, the hominid is broad medio-laterally and narrow antero-posteriorly, and the *Dryopithecus* suture is narrow in both dimensions. In lateral view, RUD 200 and hominid profiles are quite similar, excepting the substantial distance in the former between glabella (GLA) and L5. They contrast with the hylobatid profile, which slopes posteriorly immediately superior to glabella.

Perhaps the most important conclusion one can draw regarding hypotheses for *Dryopithecus* is that none of these specimens demonstrated an affinity to *Pongo*. In the whole-sample analysis, the probability of any *Dryopithecus* specimen grouping with *Pongo* was less than 0.0001; replicate analyses grouped RUD 77 with the orangutans only 5% of the time—the other specimens less than 1%. Perhaps the most convincing evidence is from the node 2 analysis. In 10,000 discriminant analyses separating *Pongo* from the hominines, all three specimens grouped with the latter in more than 90% of the replicates. This weighs heavily against the hypothesis that *Dryopithecus* shares supraorbital features with *Pongo* (Köhler et al., 2001). While it is clear that these data are robust for delineating orangutan morphology from that of the other apes, they do not reveal any similarities between *Dryopithecus* and *Pongo*.

Sivapithecus

Of the fossil specimens examined here, only GSP 15000 demonstrated strong affinities to *Pongo*. Curiously, this was not evident in scores from the first four eigenvectors. Indeed, this specimen was fairly unique on the first and third PCs (see Figures 2 and 3). Yet, in analyses designed to sort among known groups, it was well supported as a pongine; the whole-sample discriminant analysis grouped this fossil with *Pongo* at the highest probability, supported by 75% of the replicate assignments. These results bear particular significance in light of the highly autapomorphic nature of the orangutan brow (see Table 5). In node analyses, GSP 15000 showed a close affinity to *Pongo*, and secondarily to *Hylobates*. This statistically corroborates the consensus opinion about similarities between *Sivapithecus* and *Pongo*, and strongly supports a single character state for the two. Results from node 1, however, caution against drawing any further evolutionary significance from such analyses. In the shared features that distinguish extant hominids from hylobatids, GSP 15000 is demonstrably more like the latter. But, whether the node 1 and node 2 results represent symplesiomorphy in *Sivapithecus*, autapomorphy in *Pongo*, or convergence in both cannot be ascertained from this study.

Figure 6c shows the landmark configuration of GSP 15000 superimposed over the shape differences between pongine and hominine brows (node 2). In nearly all aspects of the supraorbital morphology, *Sivapithecus* bears a strong similarity to the pongine configuration. Space curves in both are inferiorly placed in the midline, arch substantially, and terminate well superolateral to the hominine morphology. Mid-torus landmarks, especially in *Sivapithecus*, are medial to those of the hominine, reflecting the narrow interorbital breadth. The fronto-malar suture (FMT-FMO) in GSP 15000 and the pongine is broad medio-laterally with a strong supero-inferior component relative to the hominine; all three are similar in their antero-posterior dimension. The lateral view illustrates further similarities between the pongine and *Sivapithecus* morphology. Neither has a well-developed glabellar region, unlike the hominine configuration. And, the overall antero-posterior dimension of the brow is substantially reduced compared to the African apes and humans. This indicates a flatter morphology across the front of the upper face.

The analyses here were capable both of distinguishing *Pongo* from the other apes and of recognizing pongine affinities in an unknown specimen (GSP 15000). Orangutan features were noticeably absent from all other fossils. These results cannot preclude the argument that *Dryopithecus* and *Graecopithecus* were too primitive in pongine ancestry to exhibit many derived features (Köhler et al., 2001). They do suggest, however, that any such derived morphology is not present in the supraorbital region (*contra* Köhler et al., 2001).

Graecopithecus

Results for XIR 1 place it rather unambiguously with the hominines. In genus-level discriminant analyses, it was overwhelmingly linked to *Gorilla*. In node analyses, it grouped closely with hominids (89%) at node 1, and hominines at node 2 (96% vs *Pongo*, 89% vs *Hylobates*); these results strongly support hypotheses placing this morphology with the African apes and humans (Begun, 1992; Benefit and McCrossin, 1995). There is further evidence here linking XIR 1 to *Gorilla* (Dean and Delson, 1992), although support at node 3 (68%) was only moderate. As with *Dryopithecus*, there is no evidence to suggest pongine affinities. There is also no support here for a

Graecopithecus–Homo group (Bonis and Koufos, 2001); indeed, in genus-level replicate discriminations, XIR 1 grouped least-often with the modern human morphology.

Figure 6d depicts *Graecopithecus* superimposed over configurations representing the discrimination between *Gorilla* and the *Pan–Homo* clade. In frontal view, XIR 1 is similar to *Gorilla* in having a flatter space curve relative to rounded arch seen in *Pan–Homo*. Its overall contour, however, is not especially similar to either extant clade. *Graecopithecus* and *Gorilla* configurations are also broader than that of *Pan–Homo*, yet both have medially placed mid-torus landmarks; this indicates a narrower interorbital breadth compared to the hominins and chimpanzees. *Gorilla* and *Graecopithecus* also share broad lateral orbital pillars in medio-lateral and antero-posterior dimensions; in the supero-inferior dimension, however, the fossil specimen is greatly reduced relative to both groups. In lateral view, XIR 1 tends to mimic the *Gorilla* morphology except at glabella and nasion (NAS). The marked inferior displacement of nasion in *Graecopithecus* resembles neither extant morphology. While the overall supraorbital morphology of *Graecopithecus* is more similar to *Gorilla* than to the *Pan–Homo* group, in many features it appears to be fairly unique.

CONCLUSIONS

This project used three-dimensional landmark-based morphometric analyses to quantify morphology and variation in the supraorbital region of extant and fossil hominoids. Based on Procrustes superimposition and a battery of statistical approaches, several results were obtained. First, it was demonstrated that supraorbital morphology is robust for distinguishing among extant hominoids. Three character states are exhibited in living apes, separating hominines, *Pongo*, and *Hylobates*; *Homo* is best placed with the African apes in brow morphology, rather than in a separate category. Second, Late Miocene hominoid specimens of *Dryopithecus*, *Sivapithecus*, and *Graecopithecus* were shown to have affinities with particular branches of the hominoid phylogeny. *Dryopithecus* from Hungary best represents stem hominid morphology; *Dryopithecus* from Spain is fairly unique, with uncertain affinities. *Sivapithecus* shows strong affinities to *Pongo* and the pongine lineage, but displays some similarity to hylobatids. Finally, *Graecopithecus* clearly groups with the hominines, and shows some affinity to the *Gorilla* lineage.

ACKNOWLEDGMENTS

I thank Dr. Slice for organizing the AAPA morphometrics symposium, for inviting me to participate in this volume, and especially for the help he has given me and other morphometricians around the world. I am grateful to all those who provided access to specimens in their care: Bob Randall, Ross MacPhee, Richard Thorington, Linda Gordon, Judith Chupasko, David Pilbeam, John Harrison, Malcolm Harman, Manfred Ade, Wim Van Neer, László Kordos, David Begun, and Jay Kelley. I especially thank Eric Delson, Terry Harrison, David Begun, Michelle Singleton, Jim Rohlf, Steve Frost, and Katerina Harvati for their support in this and many other endeavors, as well as Suzanne Hagell and Karen Baab for real-time solutions. Finally, many thanks to the late Dr. Leslie Marcus for bringing life, laughter, and enlightenment to the world of morphometrics. This project was undertaken as part of my dissertation work at the City University of New York, supported by the New York Consortium in Evolutionary Primatology and American Museum of Natural History, and partially funded by NSF grants to NYCEP (DBI 9602234) and the NYCEP Morphometrics Group (ACI 9982351). This is NYCEP Morphometrics contribution number nine.

REFERENCES

Andrews, P., 1992, Evolution and the environment in the Hominoidea, *Nature* 360:641–646.

Begun, D. R., 1992, Miocene fossil hominoids and the chimp-human clade, *Science* 257:1929–1933.

Begun, D. R., 1994, Relations among the great apes and humans: New interpretations based on the fossil great ape *Dryopithecus, Yearb. Phys. Anthropol.* 37:11–63.

Begun, D. R. and Moyà Solà, S., 1992, A new partial cranium of *Dryopithecus laietanus* from Can Llobateres, *Am. J. Phys. Anthropol.* (Suppl.) 14:47.

Begun, D. R., Ward, C. V., and Rose, M. D., 1997, Events in hominoid evolution, in: *Function, Phylogeny, and Fossils. Miocene Hominoid Evolution and Adaptations,* D. R. Begun, C. V. Ward, and M. D. Rose, eds., Plenum Press, New York, pp. 389–416.

Benefit, B. R. and McCrossin, M. L., 1995, Miocene hominoids and hominid origins, *Annu. Rev. Anthropol.* 24:237–256.

Bonis, L., de and Koufos, G. D., 2001, Phylogenetic relationships of *Ouranopithecus macedoniensis* (Mammalia, Primates, Hominoidea, Hominidae) of the late Miocene deposits of central Macedonia (Greece), in: *Phylogeny of the Neogene Hominoid Primates of Eurasia*, L. de Bonis, G. D. Koufos, and P. Andrews, eds., Cambridge University Press, Cambridge, pp. 254–268.

Bookstein, F. L., Schäfer, K., Prossinger, H., Fieder, M., Stringer, C. Weber, G. et al., 1999, Comparison of frontal cranial profiles in archaic and modern *Homo* by morphometric analysis, *Anat. Rec. (The New Anatomist)* 257:217–224.

Cameron, D. W., 1999, The single species hypothesis and *Hispanopithecus* fossils from the Vallés Penedés basin, Spain, *Z. Morph. Anthropol.* 82:159–186.

Collard, M. and Wood, B., 2000, How reliable are human phylogenetic hypotheses? *PNAS* 97:5003–5006.

Dean, D., 1993, The middle Pleistocene *Homo erectus/Homo sapiens* transition: New evidence from space curve statistics, Unpublished Ph.D. Dissertation, Department Anthropology, City University of New York.

Dean, D. and Delson, E., 1992, Second gorilla or third chimp? *Nature* 359:676–677.

Dean, D., Marcus, L. F., and Bookstein, F. L., 1996, Chi-square test of biological space curve affinities, in: *Advances in Morphometrics, NATO ASI Series*, L. F. Marcus, M. Corti, A. Loy, G. J. P. Naylor, and D. E. Slice, eds., Plenum Press, New York, pp. 235–251.

Dryden, I. L. and Mardia, K. V., 1998, *Statistical Shape Analysis*, John Wiley, New York.

Harvati, K., 2001, The Neanderthal problem: 3-D geometric morphometric models of cranial shape variation within and among species, Unpublished Ph.D. Dissertation, Department Anthropology, City University of New York.

Köhler, M., Moyà Solà, S., and Alba, D. M., 2001, Eurasian hominoid evolution in the light of recent *Dryopithecus* findings, in: *Phylogeny of the Neogene Hominoid Primates of Eurasia*, L., de Bonis, G. D. Koufos, and P. Andrews, eds., Cambridge University Press, Cambridge, pp. 192–212.

Kordos, L., 1987, Description and reconstruction of the skull of *Rudapithecus hungaricus* Kretzoi (Mammalia), *Ann. Hist. Natur. Mus. Nat. Hung.* 79:77–88.

Kordos, L. and Begun, D. R., 2001, A new cranium of *Dryopithecus* from Rudabáya, Hungary, *J. Hum. Evol.* 41:689–700.

Marcus, L. F., 1969, Measurement of selection using distance statistics in the prehistoric orang-utan *Pongo pygmaeus paleosumatrensis*, *Evolution* 23(2):301–307.

Marshall, J. and Sugardjito, J., 1986, Gibbon systematics, in: *Comparative Primate Biology, Volume 1: Systematics, Evolution, and Anatomy*, D. R. Swindler and J. Erwin, eds., Alan R. Liss, New York, pp. 137–185.

McNulty, K. P., 2003, Geometric morphometric analyses of extant and fossil hominoid craniofacial morphology, Unpublished Ph.D. Dissertation, Department Anthropology, City University of New York.

Moyà Solà, S. and Köhler, M., 1995, New partial cranium of *Dryopithecus* lartet, 1863 (Hominoidea, Primates) from the upper Miocene of Can Llobateres, Barcelona, Spain, *J. Hum. Evol.* 29:101–139.

Neff, N. A. and Marcus, L. F., 1980, A survey of multivariate methods for systematics. Numerical Methods in Systematic Mammalogy Workshop, American Society of Mammalogists.

O'Higgins, P. and Jones, N., 1998, Facial growth in *Cercocebus torquatus*: an application of three-dimensional geometric morphometric techniques to the study of morphological variation, *J. Anat.* 193:251–272.

Pilbeam, D. R. and Young, N. M., 2001, *Sivapithecus* and hominoid evolution: Some brief comments, in: *Hominoid Evolution and Climate Change in Europe, Volume 2: Phylogeny of the Neogene Hominoid Primates of Eurasia*, L. de Bonis, G. D. Koufos, and P. Andrews, eds., Cambridge University Press, Cambridge, pp. 349–364.

Rohlf, F. J., 1999, *tpsSmall* v. 1.17, Department of Ecology and Evolution, State University of New York, Stony Brook, New York.

Rohlf, F. J. and Marcus, L. F., 1993, A revolution in morphometrics, *Trends Evol. Ecol.* 8:129–132.

Ruvolo, M., 1994, Molecular evolutionary processes and conflicting gene trees: The hominoid case, *Am. J. Phys. Anthropol.* 94:89–114.

Schwartz, J. H., 1987, *The Red Ape. Orang-utans and Human Origins*, Houghton Mifflin Company, Boston.

Slice, D. E., 1998, *Morpheus et al.: Software for Morphometric Research*, Department of Ecology and Evolution, State University of New York, Stony Brook, New York.

Slice, D. E., 2001, Landmark coordinates aligned by Procrustes analysis do not lie in Kendall's shape space, *Syst. Biol.* 50:141–149.

Slice, D. E., Bookstein, F. L., Marcus, L. F., and Rohlf, F. J., 1996, Appendix I—a glossary for geometric morphometrics, in: *Advances in Morphometrics*, L. F. Marcus, M. Corti, A. Loy, G. J. P. Naylor, and D. E. Slice, eds., Plenum Press, New York, pp. 531–551.

Ward, S. C. and Kimbel, W. H., 1983, Subnasal alveolar morphology and the systematic position of *Sivapithecus*, *Am. J. Phys. Anthropol.* 61:157–171.

INDEX